W0036932

Genetic Engineering of Plants: Recent Advances

Genetic Engineering of Plants: Recent Advances

Contributors

Patrick M Boyle,Devin R Burrill et al.

AURIS
Reference

www.aurisreference.com

Genetic Engineering of Plants: Recent Advances

Contributors: Patrick M Boyle,Devin R Burrill et al.

Published by Auris Reference Limited

www.aurisreference.com

United Kingdom

Copyright 2016
Printed in 2017 for Sale in the Indian Subcontinent

The information in this book has been obtained from highly regarded resources. The copyrights for individual articles remain with the authors, as indicated. All chapters are distributed under the terms of the Creative Commons Attribution License, which permit unrestricted use, distribution, and reproduction in any medium, provided the original author and source are credited.

Notice

Contributors, whose names have been given on the book cover, are not associated with the Publisher. The editors and the Publisher have attempted to trace the copyright holders of all material reproduced in this publication and apologise to copyright holders if permission has not been obtained. If any copyright holder has not been acknowledged, please write to us so we may rectify.

Reasonable efforts have been made to publish reliable data. The views articulated in the chapters are those of the individual contributors, and not necessarily those of the editors or the Publisher. Editors and/or the Publisher are not responsible for the accuracy of the information in the published chapters or consequences from their use. The Publisher accepts no responsibility for any damage or grievance to individual(s) or property arising out of the use of any material(s), instruction(s), methods or thoughts in the book.

Genetic Engineering of Plants: Recent Advances

ISBN: 978-1-78154-955-1

British Library Cataloguing in Publication Data
A CIP record for this book is available from the British Library

Printed in the United Kingdom

Exclusively distributed by CBS Publishers & Distributors Pvt. Ltd.

Sales & Distribution Rights only for India, Pakistan, Bangladesh, Sri Lanka, Nepal and Bhutan.This book is not to be sold outside these territories.

Contents

List of Abbreviations

AFLP	Amplified Fragment Length Polymorphism
AIR	Alcohol Insoluble Residue
APHIS	Animal and Plant Health Inspection Service
ARF	Auxin Response Factor
BAC	Bacterial Artificial Chromosome
CAD	Cinnamyl Alcohol Dehydrogenase
CDP	Calcium Dependent Protein Kinases
EIQ	Environmental Impact Quotient
ERF	Element Binding Factor
EST	Expressed Sequence Tag
FDA	Food and Drug administration
FIGE	Field Inversion Gel Electrophoresis
GE	Genetic engineering
GFP	Green Fluorescent Protein
GM	Genetically Modified
GM	Genetically Modified Organisms
GOI	Gene of Interest
GS	Germinating seedlings
HFSE	High Frequency Somatic Embryogenesis
HGT	Horizontal Gene Transfer
HSP	Heat Shock Protein
HTS	High Throughput Sequencing
ICGN	International Coffee Genome Network
ISSR	Intersimple Sequence Repeat
LEA	Late Embryogenesis Abundant
LFSE	Low Frequency Somatic Embryogenesis
LR	Locus Rescue
MAP	Mitogen Activated Protein Kinases
MAS	Marker Assisted Selection
MTP	Monosaccharide Transport Protein
NGS	Next Generation Sequencing
NSF	National Science Foundation
PFGE	Pulsed Field Gel Electrophoresis
PMP	Plant Made Pharmaceutical
PRSV	Papaya Ringspot Virus
PTGS	Post Transcriptional Gene Silencing
RFLP	Restriction Fragment Length Polymorphism
ROS	Reactive Oxygen Species

SA	Salicylic Acid
SM	Scorable Marker Genes
SN	Single-Nucleotide Polymorphisms
SNF	Symbiotic nitrogen fixation
SRAP	Sequence Related Amplified Polymorphism
SSR	Simple Sequence Repeats
TF	Transcription factors
TMV	Tobacco Mosaic Virus

List of Contributors

Patrick M Boyle
Department of Systems Biology, Harvard Medical School, Boston, MA02115USA

Devin R Burrill
Department of Systems Biology, Harvard Medical School, Boston, MA02115USA

Mara C Inniss
Department of Systems Biology, Harvard Medical School, Boston, MA02115USA

Christina M Agapakis
Department of Systems Biology, Harvard Medical School, Boston, MA02115USA
Department of Chemical and Biomolecular Engineering, University of California, Los Angeles, CA 90095, USA

Aaron Deardon
Harvard College, Harvard University, Cambridge, MA 02138, USA

Jonathan G DeWerd
Harvard College, Harvard University, Cambridge, MA 02138, USA

Michael A Gedeon
Harvard College, Harvard University, Cambridge, MA 02138, USA

Jacqueline Y Quinn
Harvard College, Harvard University, Cambridge, MA 02138, USA

Morgan L Paull
Harvard College, Harvard University, Cambridge, MA 02138, USA

Anugraha M Raman
Harvard College, Harvard University, Cambridge, MA 02138, USA

Mark R Theilmann
Harvard College, Harvard University, Cambridge, MA 02138, USA

Lu Wang
Harvard College, Harvard University, Cambridge, MA 02138, USA

Julia C Winn
Harvard College, Harvard University, Cambridge, MA 02138, USA

Oliver Medvedik
Department of Molecular and Cellular Biology, Harvard University, Cambridge, MA 02138, USA

Kurt Schellenberg
The Arnold Arboretum of Harvard University, Boston, MA 02131, USA

Karmella A Haynes
Department of Systems Biology, Harvard Medical School, Boston, MA 02115 USA
School of Biological and Health Systems Engineering, Arizona State University, Tempe, AZ 85287, USA

Alain Viel
Department of Molecular and Cellular Biology, Harvard University, Cambridge, MA 02138, USA

Tamara J Brenner
Department of Molecular and Cellular Biology, Harvard University, Cambridge, MA 02138, USA

George M Church
Wyss Institute for Biologically Inspired Engineering, Harvard University, Boston, MA 02115, USA
Department of Genetics, Harvard Medical School, Boston, MA 02115, USA

Jagesh V Shah
Department of Systems Biology, Harvard Medical School, Boston, MA 02115 USA

Pamela A Silver
Department of Systems Biology, Harvard Medical School, Boston, MA 02115 USA
Wyss Institute for Biologically Inspired Engineering, Harvard University, Boston, MA 02115, USA

Laura C Roden
Broom's Barn Research Station, Higham, Bury St Edmunds, Suffolk IP28 6NP, UK
Dept. Mol. & Cell Biol., UCT, Private Bag Rondebosch, 7701, Cape Town, South Africa

Berthold Göttgens
Cambridge Institute for Medical Research, Hills Road, Cambridge, CB2 2XY, UK

Effie S Mutasa-Göttgens
Broom's Barn Research Station, Higham, Bury St Edmunds, Suffolk IP28 6NP, UK

Hari P. Singh
Agricultural Research Station, Fort Valley State University, Fort Valley, GA, USA

Bharat P. Singh
Agricultural Research Station, Fort Valley State University, Fort Valley, GA, USA

M. K. Mishra
Central Coffee Research Institute, Coffee Research Station, Chikmagalur, Karnataka 577117, India

A. Slater
The Biomolecular Technology Group, Faculty of Health and Life Sciences, De Montfort University, Gateway, Leicester LE1 9BH, UK

Hongyan Wang
Institute of Technology, Yantai Academy of China Agriculture University, Yantai, China

Honglei Wang
Institute of Technology, Yantai Academy of China Agriculture University, Yantai, China

Hongbo Shao
Jiangsu Key Laboratory for Bioresources of Saline Soils, Provincial Key Laboratory of Agrobiology, Institute of Biotechnology, Jiangsu Academy of Agricultural Sciences, Nanjing, China

Key Laboratory of Coastal Biology and Bioresources Utilization, Yantai Institute of Coastal Zone Research, Chinese Academy of Sciences, Yantai, China

Xiaoli Tang
Key Laboratory of Coastal Biology and Bioresources Utilization, Yantai Institute of Coastal Zone Research, Chinese Academy of Sciences, Yantai, China

A. Karthikeyan
Centre for Plant Molecular Biology, Department of Plant Molecular Biology and Biotechnology, Tamil Nadu Agricultural University, Coimbatore-641-003, India

R. Valarmathi
Centre for Plant Protection Studies, Department of Entomology, Tamil Nadu Agricultural University, Coimbatore-641-003, India

S. Nandini
Centre for Plant Protection Studies, Department of Entomology, Tamil Nadu Agricultural University, Coimbatore-641-003, India

M.R. Nandhakumar
Department of Agronomy, Directorate of Crop Management, Tamil Nadu Agricultural University, Coimbatore-641-003, India

Asis Datta
National Institute of Plant Genome Research, New Delhi 110067, India

Fabian Afonso-Grunz
Institute for Molecular BioSciences, Goethe University Frankfurt am Main, Frankfurt am Main, Germany
GenXPro GmbH, Frankfurt Biotechnology Innovation Center (FIZ), Frankfurt am Main, Germany

Carlos Molina
GenXPro GmbH, Frankfurt Biotechnology Innovation Center (FIZ), Frankfurt am Main, Germany
Plant Breeding Institute, Christian-Albrechts-University Kiel, Kiel, Germany

Klaus Hoffmeier
GenXPro GmbH, Frankfurt Biotechnology Innovation Center (FIZ), Frankfurt am Main, Germany

Lukas Rycak
GenXPro GmbH, Frankfurt Biotechnology Innovation Center (FIZ), Frankfurt am Main, Germany

Himabindu Kudapa
International Crops Research Institute for the Semi-Arid Tropics, Hyderabad, India

Rajeev K. Varshney
International Crops Research Institute for the Semi-Arid Tropics, Hyderabad, India

Jean-Jacques Drevon
French National Institute for Agricultural Research (INRA), Eco&Sols, Montpellier-Cedex, France

Peter Winter
GenXPro GmbH, Frankfurt Biotechnology Innovation Center (FIZ), Frankfurt am Main, Germany

Günter Kahl
Institute for Molecular BioSciences, Goethe University Frankfurt am Main, Frankfurt am Main, Germany
GenXPro GmbH, Frankfurt Biotechnology Innovation Center (FIZ), Frankfurt am Main, Germany

Vikash K. Singh
Functional and Applied Genomics Laboratory, National Institute of Plant Genome Research, New Delhi, India

Mukesh Jain
Functional and Applied Genomics Laboratory, National Institute of Plant Genome Research, New Delhi, India

Rohini Garg
Functional and Applied Genomics Laboratory, National Institute of Plant Genome Research, New Delhi, India

Keisuke Matsui
Research Institute, Suntory Global Innovation Center Limited, 1-1-1 Wakayama-dai, Shimamoto-cho, Mishima-gun, Osaka 618-8503, Japan

Junichi Togami

Safety Science Institute, Quality Assurance Division, Suntory Business Expert Limited, 57 Imaikami-cho, Nakahara-ku, Kanagawa Kawasaki 211-0067, Japan

John G. Mason
Biosciences Research Division, Department of Environment & Primary Industries, AgriBio, Centre for AgriBioscience, 5 Ring Road, La Trobe University, Bundoora, VIC 3083, Australia

Stephen F. Chandler
School of Applied Sciences, RMIT University, Bundoora, VIC 3083, Australia

Yoshikazu Tanaka
Research Institute, Suntory Global Innovation Center Limited, 1-1-1 Wakayama-dai, Shimamoto-cho, Mishima-gun, Osaka 618-8503, Japan

Vibe M Gondolf
Feedstocks Division, Joint BioEnergy Institute, Emeryville, California 94608, USA
Physical Biosciences Division, Lawrence Berkeley National Laboratory, Berkeley, California 94720, USA
Department of Plant and Environmental Sciences, University of Copenhagen, DK-1871 Frederiksberg C, Denmark

Rhea Stoppel
Feedstocks Division, Joint BioEnergy Institute, Emeryville, California 94608, USA
Physical Biosciences Division, Lawrence Berkeley National Laboratory, Berkeley, California 94720, USA

Berit Ebert
Feedstocks Division, Joint BioEnergy Institute, Emeryville, California 94608, USA
Physical Biosciences Division, Lawrence Berkeley National Laboratory, Berkeley, California 94720, USA
Department of Plant and Environmental Sciences, University of Copenhagen, DK-1871 Frederiksberg C, Denmark

Carsten Rautengarten
Feedstocks Division, Joint BioEnergy Institute, Emeryville, California 94608, USA
Physical Biosciences Division, Lawrence Berkeley National Laboratory, Berkeley, California 94720, USA

April JM Liwanag
Feedstocks Division, Joint BioEnergy Institute, Emeryville, California 94608, USA
Physical Biosciences Division, Lawrence Berkeley National Laboratory, Berkeley, California 94720, USA

Dominique Loqué
Feedstocks Division, Joint BioEnergy Institute, Emeryville, California 94608, USA
Physical Biosciences Division, Lawrence Berkeley National Laboratory, Berkeley, California 94720, USA

Henrik V Scheller
Feedstocks Division, Joint BioEnergy Institute, Emeryville, California 94608, USA
Physical Biosciences Division, Lawrence Berkeley National Laboratory, Berkeley, California 94720, USA
Department of Plant and Microbial Biology, University of California, Berkeley, California 94720, USA

Preface

Genetic engineering is the process of manually adding new DNA to an organism. Plant genetics deals with heredity in plants, specifically mechanisms of hereditary transmission and variation of inherited characteristics. The text *Genetic Engineering of Plants: Recent Advances* describes techniques in plant genetic research and the practical application of genetic engineering to plants. A biobrick compatible strategy for genetic modification of plants has been discussed in first chapter. In second chapter, we describe the basis for homologous recombination cloning in *E. coli*, the available tools and resources, together with a protocol for long range cloning and manipulation of an *Arabidopsis thaliana* gene locus, to create constructs coordinately driven by locus-specific regulatory elements. Third chapter presents a narrative on development of transgenics and their use for the improvement of field, industrial, and pharmaceuticals crops. The objective of fourth chapter is to provide an update on coffee genetic transformation over the last decade, including the *in vitro* methods used for plant generation. In fifth chapter, we review the recent progress of transcription factors (TFs) involved in plant abiotic stress responses and their potential utilization to improve multiple stress tolerance of crops in the field conditions. Sixth chapter focuses on genetically modified crops. Genetic engineering for improving quality and productivity of crops has been described in seventh chapter. Genome-based analysis of the transcriptome from mature chickpea root nodules has been presented in eighth chapter. The objective of ninth chapter is to investigate precise function of GH3 genes in legumes during development and stress conditions. In tenth chapter, we describe transgenic plants which hyperaccumulate inorganic phosphate (Pi) and which may be used to reduce environmental water pollution by phytoremediation. A gene stacking approach leads to engineered plants with highly increased galactan levels in arabidopsis has been proposed in last chapter.

Chapter 1

A BIOBRICK COMPATIBLE STRATEGY FOR GENETIC MODIFICATION OF PLANTS

Patrick M Boyle[1], Devin R Burrill[1] , Mara C Inniss[1] , Christina M Agapakis[1,7] , Aaron Deardon[2] , Jonathan G DeWerd[2] , Michael A Gedeon[2] , Jacqueline Y Quinn[2] , Morgan L Paull[2] , Anugraha M Raman[2] , Mark R Theilmann[2] , Lu Wang[2] , Julia C Winn[2] , Oliver Medvedik[3] , Kurt Schellenberg[4] , Karmella A Haynes[1,8], Alain Viel[3] , Tamara J Brenner[3] , George M Church[5,6], Jagesh V Shah[1] and Pamela A Silver[1,5]

[1]Department of Systems Biology, Harvard Medical School, Boston, MA02115USA

[2] Harvard College, Harvard University, Cambridge, MA 02138, USA

[3] Department of Molecular and Cellular Biology, Harvard University, Cambridge, MA 02138, USA

[4] The Arnold Arboretum of Harvard University, Boston, MA 02131, USA

[5] Wyss Institute for Biologically Inspired Engineering, Harvard University, Boston, MA 02115, USA

[6] Department of Genetics, Harvard Medical School, Boston, MA 02115, USA

[7] Department of Chemical and Biomolecular Engineering, University of California, Los Angeles, CA 90095, USA

[8]School of Biological and Health Systems Engineering, Arizona State University, Tempe, AZ 85287, USA

ABSTRACT

Background

Plant biotechnology can be leveraged to produce food, fuel, medicine, and materials. Standardized methods advocated by the synthetic biology community can accelerate the plant design cycle, ultimately making plant engineering more widely accessible to bioengineers who can contribute diverse creative input to the design process.

Results

This paper presents work done largely by undergraduate students participating in the 2010 International Genetically Engineered Machines (iGEM) competition. Described here is a framework for engineering the model plant *Arabidopsis thaliana* with standardized, BioBrick compatible vectors and parts available through the Registry of Standard Biological Parts (http://www.partsregistry.org). This system was used to engineer a proof-of-concept plant that exogenously expresses the taste-inverting protein miraculin.

Conclusions

Our work is intended to encourage future iGEM teams and other synthetic biologists to use plants as a genetic chassis. Our workflow simplifies the use of standardized parts in plant systems, allowing the construction and expression of heterologous genes in plants within the timeframe allotted for typical iGEM projects.

BACKGROUND

Selective breeding has long been used to modify plant characteristics such as growth rate, seed size, and flavor [1]. For much of agricultural history, the targeted traits reflected the needs of local growers and consumers, creating a vast array of crop varieties. Advances in the field of genetics and the advent of recombinant DNA technology accelerated our ability to manipulate food crops [1–5]. In particular, the introduction of multiple genes (termed *gene stacking* in plants) has made plants accessible to synthetic biology applications [6–11]. In contrast to previous developments in agricultural technology, genetic modification of plants has been primarily performed for the benefit of large-scale monocultures of agricultural crops.

This work aims to create a standardized, modular system for the production of genetically enhanced plants to facilitate their adoption by diverse users. Ideally, a plant engineering system is customizable, yet has convenient standard features that minimize the need to re-invent common steps such as transferring genetic material into the plant. We demonstrate the feasibility of small-scale engineering projects in the model organism, *Arabidopsis thaliana* (Arabidopsis), using a BioBrick-modified plant vector system (Figure 1), performed within the time constraints of the iGEM competition.

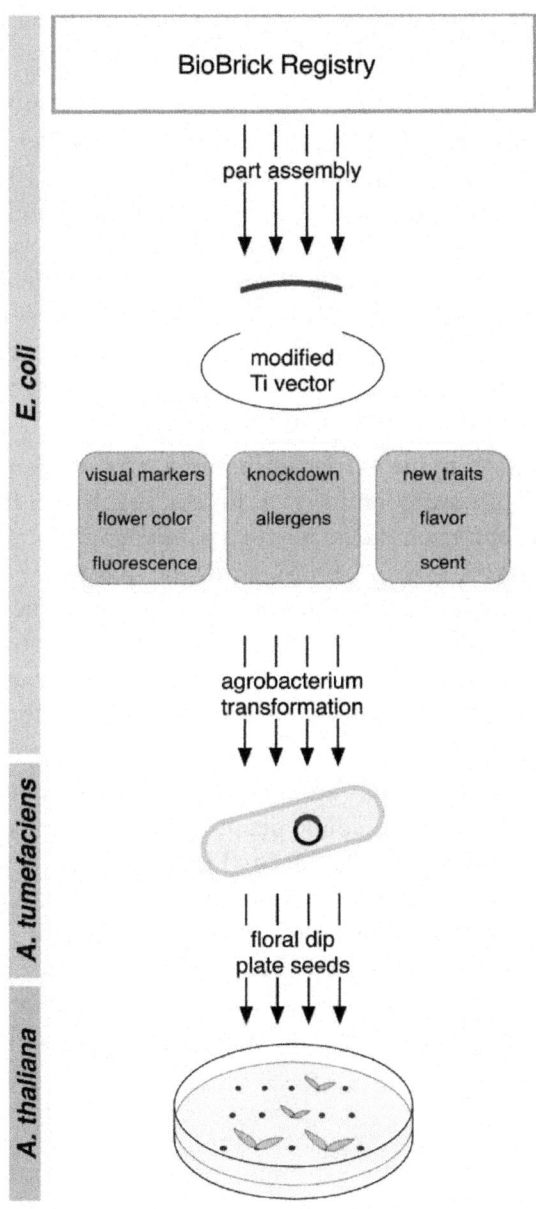

Figure 1: A standardized, modular system for the production of genetically-modified plants. Genetic parts (such as those obtained from the BioBrick Registry) were assembled and inserted into modified vectors (Open, Expression, or Reporter) in *E. coli*. These parts may be assembled to build constructs to impact a wide variety of plant phenotypes. Once assembled, these vectors were transformed into *Agrobacterium*. Using the floral dip procedure, *Agrobacterium* infected Arabidopsis, thereby transferring

the assembled construct. Once seeds were produced, they were plated on selective media to obtain transgenic plants carrying the assembled construct.

Using BioBrick compatible plant vectors, we sought to modify the taste of Arabidopsis, specifically enhancing the sweetness of a bitter plant without altering sugar content. Several naturally occurring proteins are 100–3000 times sweeter than sugar by weight [12]. Brazzein, monellin, thaumatin, pentadin, mabinlin, and curculin are sweet proteins found in a variety of African and South Asian fruits, with no sequence similarity or common features [13]. Brazzein, isolated from the West African fruit *Pentadiplandra brazzeana*, is the smallest of these proteins with only 54 amino acids. It is exceptionally heat stable and was previously expressed heterologously in *Escherichia coli* (*E. coli*) [13], *Zea mays*[14], and *Lactobacillus lactis*[15].

Miraculin, isolated from the berries of the West African plant *Synsepalum dulcificum*, does not taste sweet on its own. Rather, it acts as a flavor-inverter by binding to taste receptors on the tongue in a pH-dependent manner, causing sour foods to taste sweet [16]. A 1 μM miraculin solution is sufficient to activate this inversion, where 20 mM citrate corresponds to the sweetness of 300 mM sucrose [17]. Miraculin is a glycosylated homodimer that has been heterologously expressed in lettuce [18], tomato [19], and even *E. coli*[17], indicating that endogenous *S. dulcificum* glycosylation is not required for functional expression.

Beyond creative gastronomy, we imagine this system being used to enhance the nutritional content of edible plants or help allergy sufferers enjoy the benefits of fresh home-grown produce (Figure 1). The development of efficient transformation techniques for many plants remains a key hurdle for commercial and personal agriculture. However, flexible genetic customization of plants also requires a system of easily transferable, standardized components such as those presented here. We hope this work will lead to techniques that yield a diversity of produce tailored to individual, community, and local environmental needs.

RESULTS

Design of BioBrick Compatible Vectors for Arabidopsis Transformation

Arabidopsis is readily transformed by *Agrobacterium*: when a plant is injured, *Agrobacterium* migrates to the wound site and transfers the T-DNA region of its tumor-inducing (Ti) plasmid into the plant cell [20]. The T-DNA localizes to the nucleus and integrates into the plant's chromosomal DNA. A series of vectors (the pORE series) have been developed from*Agrobacterium*'s Ti plasmid to allow transformation of heterologous DNA into plants via *Agrobacterium*[20]. pORE vectors come equipped with a multiple cloning site (MCS) containing twenty-one unique restriction endonuclease sites. Reporters or promoters are included to create expression vectors, reporter vectors, or vectors that can carry an exogenous promoter or open reading frame. This vector series offers either glufosinate resistance via the *pat* gene, or kanamycin resistance via the *nptII* gene, to enable the selection of successfully transformed plants.

We developed a new set of six BioBrick DNA assembly compatible plant transformation vectors based on the pORE series (Table 1). Vectors V1 and V2 (modified Open vectors) contain no promoter or reporter gene, allowing integration of constructs under the control of a chosen promoter (Figure 2A, Table 1). Vectors V3 and V4 (modified Expression vectors) contain the constitutive pENTCUP2 promoter upstream of the MCS (Figure 2B, Table 1), while V5 and V6 (modified Reporter vectors) contain no promoter but have either the reporter gusA or soluble modified GFP (smGFP) downstream of the cloning site (Figure 2C, Table 1). Each vector contains an MCS that is compatible with three widely used BioBrick standards (RFC 10, 20, 23, http://www.partsregistry.org).

Table 1: Features of BioBrick plant vectors

Vector	BioBrick Registry ID	Bacterial Resistance	Plant Resistance	Promoter	Reporter	Original pORE vector
V1	BBa_K382000	kan	pat	none	none	pORE O1
V2	BBa_K382001	kan	nptII	none	none	pORE O2
V3	BBa_K382002	kan	pat	pENT-CUP2	none	pORE E3

V4	BBa_K382003	kan	nptII	pENT-CUP2	none	pORE E4
V5	BBa_K382004	kan	nptII	none	gusA	pORE R1
V6	BBa_K382005	kan	nptII	none	smGFP	pORE R3

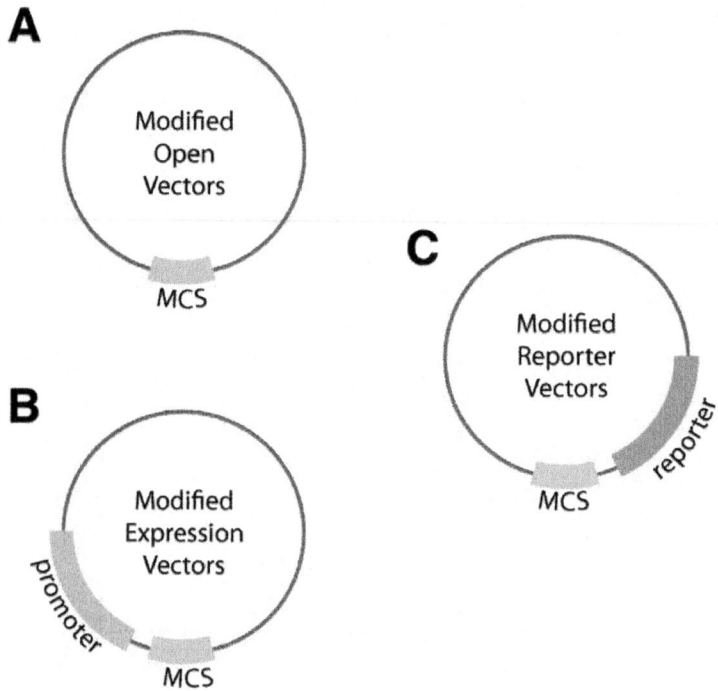

Figure 2: Schematic of BioBrick plant vectors. (A) Modified Open vectors are based on vectors pORE O1 and O2 [14]. They are designed for general insertion of a construct. **(B)** Modified Expression vectors are based on vectors pORE E3 and E4 [14]. They contain an inducible promoter preceding the BioBrick MCS, to permit user-controlled expression of the inserted construct. **(C)** Modified Reporter vectors are based on vectors pORE R1 and R2 [14]. They contain a reporter gene following the BioBrick MCS, such that expression of the reporter follows that of the inserted construct.

Expression of Standardized Flavor Protein Genes in Industrial Microorganisms

We first tested the expression of standardized miraculin and brazzein genes in *E. coli* and the yeast *Saccharomyces cerevisiae* (*S. cerevisiae*), since the introduction of exogenous genes is faster in these organisms. Full-length miraculin and brazzein genes were commercially synthesized and codon-optimized for expression in Arabidopsis. BioBrick compatible restriction enzyme sites bracketed each open reading frame. Constructs were tagged at either the N- or C-terminus with the Strep-II tag [21] for western blot analysis. Miraculin (Figure 3A) and brazzein (Figure 3B) were expressed from an IPTG-inducible T7 promoter in *E. coli*. Monomeric miraculin was expressed at very low levels at approximately 24 kDa regardless of tag location, which is consistent with previous work [17]. Brazzein was highly expressed in the same system at about 12 kDa, regardless of tag location, as has been previously observed [13]. Brazzein was also highly expressed from the constitutive TEF and copper-inducible CUP1 promoters in *S. cerevisiae* (Figure 3C). The higher molecular weight of the Strep-II tagged brazzein observed by western blot in yeast, compared to *E. coli* (~35 kDa versus 12 kDa) is likely due to yeast-specific glycosylation of the brazzein protein [22]. While expression of the miraculin gene was not verified in yeast, integration of both miraculin and brazzein constructs in Arabidopsis was attempted.

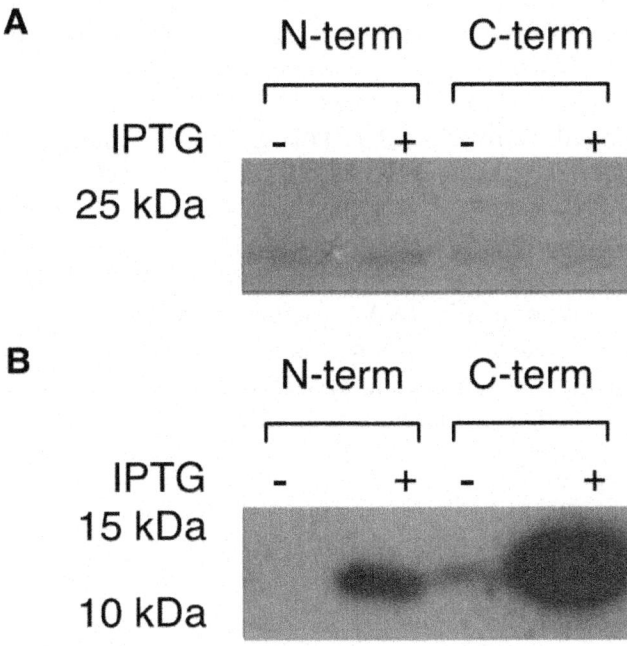

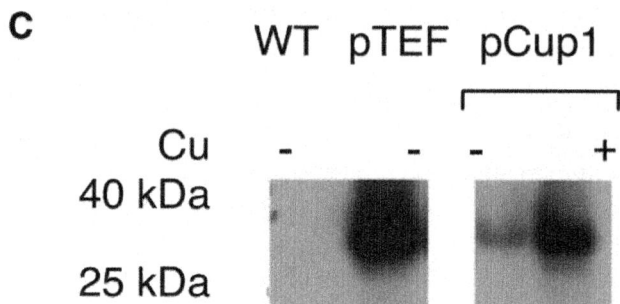

Figure 3: BioBrick miraculin and brazzein protein expression in bacteria and yeast. (A) Miraculin and **(B)** brazzein BioBricks were expressed from an IPTG-inducible promoter in *E. coli* with an N- or C-terminal Strep-II tag. Miraculin was expressed at low levels, with only a faint band appearing at 24 kDa. Brazzein was well expressed in an IPTG-dependent manner. **(C)** Brazzein BioBrick was expressed in yeast from pTEF or pCup1 promoters with a C- terminal Strep-II tag. Brazzein appeared larger in yeast vs. *E. coli*, most likely due to glycosylation of brazzein in yeast.

Expression of Flavor Proteins in Arabidopsis

We successfully introduced two different BioBrick plant vectors into Arabidopsis and selected for seeds carrying genomically-integrated miraculin and brazzein transgenes. Miraculin- or brazzein-encoding DNA under control of the pENTCUP2 promoter and NosT transcriptional terminator on either the V3 (glufosinate resistance) or V4 (kanamycin resistance) BioBrick vector was introduced into Arabidopsis via *Agrobacterium*-mediated transformation [23]. Transformed seeds were selected on MS-agar, and resistant plants were moved to soil and allowed to produce seeds. T1 generation seeds were collected and re-plated on selective plates. Resistant plants were once again moved to soil and allowed to produce T2 generation seeds. While integration of both the miraculin and brazzein genes into the plant genome was verified by PCR (Figure 4A), only miraculin RNA expression was detected by end-point PCR (Figure 4B). Miraculin expression could not be verified by western blot as the antibody showed significant background binding. However, the RNA expression data indicates that miraculin mRNA is expressed in our transgenic plants.

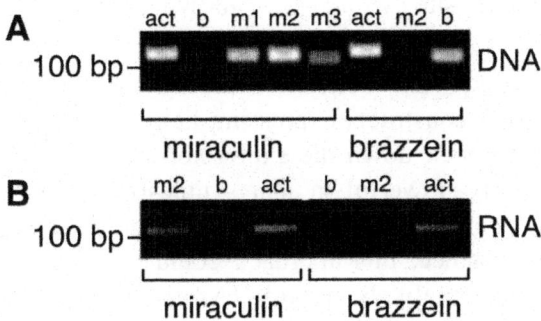

Figure 4: BioBrick miraculin DNA and RNA expression in Arabidopsis. (A) Integration of the miraculin and brazzein genes in the Arabidopsis genome was confirmed. Primer sets for miraculin (m) and brazzein (b) demonstrated that only the desired gene was integrated. **(B)** Miraculin mRNA was constitutively expressed in Arabidopsis however brazzein expression was not detected. act: actin control; b: brazzein primer set; m1-m3: miraculin primer sets.

DISCUSSION AND CONCLUSIONS

Genetic engineering of plants at the industrial and laboratory scale is well established. Technological advances have yielded crops that reduce food production costs through resistance to pests, herbicide, drought, and flood [24]. Additionally, modification of crops (e.g., rice) to contain pro-vitamins can help treat health issues such as vitamin A deficiency in countries where staple foods do not provide the necessary nutrients [24, 25]. However, advances in small-scale experimental horticulture, farming, and gardening have been impeded by the lack of readily available modular parts for the genetic modification of plants. Access to standardized plant vectors in the Registry of Standard Biological Parts will facilitate the design of small-scale plant engineering projects.

We have modified existing plant integration vectors to make them compatible with the BioBrick assembly standard 23 [26], demonstrated that they can be used to integrate transgenes in Arabidopsis, and showed successful integration and expression of the taste modifying gene miraculin. All constructs have been submitted to the Registry of Biological Parts (http://www.partsregistry.org) and are available as a resource for the synthetic biology and plant engineering communities. These include vectors modified from the pORE vector series [20], plant-specific regulatory elements (e.g., promoters, terminators), resistance markers, and the coding sequence of miraculin. The vector series features variations containing the constitutive promoter pENTCUP2 (V3 and V4), visible reporters gusA or smGFP (V5 and V6), or a simple multiple cloning site (V1 and V2), allowing expression of a gene from a promoter of choice.

In addition to using these vectors to express exogenous proteins, we have considered integrating constructs expressing hairpin RNAs [27] or artificial microRNAs [28] to knock down the expression of endogenous genes. This strategy is particularly powerful in that synthesizing a DNA sequence to match any gene transcript of choice allows the regulation of potentially any plant protein. For instance, this approach could be used reduce allergenic protein levels [29]. Alternatively, microRNAs could be targeted to metabolic regulators so that key metabolites, such as pigments or nutrients, are allowed to accumulate [30] and enhance the color or nutritional content of the plant. Modification of existing vectors to conform to a BioBrick assembly standard allows them to be integrated into a BioBrick cloning based workflow. In addition to simplifying the construction of more complex genetic devices, adhering to an assembly standard allows for the possibility of automation of the assembly process.

We hope that availability of plant integration vectors compatible with a common assembly standard will facilitate the use of plants as a chassis in synthetic biology. Local-scale design of food plants, in which the grower selects traits desired in their community, can be made possible through the availability of standardized, modular genetic parts. Personalized engineering of plants to modify flavor, nutritional value, or allergenicity could create a new class of designed foods that are grown and consumed at a local scale. We encourage the iGEM community to continue to explore these concepts via plant engineering.

A significant barrier to the adoption of local-scale plant engineering is the uncertain regulatory landscape for the deployment of genetically modified organisms. This landscape has been defined by large-scale commercial agriculture. As the tools of synthetic biology become more accessible, efforts by small groups such as our own will continue to challenge existing regulatory frameworks. We hope that our concept of genetic engineering tools in the hands of local growers will spur discussion and debate on how to responsibly regulate the synthetic biology scenarios of the near future.

MATERIALS AND METHODS

Plasmids and Cloning

Gene assembly was performed in *E. coli* DH5α using BioBrick assembly standard 23 [26], and all described parts were submitted to the BioBrick

Registry. Arabidopsis pORE series vectors were provided by The Arabidopsis Information Resource (TAIR) and engineered to support BioBrick cloning through PCR-based methods (see Additional file 1: Table S1). pORE Open Series vectors O1 and O2 were digested with SpeI and SacII and ligated with an annealed oligonucleotide insert with NheI and SacII overhangs containing the BioBrick Multiple Cloning Site (MCS) to create vectors V1 and V2. pORE Expression Series vectors E3 and E4 were digested with HindIII and SpeI and ligated with an insert PCR-amplified from the expression vectors containing a HindIII site upstream of the pENTCUP2 promoter and the BioBrick MCS and an NheI site downstream to create vectors V3 and V4. pORE Reporter Series vectors R1, containing the gusA reporter, and R3, containing the smGFP reporter, were digested with HindIII and SpeI and ligated with inserts containing the reporter gene PCR-amplified with primers containing a HindIII site followed by the BioBrick MCS upstream and NheI downstream, yielding vectors V5 and V6.

Brazzein and miraculin were codon-optimized for expression in Arabidopsis, commercially synthesized (Mr. Gene, Regensburg, Germany), and assembled with the pENTCUP2 promoter and NosT transcriptional terminator. Completed constructs were subcloned from BioBrick assembly vector V0120 to BioBrick modified pORE vectors through digestion with EcoRI and PstI.

Plant Maintenance

Wild-type Col-0 *Arabidopsis thaliana* seeds were sterilized by washing with 70% ethanol, 0.1% Triton X-100, followed by two 95% ethanol washes and two sterile dH$_2$O washes. Seeds were then plated on 1X Murashige & Skoog (MS) media with 0.7% agar supplemented with 150 uM carbenicillin and placed in the dark at 4°C for three days before moving to an incubator with 16 h illumination at 20°C and 8 h dark at 15°C per day to allow seeds to germinate. Once plants produced secondary leaves, they were moved to soil and allowed to mature and produce seeds. Seeds were collected and stored at 4°C.

Plant Transformation

Agrobacterium-mediated transformation was performed according to previously reported techniques [23]. Briefly, *Agrobacterium* was made electro-competent by washing in cold sterile water and resuspending in 10% glycerol. Vector DNA was dialyzed to remove excess salt, and electroporated into *Agrobacterium*. Kanamycin-resistant colonies were grown in YEB media, spread on YEB plates, and allowed to form a lawn. Lawns were scraped and suspended in a solution of 20% YEB, 4% sucrose (w/v), and 0.024% Silwet L-77 surfactant (Helena Chemical Company, Collierville, TN). Wild-type

Col-0 Arabidopsis flowers were dipped in the *Agrobacterium* solution and allowed to grow and develop seed pods. Seeds were collected from mature plants and selected on 1x MS media with 0.7% agar supplemented with 5 mg/L glufosinate or 50 µg/ml kanamycin.

E. coli and Yeast Protein Expression

In BL21(DE3) *E. coli*, StrepII-tagged brazzein and miraculin were inserted at multiple cloning site 1 of a BioBrick-modified pET-duet vector [31]. Cells were grown to mid-log phase and induced with a final concentration of 1 mM IPTG. Protein expression was measured by western blot.

In PSY580a yeast *S. cerevisiae*, StrepII-tagged brazzein was cloned with the constitutive TEF promoter or the copper-inducible CUP1 promoter and integrated at the LEU2 locus. Transformants were grown in YEPD media with 0.3 mM $CuSO_4$ to induce protein expression, which was measured by western blot.

Verification of Genomic Transgenes

Genomic DNA was extracted from Arabidopsis using the DNEasy kit (Qiagen) and amplified by PCR (see Additional file 1: Table S2). Whole cell RNA was collected using the plant RNEasy kit (Qiagen). cDNA was synthesized with the SuperScript III First-Strand synthesis kit (Invitrogen). qPCR was performed with primer pairs (see Additional file 1: Table S2) amplifying 100 base pair amplicons within target genes to identify expression of heterologous genes or endogenous gene knockdown.

SDS-Page and Western Blotting

Protein samples were extracted from Arabidopsis, *E. coli*, and yeast and normalized using the Bradford assay (Bio-Rad, Hercules, CA). Samples were diluted into SDS-PAGE loading buffer and loaded onto a 4–20% Tris/glycine/SDS acrylamide gel. α-Strep-tag II antibody (HRP-conjugated, Novagen, Gibbstown, NJ) was used to measure brazzein and miraculin protein expression in *E. coli* and yeast, and α-miraculin antibody [18] (provided by Tadayoshi Hirai, Graduate School of Life and Environmental Sciences, University of Tsukuba, Japan) was used to detect levels of miraculin expression in Arabidopsis. Monoclonal Anti-β-Tubulin antibody (Sigma-Aldrich, St-Louis, MO) was used to detect tubulin in Arabidopsis.

SUPPLEMENTARY MATERIAL

Supplementary Data for A BioBrick Compatible Strategy for Genetic Modification of Plants. Contains tables of primer sequences describing primers used to modify pORE vectors and verify integration and RNA expression of transgenes.

Supplmentary Table 1: Primers used to engineer Arabidopsis pORE series transformation vectors (TAIR*)

Primer Name	Sequence	Purpose
O_5'	GGGAATTcgcggccgcttctagaactagtagcggccgctgcaga	Add MCS** to pORE Open Series vectors O1 and O2 to create V1 and V2
O_3'	CTAGTctgcagcggccgctactagttctagaagcggccgcgAATTCccgc	Add MCS** to pORE Open Series vectors O1 and O2 to create V1 and V2
E_5'	cgtgatAAGCTTgggatcttctgcaagcatctc	Add MCS** to pORE Expression Series vectors E3 and E4 to create V3 and V4
E_3'	caagaGCTAGCctgcagcggccgctactagtcctctaattctagaagcggccgcgaattctccggtgggtttgaggtgag	Add MCS** to pORE Expression Series vectors E3 and E4 to create V3 and V4
R1_5'	actacgaagcttgaattcgcggccgcttctagaattagaggactagtagcggccgctgcagAgagctcatgttacgtcctgtagaaacc	Add MCS** to pORE Reporter Series vector R1 to create V5
R1_3'	caagaGCTAGCAaaaggtacctcattgtttgc	Add MCS** to pORE Reporter Series vector R1 to create V5
R3_5'	actacgaagcttgaattcgcggccgcttctagaattagaggactagtagcggccgctgcagAgagctcatggcgagtaaaggagaagaac	Add MCS** to pORE Reporter Series vector R3 to create V6
R3_3'	atctGCTAGCtttggtaccttatttgtatag	Add MCS** to pORE Reporter Series vector R3 to create V6

*Original vectors were provided by The Arabidopsis Information Resource (TAIR)

**MCS: BioBrick multiple cloning site

Supplmentary Table 2: Primers for testing DNA and RNA expression of miraculin and brazzein in Arabidopsis

Primer Name	Sequence
BRAZ_5'	AATTGGCAAACCAGTGCAA
BRAZ_3'	TCACAGATACACTGGAGGTTCC
MIR1_5'	CGTGACTATCGGAGGAGTGAAGGGT
MIR1_3'	CCACACACAGTTGGGCAAAACACA
ACTIN_5'	TGTGCCAATCTACGAGGGTTT
ACTIN_3'	TTTCCCGCTCTGCTGTTGT

AUTHORS' CONTRIBUTIONS

Cloning schemes were designed by PMB, DRB, MCI, CMA, AD, JGdW, MAG, JYQ, MLP, AMR, MRT, LW, JCW, and OM. KS provided technical assistance with Arabidopsis culture and *Agrobacterium*-mediated transformation. PMB, DRB, MCI, and CMA performed PCR, qPCR and western blots; all other cloning and experiments were performed by AD, JGdW, MAG, JYQ, MLP, AMR, MRT, LW, and JCW. KAH, AV, TJB, GMC, JVS, and PAS provided general advising throughout the project. The manuscript was drafted by PMB, DRB, MCI, and CMA. All authors read and approved the final manuscript.

ABBREVIATIONS

MCS: Multiple cloning site.

ACKNOWLEDGEMENTS

We would like to thank the people at iGEM for supporting this project and allowing us the opportunity to present it at the 2010 iGEM competition. Further thanks are owed to Sarah Mathews of the Harvard Arnold Arboretum for providing advice, access to her lab, and reagents. PMB was supported by the Harvard University Center for the Environment Graduate Consortium. PMB and DRB were supported by the National Science Foundation (NSF) Synthetic Biology Engineering Research Center (SynBERC). MCI was supported by the Natural Science and Engineering Research Council of Canada. CMA was supported by a NSF Graduate Research Fellowship. AD, JGdW, MAG, JYQ, MLP, AMR, MRT, LW, JCW, and OM were generously supported by the Wyss Institute for Biologically Inspired Engineering, Harvard's Office of the Provost, and a grant from the Howard Hughes Medical Institute Undergraduate Education Program awarded to Robert A. Lue, Department of Molecular and Cellular Biology, Harvard University. KS was supported by the Arnold Arboretum.

REFERENCES

1. Kingsbury N: *Hybrid: the history and science of plant breeding - Noël Kingsbury - Google Books*. University of Chicago Press; 2009.

2. Weiling F: Historical study: Johann Gregor Mendel 1822–1884. 1991, 40: 1-25.

3. Jackson DA, Symons RH, Berg P: Biochemical method for inserting new genetic information into DNA of Simian Virus 40: circular SV40 DNA molecules containing lambda phage genes and the galactose operon of Escherichia coli. *Proc Natl Acad Sci USA* 1972, 69:2904-2909. 10.1073/pnas.69.10.2904

4. Lobban PE, Kaiser AD: Enzymatic end-to end joining of DNA molecules. *J Mol Biol* 1973, 78: 453-471. 10.1016/0022-2836(73)90468-3

5. Cohen SN, Chang AC, Boyer HW, Helling RB: Construction of biologically functional bacterial plasmids in vitro. *Proc Natl Acad Sci USA* 1973, 70: 3240-3244. 10.1073/pnas.70.11.3240

6. Antunes MS, Ha S-B, Tewari-Singh N, Morey KJ, Trofka AM, Kugrens P, Deyholos M, Medford JI: A synthetic de-greening gene circuit provides a reporting system that is remotely detectable and has a re-set capacity. *Plant Biotechnol J* 2006, 4: 605-622. 10.1111/j.1467-7652.2006.00205.x

7. Herrera-Estrella L, Depicker A, Van Montagu M, Schell J: Expression of chimaeric genes transferred into plant cells using a Ti-plasmid-derived vector. *Nature* 1983, 303: 209-213. 10.1038/303209a0

8. Bevan MW, Flavell RB, Chilton M-D: A chimaeric antibiotic resistance gene as a selectable marker for plant cell transformation. *Nature* 1983, 304: 184-187. 10.1038/304184a0

9. Fraley RT, Rogers SG, Horsch RB, Sanders PR, Flick JS, Adams SP, Bittner ML, Brand LA, Fink CL, Fry JS, Galluppi GR, Goldberg SB, Hoffmann NL, Woo SC: Expression of bacterial genes in plant cells. *Proc Natl Acad Sci USA* 1983, 80: 4803-4807. 10.1073/pnas.80.15.4803

10. Murai N, Kemp JD, Sutton DW, Murray MG, Slightom JL, Merlo DJ, Reichert NA, Sengupta-Gopalan C, Stock CA, Barker RF, Hall TC:Phaseolin gene from bean is expressed after transfer to sunflower via tumor-inducing plasmid vectors. *Science* 1983, 222: 476-482. 10.1126/science.222.4623.476

11. Shewry PR, Jones HD, Halford NG: *Advances in Biochemical Engineering/Biotechnology.* Berlin, Heidelberg: Springer Berlin Heidelberg; 2008:149-186.

12. Kant R: Sweet proteins–potential replacement for artificial low calorie sweeteners. *Nutr J* 2005, 4: 5. 10.1186/1475-2891-4-5

13. Assadi-Porter F: Efficient Production of Recombinant Brazzein, a Small, Heat-Stable, Sweet-Tasting Protein of Plant Origin. *Arch Biochem Biophys* 2000, 376: 252-258. 10.1006/abbi.2000.1725

14. Lamphear BJ, Barker DK, Brooks CA, Delaney DE, Lane JR, Beifuss K, Love R, Thompson K, Mayor J, Clough R, Harkey R, Poage M, Drees C, Horn ME, Streatfield SJ, Nikolov Z, Woodard SL, Hood EE, Jilka JM, Howard JA: Expression of the sweet protein brazzein in maize for production of a new commercial sweetener. *Plant Biotechnol J* 2004, 3: 103-114. 10.1111/j.1467-7652.2004.00105.x

15. Berlec A, Jevnikar Z, Majhenič AČ, Rogelj I, Štrukelj B: Expression of the sweet-tasting plant protein brazzein in Escherichia coli and Lactococcus lactis: a path toward sweet lactic acid bacteria. *Appl Microbiol Biotechnol* 2006, 73: 158-165. 10.1007/s00253-006-0438-y

16. Koizumi A, Tsuchiya A, Nakajima K-I, Ito K, Terada T, Shimizu-Ibuka A, Briand L, Asakura T, Misaka T, Abe K: Human sweet taste receptor mediates acid-induced sweetness of miraculin. *Proc Natl Acad Sci USA* 2011, 108: 16819-16824. 10.1073/pnas.1016644108

17. Matsuyama T, Satoh M, Nakata R, Aoyama T, Inoue H: Functional Expression of Miraculin, a Taste-Modifying Protein in Escherichia Coli. *J Biochem* 2009, 145: 445-450. 10.1093/jb/mvn184

18. Sun H-J, Cui M-L, Ma B, Ezura H: Functional expression of the taste-modifying protein, miraculin, in transgenic lettuce. *FEBS Lett* 2006, 580: 620-626. 10.1016/j.febslet.2005.12.080

19. Hirai T, Fukukawa G, Kakuta H, Fukuda N, Ezura H: Production of Recombinant Miraculin Using Transgenic Tomatoes in a Closed Cultivation System. *J Agric Food Chem* 2010, 58: 6096-6101. 10.1021/jf100414v

20. Coutu C, Brandle J, Brown D, Brown K, Miki B, Simmonds J, Hegedus DD: pORE: a modular binary vector series suited for both monocot and

dicot plant transformation. *Transgenic Res* 2007, 16: 771-781. 10.1007/s11248-007-9066-2

21. Schmidt TG, Koepke J, Frank R, Skerra A: Molecular interaction between the Strep-tag affinity peptide and its cognate target, streptavidin. *J Mol Biol* 1996, 255: 753-766. 10.1006/jmbi.1996.0061

22. Carlson A, Armentrout RW, Ellis TP: Enhanced Production and Purification of a Natural High Intensity Sweetener. 2010, 1-34.

23. Logemann E, Birkenbihl RP, Ulker B, Somssich IE: An improved method for preparing Agrobacterium cells that simplifies the Arabidopsis transformation protocol. *Plant Methods* 2006, 2: 16. 10.1186/1746-4811-2-16

24. Ronald P: Plant Genetics, Sustainable Agriculture and Global Food Security. *Genetics* 2011, 188: 11-20. 10.1534/genetics.111.128553

25. Ye X, Al-Babili S, Klöti A, Zhang J, Lucca P, Beyer P, Potrykus I: Engineering the provitamin A (beta-carotene) biosynthetic pathway into (carotenoid-free) rice endosperm. *Science* 2000, 287: 303-305. 10.1126/science.287.5451.303

26. Phillips IE, Silver PA: A New Biobrick Assembly Strategy Designed for Facile Protein Engineering. 2006, 1-6.

27. Helliwell C, Waterhouse P: Constructs and methods for high-throughput gene silencing in plants. *Methods* 2003, 30: 289-295. 10.1016/S1046-2023(03)00036-7

28. Ossowski S, Schwab R, Weigel D: Gene silencing in plants using artificial microRNAs and other small RNAs. *Plant J* 2008, 53: 674-690. 10.1111/j.1365-313X.2007.03328.x

29. Singh MB, Bhalla PL: Genetic engineering for removing food allergens from plants. *Trends Plant Sci* 2008, 13: 257-260. 10.1016/j.tplants.2008.04.004

30. Kennedy CJ, Boyle PM, Waks Z, Silver PA: Systems-level engineering of nonfermentative metabolism in yeast. *Genetics* 2009, 183:385-397. 10.1534/genetics.109.105254

31. Agapakis CM, Ducat DC, Boyle PM, Wintermute EH, Way JC, Silver PA: Insulation of a synthetic hydrogen metabolism circuit in bacteria. *J Biol Eng* 2010, 4: 3. 10.1186/1754-1611-4-3

Chapter 2

PROTOCOL: PRECISION ENGINEERING OF PLANT GENE LOCI BY HOMOLOGOUS RECOMBINATION CLONING IN ESCHERICHIA COLI

Laura C Roden[1,3], Berthold Göttgens[2] and Effie S Mutasa-Göttgens[1]

[1]Broom's Barn Research Station, Higham, Bury St Edmunds, Suffolk IP28 6NP, UK

[2]Cambridge Institute for Medical Research, Hills Road, Cambridge, CB2 2XY, UK

[3]Dept. Mol. & Cell Biol., UCT, Private Bag Rondebosch, 7701, Cape Town, South Africa

ABSTRACT

Plant genome sequence data now provide opportunities to conduct molecular genetic studies at the level of the whole gene locus and above. Such studies will be greatly facilitated by adopting and developing further the new generation of genetic engineering tools, based on homologous recombination cloning in *Escherichia coli*, which are free from the constraints imposed by the availability of suitably positioned restriction sites. Here we describe the basis for homologous recombination cloning in *E. coli*, the available tools and resources, together with a protocol for long range cloning and manipulation of an *Arabidopsis thaliana* gene locus, to create constructs co-ordinately driven by locus-specific regulatory elements.

INTRODUCTION

Plant bacterial artificial chromosome (BAC) resources are being generated for ever increasing numbers of species, providing scientists with long-range physical maps and associated sequence data for both model and crop plants. This provides opportunities for reverse genetics and functional studies at the level of the gene locus and above. The latter requires methods for the cloning and manipulation of large DNA fragments, without the limitations imposed by the need for suitably positioned restriction enzyme sites. Significant advances in this respect arose from the development of homologous recombination

(HR) cloning in *Escherichia coli*, based on RecE/RecT (ET) [1, 2] and λ RED operon gene products [3, 4]. Essentially, in ET-based strategies, PCR-amplified linear DNA fragments with short regions of homology (~50 bp to 60 bp) are precisely targeted into any DNA sequence including high copy number plasmids, the *E. coli* chromosome and BACs. RED-based protocols rely on a defective λ prophage to provide functions that protect and recombine the linear DNA fragments, under the control of a temperature sensitive λ cl-repressor, with recombinogenic functions switched on at 42°C and off at 32°C. This fixed induction window helps to reduce unwanted rearrangements, allowing DNA to be stably cloned.

HR-cloning in *E. coli* is widely used in the biomedical research field and is becoming an established tool for BAC engineering in functional genomic studies [5]. Its applications include recombinogenic targeting for gene disruption or replacement and subcloning of BAC DNA by direct isolation of specific genomic regions. A general schematic of HR cloning is given in Fig. 1. Thus, the construction of transgenes for plant functional genomics or the next generation of genetically modified crop plants may benefit from the level of precision engineering offered by HR-cloning.

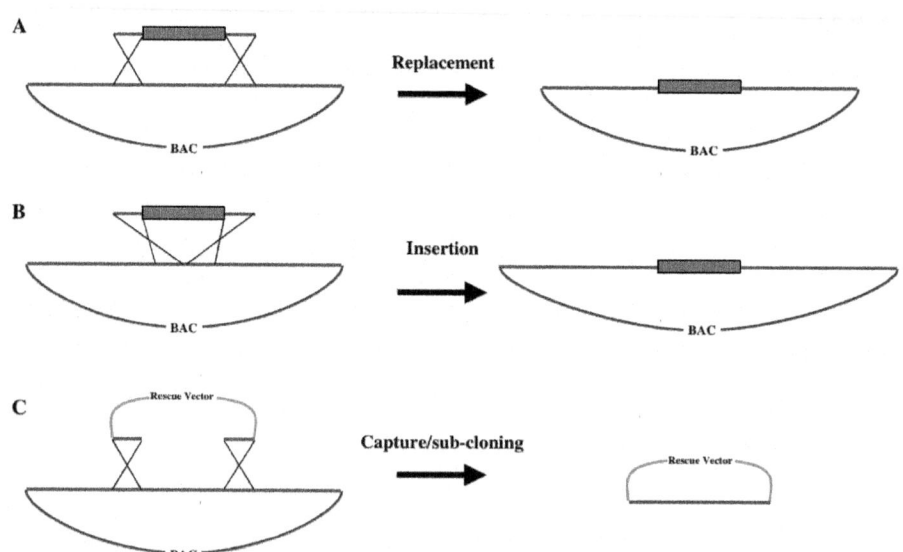

Figure 1: Schematic representation of the basic applications of homologous recombi-nation cloning in *E. coli* for genetic engineering. Homologous recombination cloning in *E. coli* can be used for gene replacement (A), insertion (B) or sub-cloning of target sequences into alternative plasmid vectors. The recombination is mediated by linear DNA fragments (usually generated by PCR), including target site-specific homology arms and a counter selectable antibiotic resistance gene marker.

Our interest in long-range HR cloning was driven by a desire to create plant-specific tools and transgene constructs that target expression to the shoot apical meristem. We wanted to express the bean (*Phaseolus coccineus*) *GAPc2ox1* (encoding GA 2-OXIDASE 1, which degrades bioactive gibberellin) in the shoot apex of sugar beet (*Beta vulgaris*) plants and study the effect on flowering. We present details of our constructs and the molecular tools (plasmids) developed to create these constructs by RED cloning.

MATERIALS

Reagents

- *E. coli* strain EL250 (genotype DH10B [λ*cI857(cro-bioA)*<> *ara* C-P$_{BAD}$*flpe*] where <> indicates that *cro-bioA* has been substituted with *ara* C-P$_{BAD}$*flpe*) available from the authors of [3] who have developed a number of different strains including EL350 (with inducible araC-P$_{BAD}$*cre*). These strains carry a defective λ prophage with *red* and *gam* recombination genes under the control of the λP$_L$ promoter and *exo* and *bet* tightly controlled by the temperature sensitive cI857 repressor. Exo and Beta provide recombinogenic function while Gam inhibits the *E. coli* RecBCD nuclease from degrading electroprated linear DNA fragments. The promoter of the *ara* BAD operon (P$_{BAD}$) is induced by L-arabinose for *flpe* and cre expression enabling removal of sequences between *FRT* and *Lox* P sites respectively. We used EL250 to enable removal of the kanamycin gene in our FRT-mPGK-Tn5-neo-FRT cassette. **OUR RESULTS**: The marker gene was removed as described [3] and worked with 90%–100% efficiency and we were able to recover 100 s of colonies which had become kanamycin sensitive.

- Luria Bertani (LB) broth and plates supplemented with antibiotics as required

- Fully sequenced BAC, PAC or other clones with desired gene locus. Plant BAC and PAC clones are widely available from a number of different sources, including individual labs and organizations e.g. The Arabidopsis Biological Resource Centre (ABRC) http://www. biosci.ohio-state.edu/~plantbio/Facilities/abrc/abrchome.htm or the Nottingham Arabidopsis Stock Centre (NASC) at http://arabidopsis. info/ and, http://www.dna.affrc.go.jp/ for rice genes and others. The *AtSTM* locus used in our experiments is cloned in BAC F24o1, sourced from the Arabidopsis Biological Resource Centre, Columbus, Ohio.

- pUC-based vectors to be used for making (i) the locus rescue gap-repair construct (must be counter selectable to the BAC/PAC), and (ii) the gene of interest (GOI) targeting cassette construct (must contain a counter selectable marker to the gap-repair construct).

- High fidelity *Taq* DNA polymerase. Preferably one which retains A-tails for TA cloning, e.g. the Expand High Fidelity PCR system (Roche Diagnostics).

- PCR primers – four locus-specific primers to amplify DNA fragments at the locus border flanks for the gap-repair rescue construct and two target site-specific primers (minimum 70 bp long) to generate GOI targeting products with destination site specific 5› and 3› homology arms.

- PCR product and gel purification kits e.g. the Qiagen QIAquick™ range and *Dpn* I restriction enzyme – used to remove plasmid templates from PCR reactions because it only cleaves methylated sites.

- General reagents for standard gene cloning and gel electrophoresis

Equipment

- Orbital shaking incubator
- Orbital shaking water bath e.g. Grant OLS 200 – essential for induction of recombination functions in bacterial cells.
- Electroporator e.g. Bio-Rad *E. coli* Pulser
- PCR Machine
- Long wave UV transilluminator – long wave ultra violet light is less damaging to DNA during excision of bands from gels. UV-damaged DNA will not recombine efficiently.
- Electrophoresis equipment capable of field inversion gel electrophoresis (FIGE) or pulsed field gel electrophoresis (PFGE) e.g. BioRad CHEF DR-II, DR-III or Mapper™ XA, for efficient resolution of large DNA fragments
- Spectrophotometer for cell density quantification
- Temperature controlled centrifuge able to run at 4°C

PROTOCOL

The protocols outlined below describe the development of (i) an *AtSTM*-locus specific gap-repair rescue vector, (ii) a plant gene targeting construct with a removable kanamycin resistance marker cassette from pGK-FRT [6], under the control of both the bacterial *Tn5* promoter and the mouse phosphoglycerate kinase (mPGK) promoter for selection in prokaryotes and eukaryotes respectively. This provides templates for PCR amplification of selectable gene fragments that can be precisely targeted into any desired gene locus; and (iii) a bean (*Phaseolus coccineus*) *GAPc2ox1* transformation construct, co-ordinately driven by «all» *AtSTM* locus elements, designated p*STM* 17::*GAPc2ox1*. We have also constructed a pENTR4-based *AtSTM* gap-repair rescue vector for the production of a Gateway™ (Invitrogen) compatible entry clone and generic T-DNA transformation constructs as well as an *mgfp5-ER* targeting cassette. The p*STM* 17::*GAPc2ox1* was successfully transformed into sugar beet, demonstrating for the first time that the mouse PGK promoter is fully functional in transgenic plants, thus enabling the direct exploitation of existing mammalian tools.

Key Steps in the EL250 RED-HR Locus Rescue and Engineering Procedure

- Design of PCR primers for amplification of locus rescue (retrieval) homology arms and also for GOI targeting
- Construction of a gap-repair locus rescue vector.
- Construction of a targeting vector containing the GOI upstream of a counter selectable marker (different from that in the gap-repair construct).
- Electroporation of EL250 cells with the BAC or clone containing the desired gene locus and preparation of electrocompetent BAC/EL250 cells induced for Exo, Beta and Gam functions.
- Performance of gap-repair locus rescue, in cells treated as above; selection of recombinants and confirmation by restriction digestion analysis and sequencing. Transformation of the rescued locus plasmid into fresh EL250 cells.
- PCR amplification, purification and quantification of the GOI targeting cassette and its site-specific recombination into the rescued locus plasmid in EL250 cells. Selection and confirmation of recombinants as above.

The recombineered plasmid is now ready for application in functional analyses as desired.

Primer Design and Plasmid Constructs

Primers

Primer sequences for the *AtSTM* HR rescue protocol described here are given in Additional file 1. Careful attention must be paid to the design of primers for generating locus rescue (LR) homology arms (HA) to ensure that their orientation in the resultant gap-repair vector is correct for DNA double stranded break repair homologus recombination. A total of four short (18 to 20 bp) primers will be required and can if necessary, include restriction sites to enable cloning into the gap-repair vector so that the gap-repair construct can be linearised between the LR-HAs. Fig. 2A shows our *AtSTM* gap-repair construct.

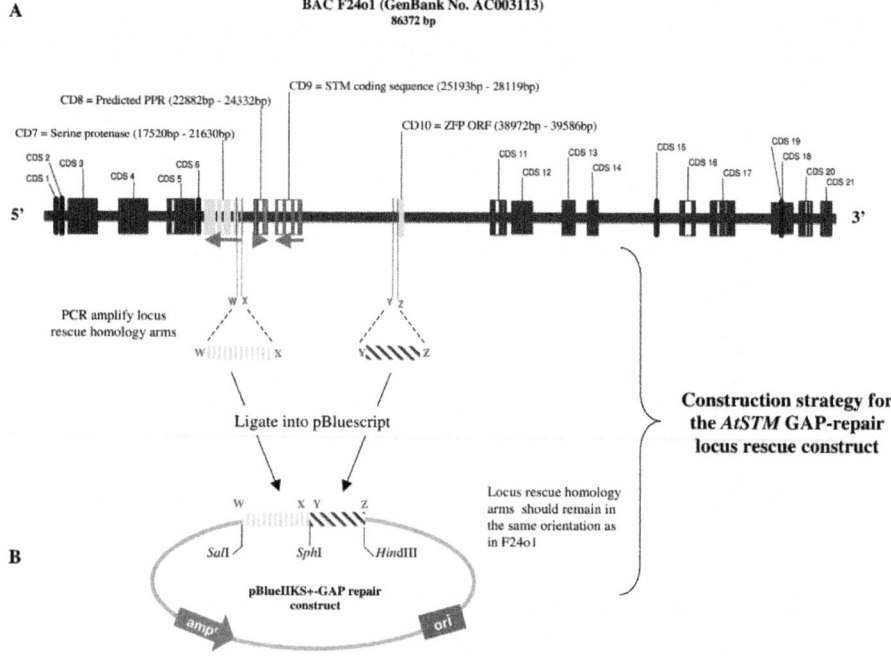

Figure 2: Maps of the *Arabidopsis* BAC F24o1 and the *AtSTM* gap-repair construct. A: A physical map of BAC F24o1, showing the relative positions of the 21 coding sequences (CDS), including *STM* (CDS 9) and its immediate neighbours (CDS 7, 8 and 10) which, are illustrated in different colours and greater detail to show the exon blocks making up each open reading frame. Grey arrows show the orientations of the serine protease, predicted pentatricopeptide repeat protein (PPR), *SHOOTMERI-STEMLESS (STM)* and Zinc Finger Protein (ZFP) open reading frames. B: The gap-repair construct and a schematic representation of the basic protocol used to generate the *AtSTM* downstream (yellow vertical dashed line: W - X) and upstream (blue diagonal dashed line: Y - Z) homology arms to create it in the pBluescriptII KS+ vector backbone. W, X, Y and Z are the PCR primers used to generate each homology arm fragment The *Sal* I, *Sph* I, and *Hind* III cloning sites were incorporated into the PCR primers, the sequences of which are given in Additional file 1.

For the GOI targeting cassette, primers must include at least 30 to 50 bp at the 5› end, to provide homology arms for site-specific recombination. The target site sequence must not have any mismatches as this will inhibit recombination. It is therefore essential to source primers from suppliers able to guarantee sequences of long primers. Our primers were custom made by Sigma Genosys.

Plasmid Constructs

IN OUR HANDS: Creation of these basic plasmids was the key limiting step as it is dependent on conventional cloning and therefore, on the availability of suitably placed restriction sites. However, once constructed, the gene targeting constructs can be used to target the expression cassette into any desired site whereas the gap-repair construct is suitable only for subcloning the specific gene fragment/locus. Our targeting construct backbone has therefore been designed to include a plant-specific polyA signal (Nopaline synthase (nos) termination sequence), for generic use with any plant cDNA sequence.

AtSTM Gap-Repair Construct

Using BAC F24o1 DNA template (represented in Fig. 2A), and the Expand High Fidelity DNA polymerase PCR system (Roche Diagnostics) we amplified 564 bp (incorporating 5'*Sal* I and 3'*Sph* I sites) and 479 bp (incorporating 5›*Sph* I and 3'*Hind* III sites) homology arms respectively at the downstream and upstream flanks of the *AtSTM* locus (Fig. 2B). At the start of our project, the pentatricopeptide repeat protein (PPR) downstream of the STM coding sequence was annotated as a predicted ORF and we therefore opted to include it in the STM locus fragment. Now, it would be excluded as the locus boundary. PCR reactions included primers *Hind* III 3' hyp/*Sph* I 5' hyp or Sprot H1 *Sal* I/Sprot H1 *Sph* I at 0.3 µM each and were incubated for 1 cycle at 94°C for 2 min. followed by 30 cycles of 94°C for 15 sec; 64°C for 30 sec; 72°C for 1.5 min; and 1 cycle of 72°C for 5 min. The individual products were then subcloned into pGEM-T Easy (Promega), recovered and cloned into pBluescript II SK+ (Stratagene) in a three-way ligation reaction, to create the gap-repair construct.

   NOTE : *In our hands, cloning of PCR products is more efficient if we shuttle them via a PCR cloning vector. Any TA cloning vector is suitable. In this case, it is important to ensure that the proof reading activity of the Taq Polymerase used does not remove A-tails.*

Targeting Construct Backbone

The 1.8 kb *FRT-mPGK-Tn5-neo-FRT* cassette was PCR amplified with *Sac* II in the 5' primer (PGK-FRT upper) and *Apa* I in the 3' primer (PGK-FRT lower) from pPGK-FRT (obtained under a Material Transfer Agreement from Dr Francis Stewart, EMBL, Heidelberg – now at University of Technology, Dresden) and cloned into *Sac* II/*Apa* I cloning sites downstream of the Green Fluorescent Protein (*GFP*) gene of polyGFP3 (a kind gift from Dr E. Amaya, Gurdon Institute, Cambridge). The *GFP* gene was then replaced with the Nopaline synthase Terminator (NosTer) from pAL69 (pFC6 with NosTer in the multiple cloning site. A kind gift from Dr Dave Lonesdale at the John Innes Centre, Norwich, UK), to create the basic plant gene targeting vector pNosTerFRT-neo (Fig. 3A), which can be used to receive any GOI.

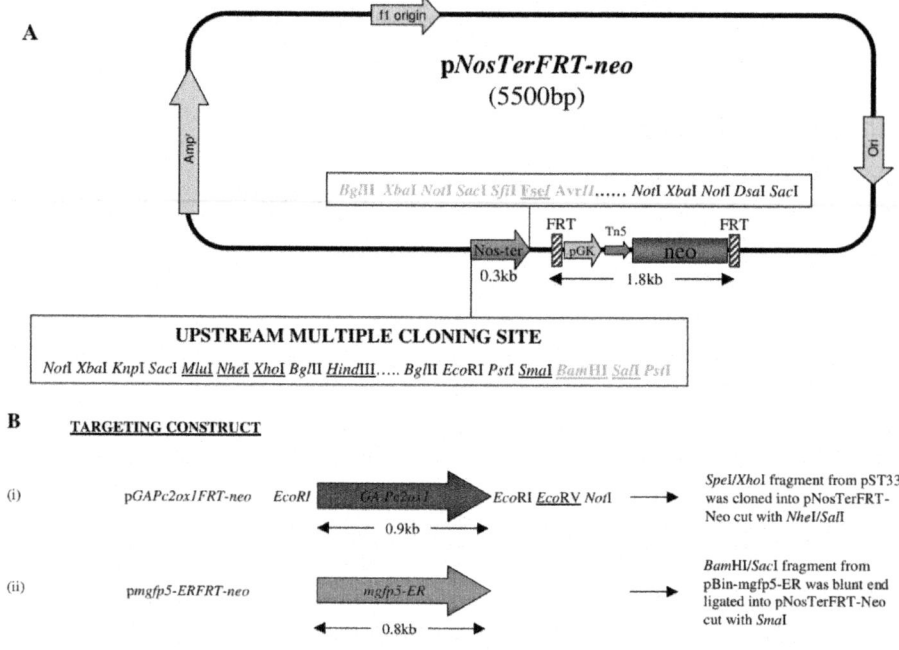

Figure 3: Map of the targeting plasmid backbone p *NosTerFRT-neo* and *GAP-c2ox1/mgfp5-ER* targeting cassettes. A: Structure of the targeting plasmid backbone pNosTerFRT-neo, showing the relative positions of the Nopaline synthase polyA signal sequence (NosTer), the FRT-mPGK-Tn5-neo-FRT selection cassette and the restriction sites mapped up and downstream of the NosTer sequence. Upstream sites may be used for cloning any GOI cDNAs to create targeting constructs. Sites shown in green were introduced with the NosTer fragment; all other sites originate from PolyG-FP3. Unique sites are underlined. B: Representations of GOI sequences with available upstream and downstream restriction sites as appropriate, together with the cloning

strategies employed to clone them into pNosTerFRT-neo to make the targeting cassette constructs.

GAPc2ox Targeting Construct

The full length *GAPc2ox1* cDNA in plasmid pST33 (a kind gift from Drs Andy Philips and Peter Hedden, Rothamsted Research) was excised with *Spe* I/*Xho* I and cloned into the *Nhe* I/*Sal* I site of pNosTerFRT-neo (Fig. 3Bi).

mgfp5-ER Targeting Cassette

The *mgfp5-ER* gene (GenBank U87974) was isolated from pBIN35S-mgfp5-ER (a kind gift from Dr Jim Haseloff, Department of Plant Sciences, University of Cambridge, UK) as a *Bam* HI/*Sac* I fragment and cloned into the *Sma* I site of pNosTerFRT-neo (Fig. 3Bii)

WARNING! *GUS* reporter cassettes are not suitable as they are able to recombine with the endogenous (chromosomal) *E. coli* gene during HR cloning.

RED Cloning Protocol

Original methods and information on how to obtain host cells can be found at the recombineering website http://recombineering.ncifcrf.gov/

A: Preparation of Electrocompetent EL250 cells

These cells will be periodically used to receive new plasmids as they are constructed and required for recombineering. It is therefore advisable to make and store a sizable batch.

- Streak out cells on LB plates and grow at 32°C. The cells are temperature sensitive and will die at 37°C
- Inoculate a single colony into 5 ml LB and grow overnight.
- Inoculate 1 ml of the overnight culture into 50 ml LB in a 500 ml flask and grow at 32°C with shaking at 200 revolutions per minute (rpm) until the cells density has reached $OD_{600} = 0.5 - 0.8$.
- Spin cells at 4°C (rotor must be pre-cooled) and wash with 5 ml ice cold sterile distilled water. Spin and discard supernatant. Repeat wash with 5 ml aliquots of ice cold SDW two more times.
- Finally resuspend cells in 500 µl of ice cold sterile distilled water and aliquot 100 µl lots into cooled 1.5 ml microfuge tubes.

NOTE : *Cells can be used immediately or re-suspended in ice cold sterile 10% (v/v) glycerol and stored at -80°C until required.*

B: Transformation of BAC F24o1 and Induction of Recombinogenic Function in EL250

- On ice, add 10 ng-100 ng of F24o1 DNA to 50 µl of competent EL250 cells. Mix by gentle pipetting and transfer to a pre-cooled 0.1 cm electroporation cuvette.

- Pulse at 1.75 kV in a Bio-Rad E. coli Gene Pulser. Immediately add 1 ml LB broth and incubate at 32°C for 1–1.5 h in a shaking incubator set at 200 rpm

- Plate cells on LB kanamycin and select for transformants. **NOTE:** *It is advisable at this stage to check the integrity of the BAC clone by restriction digestion analysis, to ensure that there have been no rearrangements.*

- Grow F24o1/EL250 cells as described in A: steps 1 – 3 except that all LB media must be supplemented with kanamycin (or relevant antibiotic) to select for the BAC. **NOTE:** *before the next step, ensure that the shaking water bath is switched on early and stabilised at 42°C ready for use and pre-warm the conical flask. An ice slurry bath must also be made ready – DO NOT USE JUST ICE – it will not cool cells fast enough.*

- For induction, transfer 10 ml of the growing culture into a pre-warmed 250 ml conical flask and incubate in the water bath at 42 °C with shaking at 200 rpm for a total of 15 min. **NOTE** : *Keep the remaining 40 ml of culture at 32 °C to act as a non-induced control.*

- Immediately after 15 min. place the flask in the ice slurry bath and swirl by hand to quickly cool down the cells. Include a similar flask with 10 ml of non-induced control cells – this will be cooled down and treated in the same way as the test cells from now on. **NOTE:** (i) *Induced cells must be used immediately as they will lose activity above 0°C. Therefore it is important to work quickly from now on. However, cells may be kept on ice for a total of 40 min. without significant loss of activity. (ii) Ensure that the centrifuge and rotor are pre-cooled to 4 °C before the next step.*

- Centrifuge the 10 ml aliquots of induced and control cells for 8 min at 5500 g and at 4 °C. Retrieve pellets and wash three times in 1 ml ice cold sterile distilled water and centrifuge as above. **NOTE**: *To save time, washing steps can be carried out in 1.5 ml microfuge tubes keeping everything ice cold and centrifuging at 4 °C for 20 seconds each time.*

- After final wash, re-suspend the cell pellet in 100 µl of ice cold sterile distilled water. This is enough for two electroporation transformation reactions.

C: AtSTM Locus Rescue from BAC F24o1 by Gap-Repair HR

Before starting: Ensure that purified and linearised gap-repair vector is available at concentrations suitable to deliver 10 to 100 ng in volumes up to 10 µl. We strongly recommend gel quantification with known standards as we find this more accurate than $OD_{260 \text{ nm}}$ measurements.

N.B. All HR experiments should be carried out with the induced and un-induced control cells in parallel.

- Using linearized gap-repair construct DNA, electroporate induced competent F24o1/EL250 cells as described in B: steps $1 - 2$.

- Select recombinants on LB supplemented with antibiotic marker for the gap-repair vector. We used pBluescriptII KS+ and therefore selected on LB ampicillin. The use of pBluescript also limits the size of insert which can be rescued and 17 kb was the largest fragment we were able to retrieve by gap repair HR cloning.

- Recover recombinant plasmids and confirm correct recombination events by restriction digestion analysis and sequencing. This is important since incorrect events may still be selectable with the antibiotic marker. **OUR RESULTS**: The number of colonies recovered was typically small $(2 - 4)$ but of these, 50% were correct. The remainder were the result of illegitimate recombination events. See Fig. 4 for our HR strategy and the result of our *AtSTM* gap-repair rescue to give the plasmid pBlueAtSTM17. **NOTE**: *Because of the large DNA fragments involved, it is advisable to use Field Inversion Gel Electrophoresis or Pulsed Field Gel Electrophoresis as appropriate for clarity in resolution.*

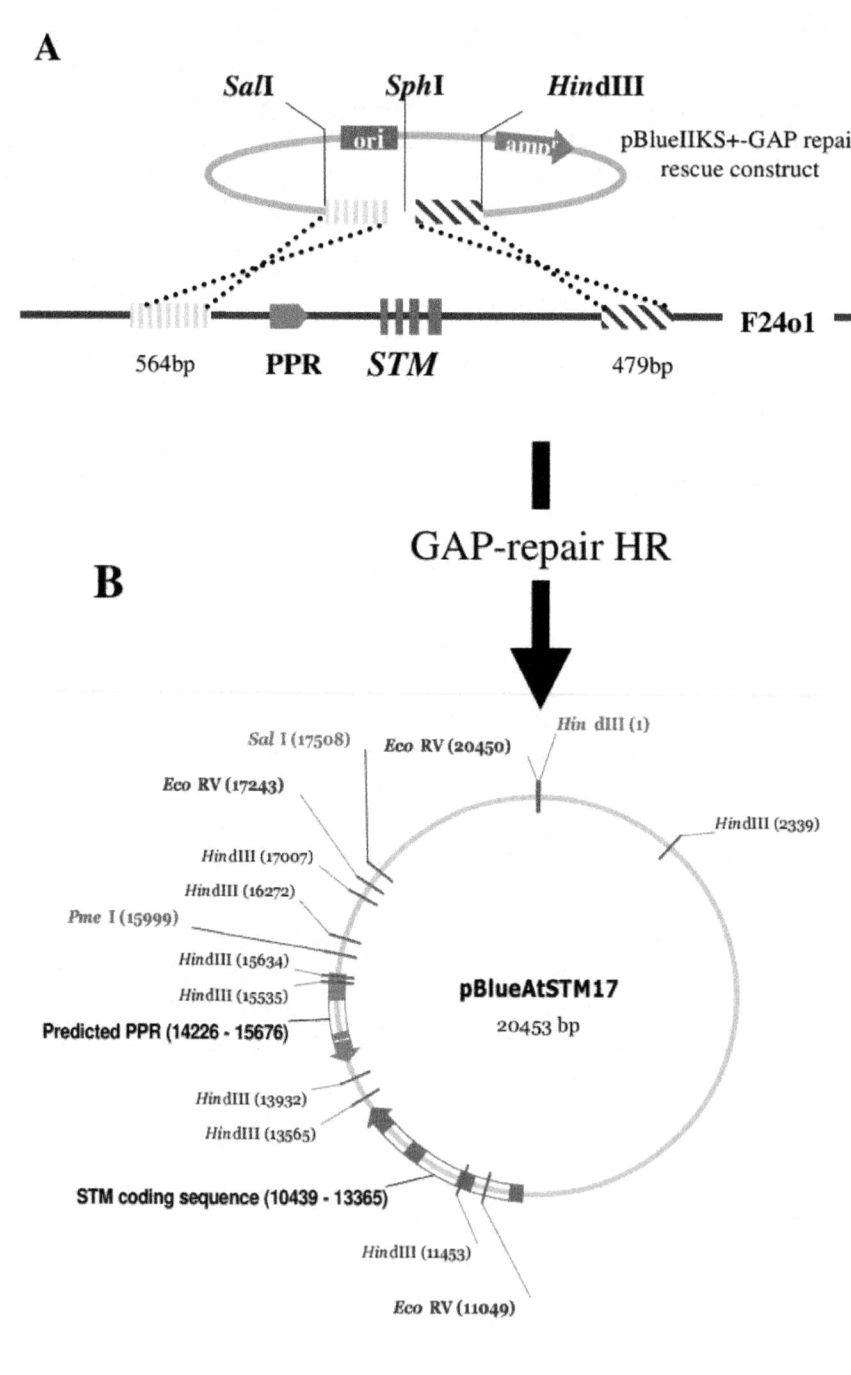

C   Restriction digestion analysis of independent GAP repair recombinations into the pBlueIIKS+ construct.

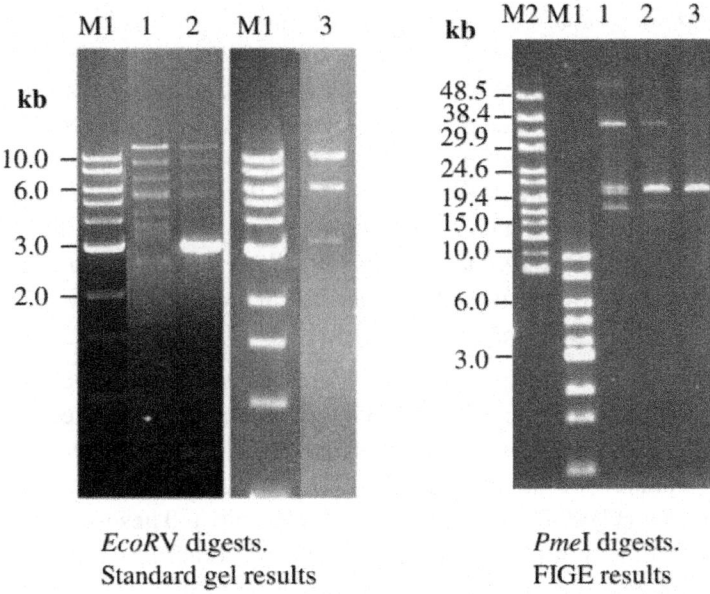

*Eco*RV digests.
Standard gel results

*Pme*I digests.
FIGE results

Lane 3 shows the restriction patterns as expected from the correct recombination event in pBlueAtSTM17 as in B

Figure 4: gap repair HR rescue of the *AtSTM* locus. Schematic representation of the HR rescue of the *AtSTM* locus into the pBluescript gap repair vector (A) and the resultant pBlueAtSTM17 construct as expected from the correct recombination event. Diagnostic *Eco* RV sites are shown in blue, *Pme* I site in red, together with their co-ordinates within pBlueAtSTM17. *Sal* I and *Hind* III sites originally engineered in the homology arms of the gap repair vector are shown in green. (B). The results of *Eco* RV and *Pme* I restriction digestion of three plasmids resulting from independent recombination events from a single experiment, showing that only the plasmid in lane 3 resulted from the correct recombination. M1 = 1 kb ladder (New England Biolabs); M2 = High Molecular Weight Marker (Invitrogen).

TROUBLE SHOOTING: Growth of un-induced colonies on selective plates suggests incomplete digestion of the gap-repair construct during linearization. However, the number of colonies should be low (we typically recovered 5 – 10 colonies from un-induced cells). Otherwise repeat with improved digestion and/or gel purification of the linearised gap repair construct.

4.   Transform the rescued plasmid into fresh EL250 cells and prepare induced competent cells as described, ready for the locus targeting experiment. We designated these cells STM17/EL250 because they contained rescued ~17.5 kb of the *AtSTM* locus.

NOTE: *For our application, we use a protoplast based direct transformation method and therefore opted to use pBluescript as the backbone for our gap repair and eventual transformation construct. However, for Agrobacterium-based systems, we recommend using a Gateway compatible Entry vector (available from Invtrogen:) as this will enable subsequent transfer of the captured, manipulated locus into a T-DNA binary destination vector for example the ones available from Plant Systems Biology* http://www.invitrogen.com(*VIB-Gent University:* http://www.psb.rug.ac.be/gateway) *or the pEarlyGates vectors, details of which can be found at the website:* http://www.biology.wustl.edu/pikaard/pEarleyGate%20plasmid%20vectors/ Table%20of%20vectors.html.

Recently, we have successfully created an *AtSTM* locus rescue vector based on the Invitrogen pENTR4 Gateway™ compatible vector in which we plan to capture/manipulate the locus as described and determine the success rate of transfer into a promoterless T-DNA destination vector pB7WG2Δ35S (based on pB7WG2 from VIB-Gent University). These are newly available resources that should enable the creation of constructs for the more generic *Agrobacterium*-mediated plant transformation systems.

WARNING!: Direct use of T-DNA vectors as gap-repair constructs in RED cloning although attractive, may prove problematic because of the common use of a limited number of identical or very similar promoter and polyA signal sequences, which if also present in the targeting cassette will result in illegitimate recombination events. For this reason, we did not attempt any experiments with T-DNA vectors, opting instead to go via the Gateway™ system as detailed above.

D: Replacement of AtSTM exon1 by In-Frame Fusion of the Promotorless GAPc2ox-FRT-neo-FRT Targeting Cassette

Before starting: The purified, *Dpn* I treated and quantified PCR amplified GOI targeting cassette should be made ready for this experiment.

1.   Using up to 100 ng of the PCR amplified *GAPc2ox1-FRT-neo-FRT* targeting cassette, electroporate induced STM17/EL250 cells as described in B: steps 1 – 2. The targeting cassette was amplified with HotStar Taq DNA polymerase (Qiagen) and primers 28001 frtlow and 2oxexon1 (0.3 µM each) in a 50 µl reaction volume. Incubation

conditions were 1 cycle 95°C for 1 min followed by 20 cycles of 94°C for 15 sec; 68°C for 4.5 min. (with a 5 sec. time increment in each cycle); followed by 1 cycle of 68°C for 10 min.

2. Select recombinants on LB kanamycin. Recover plasmids and confirm by restriction digestion analysis and sequencing. Plasmids are now ready for application in functional assays. **OUR RESULTS**: We recovered many colonies at this stage (100 s), of which ~6 % were correct by restriction digestion analysis and sequencing. For example, we typically screened between 30 and 35 colonies from which two were correct. See Fig. 5 for the results of our GOI targeting experiment. **NOTE**: *It is advisable at this stage to at least sequence across the recombination site into the GOI cassette to confirm the integrity of the gene cassette before proceeding with functional assays in transgenic plants.*

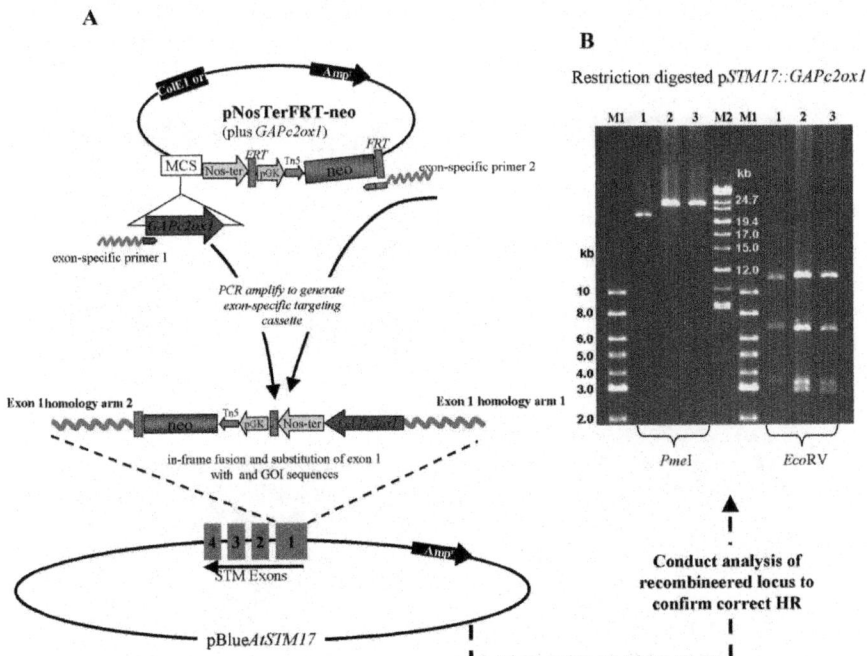

Figure 5: Replacement of *AtSTM* exon 1 by in-frame fusion of *GAPc2ox1*. A: Schematic representation of the strategy used to generate the *GAPc2ox1* cassette from pNosTerFRT-neo (plus *GAPc2ox1*) plasmid and target it into the *AtSTM* locus (captured in pBlueAtSTM17) by in-frame substitution of exon 1 sequences. B: FIGE gel results of *Pme* I and *EcoR* V digested p*STM17::GAPc2ox1* from two independent recombination events (lanes 2 and 3), clearly showing the increased size of the lin-earised construct (*Pme* I digest) and the additional fragment (*Eco* RV digest) due to

the recombination of the *GAPc2ox1* cassette. Lane 1 shows the control result with the original pBlue*AtSTM17*. M1 = 1 kb ladder (New Endgland Biolabs); M2 = High Molecular Weight Marker (Invitrogen).

TROUBLE SHOOTING: High un-induced colony numbers on selective plates suggest targeting cassette template contamination instead of recombination. Check *Dpn* I digests and use this in combination with gel purification to remove template DNA from the target cassette PCR product prior to electoporating for HR. In our experience, if the number of colonies from un-induced cells is at least 50 – 100 fold less than from the induced cells, then it was worth screening colonies from induced cells.

Our final recombineered construct was designated *AtSTM17:: GAPc2ox1* and was transformed into sugar beet guard cell protoplasts [7] from which we successfully selected transgenic callus and shoots under kanamycin selection driven by the mouse PGK promoter in the FRT-mPGK-Tn5-neo-FRT cassette (Fig. 6). We are now in the process of conducting phenotypic analyses of our *AtSTM17:: GAPc2ox1* plants.

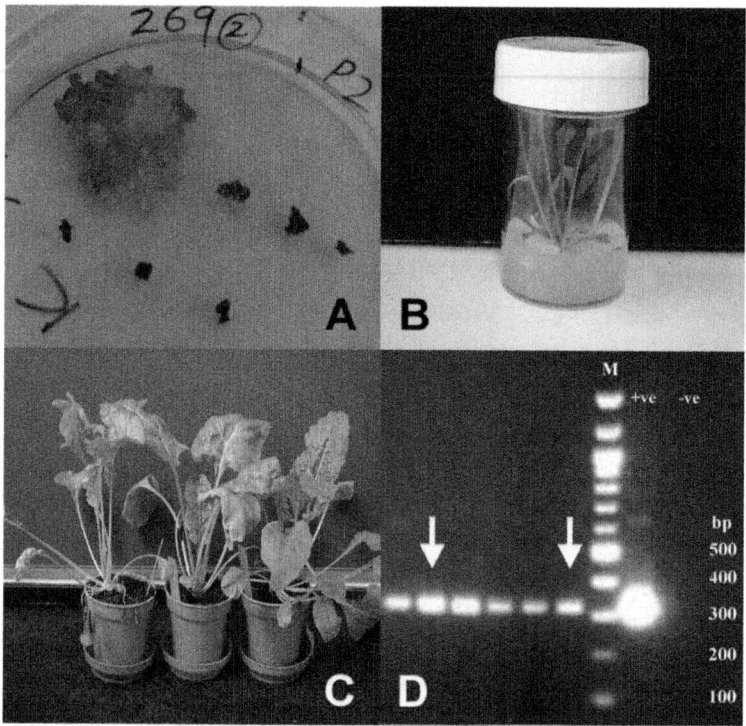

Figure 6: Kanamycin selection and regeneration of transgenic sugar beet guard cells transformed with the *pSTM17::GAPc2ox1*constructs containing the neomycin gene

driven by the mouse PGK **promoter**. A: Transgenic sugar beet callus grown from transformed sugar beet guard cell protoplasts cultured on medium supplemented with kanamycin at 100 μg ml^{-1}, clearly showing a difference between the healthy (green) callus and the dead (brown) non-transgenic callus. B: Shoots regenerated from transgenic sugar beet callus as above. C: Resultant transgenic seedlings in compost. D: Agarose gel showing the results of neomycin gene-specific PCR conducted on genomic DNA template extracted from sugar beets generated from two independent transformation events with *pSTM17::GAPc2ox1* including the kanamycin driven by the mouse PGK promoter (arrows) compared with individuals from lines transformed with constructs containing a kanamycin cassette driven the CaMV35S promoter (the remaining PCR product bands). Sequences from other regions of the *pSTM17::GAPc2ox1* construct were also detected by PCR and Southern blot hybridisation (results not shown). +ve = DNA plasmid template control, -ve = DNA template-free control; M = New England Biolabs 100 bp ladder. Diagnostic PCR for the neomycin gene sequences was performed using the primers Neo-For 5› CAG GAT GAT CTG GAC GAA GA 3› (Tm = 57.3 °C) and Neo-Rev 5› AAG AAG GCG ATA GAA GGC GA 3› (Tm = 57.3°C). The reactions (Qiagen Master Mix) contained 10 μM of each primer, and 50–100 ng genomic DNA template and were incubated at 94°C for 3 min, followed by 30 cycles of 94°C for 15 sec.; 55°C for 30 sec.; 72°C for 1 min and one cycle of 72 °C for 2 min.

COMMENTS

Manipulation of large DNA fragments to make complex constructs for functional genomics or genetic engineering for crop improvement is possible using HR cloning in *E. coli*. We have successfully used HR cloning in *E. coli* to sub-clone the *Arabidopsis thaliana SHOOTMERISTEMLESS* (*STM*) gene locus from a BAC clone into pBluescript and to replace exon 1 sequences with a Gibberellin 2-oxidase cDNA gene-of-interest cassette tightly linked to an FRT-flanked kanamycin selection marker gene. This cassette is of generic use because firstly, it can be targeted/recombineered into any locus/ destination site. Secondly, the kanamycin resistance gene is under the control of both the bacterial Tn5 promoter and the mouse phosphoglycerate kinase promoter (mPGK), which respectively allow for selection in prokaryotes and eukaryotes. We have now demonstrated the utility of the mPGK promoter for driving expression in transgenic plants and this suggests that there may well be increased scope for plant scientists to directly benefit from existing molecular genetic tools developed for application in the biomedical field.

E. coli ET- and RED-HR cloning are well established technologies within the biomedical field and they have many uses besides the creation of transformation constructs with long-range regulatory elements. The identification of regulatory elements or locus control regions located at a distance from the gene sequence can be assisted by this strategy. Point mutations, deletions or insertions, gene fusions and antisense constructs can be engineered on any BAC for functional genomics studies. The scope for plant science is further enhanced by the recently reported application of HR to convert BACs into binary vectors [8] together with (i) the availability of a BAC-based physical map of *A. thaliana*, (ii) freely available genome sequence information through the *Arabidopsis* Genome Initiative, (iii) access to rice sequence data and BAC resources through The Institute for Genomic Research (TIGR) and the Rice Genome Resource Center (RGP).

Resources for RED/ET cloning are available from Neal Copeland and Nancy Jenkins for both profit and non-profit organisations. Details can be found at the following website: http://recombineering.ncifcrf.gov/reagent_request. asp. The commercial company GeneBridges http://www.genebridges.com/ web/company/index.htm also offers reagents and a DNA engineering service.

ABBREVIATIONS

BAC: BAC = Bacterial artificial chromosome

bp = base pairs

FIGE = field inversion gel electrophoresis

GA 2ox = gibberellin 2-oxidase

GFP = gree fluorescent protein

GOI = gene of interest

HA = homology arm(s)

HR = homologous recombination

LB = Luria Bertani medium

LR = Locus rescue

mPGK = mouse phosphoglycerate kinase promoter

OD = optical density

PCR = polymerase chain reaction

ORF = open reading frame

rpm = revolutions per minute

UV = ultraviolet.

ACKNOWLEDGEMENTS

Ann Mathews, Roz Williamson and Sarah Yallop for technical assistance molecular analyses and sugar beet transformation.

The project was funded by the Biotechnology and Biological Sciences Research Council of the UK as part of the ROPA scheme.

SUPPLEMENTARY MATERIAL

Additional File 1: PCR Primer Sequences. Details of the primers used isolate and manipulate the *AtSTM*gene locus by homologous recombination in *E. coli* EL250.

Primer Sequences

STM locus-specific nucleotides are shown in red, GOI specific nucleotides in black, incorporated restriction sites in bolt italics, and nucleotides to protect the restriction site are in bold

Name	Sequence	Tm
prot H1 *Sal*I	5' **AGCT*GTCGAC*CAGACTTGTT-GAGGAAGTTCCA** 3' - primer W	69.5 °C
Sprot H1 *Sph*I	5' **AGCT*GCATGC*CAAAGCTATGGC-GTTAGAAGCA** 3' - primer X	69.5 °C
*Sph*I 5' hyp	5' **AGCT*GCATGC*TTGTATCGGATC-CGAAACTA** 3' - primer Y	66.8 °C
*Hind*III 3' hyp	5' **AGCT*AAGCTT*GTCCGT-TAGGGAAGACATCA** 3' - primer Z	66.8 °C
PGK-FRT upper	5' CCTATGCTACTCCGTCG 3'	56.7 °C
PGK-FRT lower	5' TCCCGGCGGATTTGTCCTACT-CAGGAGAGCG 3'	82.5 °C
28001 frtlow	5'TCATAGGACACATCGGACCAT-CACTATTATCCCCGGCAAAAGC-CATTGGACGGATTTGTCCTACTCAG-GAGAGCG 3'	92.3 °C
2oxexon1	5'TATAGCAAAGCCAAAGTGAATA-ATAATACTAGTGAGAGAAAGAGAA-**GATG**GTTGTTCTGTCTCAGCCAGC 3'	82.4 °C

AUTHORS' CONTRIBUTIONS

ESM-G and BG conceived of the study and participated in its design. ESM-G and LCR drafted the manuscript. LCR participated in the design of the study and carried out most of the experimental. ESM-G directed the work, participated in experimental work. All authors read and approved the final manuscript.

REFERENCES

1.   Zhang Y, Buchholz F, Muyrers JP, Stewart AF: A new logic for DNA engineering using recombination in Escherichia coli. Nat Genet. 1998, 20: 123-128.

2.   Zhang Y, Muyrers JP, Testa G, Stewart AF: DNA cloning by homologous recombination in Escherichia coli. Nat Biotechnol. 2000, 18: 1314-1317.

3.   Lee EC, Yu D, Martinez de Velasco J, Tessarollo L, Swing DA, Court DL, Jenkins NA, Copeland NG: A highly efficient Escherichia coli-based chromosome engineering system adapted for recombinogenic targeting and subcloning of BAC DNA. Genomics. 2001, 73: 56-65.

4.   Yu D, Ellis HM, Lee EC, Jenkins NA, Copeland NG, Court DL: An efficient recombination system for chromosome engineering in Escherichia coli. Proc Natl Acad Sci U S A. 2000, 97: 5978-5983.

5.   Copeland NG, Jenkins NA, Court DL: Recombineering: a powerful new tool for mouse functional genomics. Nat Rev Genet. 2001, 2: 769-779.

6.   Angrand PO, Daigle N, van der Hoeven F, Scholer HR, Stewart AF: Simplified generation of targeting constructs using ET recombination. Nucleic Acids Res. 1999, 27: e16-

7.   Hall RD, Riksen-Bruinsma T, Weyens GJ, Rosquin IJ, Denys PN, Evans IJ, Lathouwers JE, Lefebvre MP, Dunwell JM, van Tunen A, Krens FA: A high efficiency technique for the generation of transgenic sugar beets from stomatal guard cells. Nat Biotechnol. 1996, 14: 1133-1138.

8.   Takken FL, Van Wijk R, Michielse CB, Houterman PM, Ram AF, Cornelissen BJ: A one-step method to convert vectors into binary vectors suited for Agrobacterium-mediated transformation. Curr Genet. 2004, 45: 242-248.

Chapter 3

GENETIC ENGINEERING OF FIELD, INDUSTRIAL AND PHARMACEUTICAL CROPS

Hari P. Singh, Bharat P. Singh

Agricultural Research Station, Fort Valley State University, Fort Valley, GA, USA

ABSTRACT

Ability to modify plants at the genomic level by advanced molecular technology has enhanced the scope of improvements in plant traits attempted earlier through conventional breeding methods. Techniques such as genetic transformation have opened new vistas whereby functional genes, not commonly present in a particular species can be added from other species. The traits incorporated into the genetically engineered plants in the beginning were confined to those governed by dominant genes, e.g. insecticide resistance and herbicide tolerance but advancements with time now also permit the transfer of complexly inherited traits such as drought and cold tolerance. Transgenic technology is also useful in understanding gene expression and metabolic pathways which can then be used to harness the full genomic potential of the plant. This review presents a narrative on development of transgenics and their use for the improvement of field, industrial and pharmaceuticals crops. In addition, discussions are made on current status on genetically modified crops, hurdles to genetic engineering, overcoming strategies and future scope.

INTRODUCTION

The alteration or modification in an organism's genome using modern DNA technology is called genetic engineering or genetic modification. Since it involves the introduction of foreign DNA or synthetic genes into the organism of interest the resulting artifact is often referred as transgenic and or genetically modified (GM). The ability to introduce alien genes from distant species or life forms into plants has made available an entirely new and novel gene resource pool to breeders in their pursuit to improve crops for survival, productivity, and products. Transgenic crops can be generated with the use of recombinant DNA techniques which alter the crop's genetic makeup by manipulating the genome—either by introduction, deletion, substitution, or silencing of an individual gene or group of genes of interest. The functionality of transgenes has expanded with time. In the beginning, only traits that exhibited complete dominance, free of the interaction from the native plant genome or the environment, were targeted. For such traits, only one copy of the trait introduced into one of the inbred parents was required. Fortunately, some of these dominant traits such as insect resistance and herbicide tolerance provided solution to major production hurdles encountered by farmers in producing major crops of food and fiber.

Cultural practice modification offered by early transgenic varieties not only enhanced economic return to the farmer because of higher crop yields but also had multiplier effects of soil erosion prevention from reduced tillage, and reduction in environmental pollution from the residual herbicides and insecticides. Now, the scope of transgenic technology has expanded to include quantitative traits such as stress tolerance and yield improvement that necessitate the interaction of the introgressed genes with native genes engaged in the metabolic pathway for the phenotypic trait expression where environment may also influence considerably the final phenotypic expression. Stacking of introgressed genes in hybrids to create value combination of traits is also receiving considerable attention. Genetic engineering (GE) is usually resorted when improvement through conventional breeding and mutagenesis have been exhausted. Such situations arise if a desired trait is not present in the crop germplasm, the trait has proven difficult or very time consuming to improve through conventional breeding, or there is a need to remove or switch off particular genes. Commercialization of first genetically engineered crop started back in 1996 and since then it has reached new heights in its application and wide adaptability to various sectors of modern agriculture. Since 1996 to 2013 there has been tremendous increase in the acreage of genetically engineered crops. Between 1996 and 2013 there has been more than 100 fold increase in the acreage of genetically engineered crops [1] .

DEVELOPMENT OF TRANSGENIC

The ability of genetic engineering to incorporate foreign traits into plants was first exhibited in 1970s. Although approved earlier for limited sale, it was not until 1996 that the Monsanto Company (USA) got the approval to commercially market European corn borer (Ostrinia nubilalis) resistant corn (Zea mays L.), Colorado potato beetle (Leptinotarsa decemlineata) resistant potato (Solanum tuberosum L.), cotton bollworm complex (tobacco budworm—Heliothis virescence, bollworm—Helicoverpa zea, and pink bollworm—Pectinophora gossypiell) resistant cotton (*Gossypium hirsutum* L.), and a non-selective, broad spectrum herbicide glyphosate (N-phospho- nomethylglycine) tolerant soybean (*Glycine max* (L.) Merr.). The genes for all insecticidal proteins, modified cry1Ab in corn [2] , modified cry1Ac in cotton [3] , and modified cry3Ab in potato [4] derived from commonly found soil bacteria—Bacillus thuringiensis (Bt). Glyphosate tolerant soybean was developed by incorporating a bacterium glyphosate resistant EPSP (5-enolpyruvyl shikimate-3-phosphate) synthase gene [5] . The transgenic crop research developed over years can be grouped into three generations (Figure 1). Each generation represents unique thrust areas for developing transgenic crops and has contributed to the present pool of commercially adopted transgenic crops [6] .

Modern agriculture has been quick to adopt the commercial first generation transgenic crops expressing herbicide tolerance and insect resistance since these effects were clearly visible in the crop production systems. The second generation transgenic crops were designed for product quality characteristics but they did not lived up to expectations, and no commercial crop with these specific characteristics is presently in the market. However, the first approved transgenic food, Calgene's Flavr Savr tomato which reached the market in 1994, succeeded in delaying softening of the ripe fruit after harvesting but was a complete commercial failure and was withdrawn from the market in 1997 out of safety concerns. The third generation of transgenic crops has been engineered for use as biofactories or living reactors in the production of pharmaceuticals and industrial chemicals and is often referred as "molecular farming".

GENETIC ENGINEERING OF CROPS

The early and most cost-reward producing use of GE has been in the development of insecticide and pesticide resistance in field crops. A great deal of interest has currently been shown in incorporating tolerance to environmental stresses in crop cultivars in order to stabilize the yield under fluctuating environmental conditions. In addition, as enhanced nutritive value of crop has gathered much interest to combat malnutrition in developing countries and to meet the food preference of naturalists, several transgenic cultivars with fortified nutritive values have been released. Some degree of success has also been accomplished in developing crops with chemical constituent of industrial value and the use of plants as hosts for pharmaceutical products.

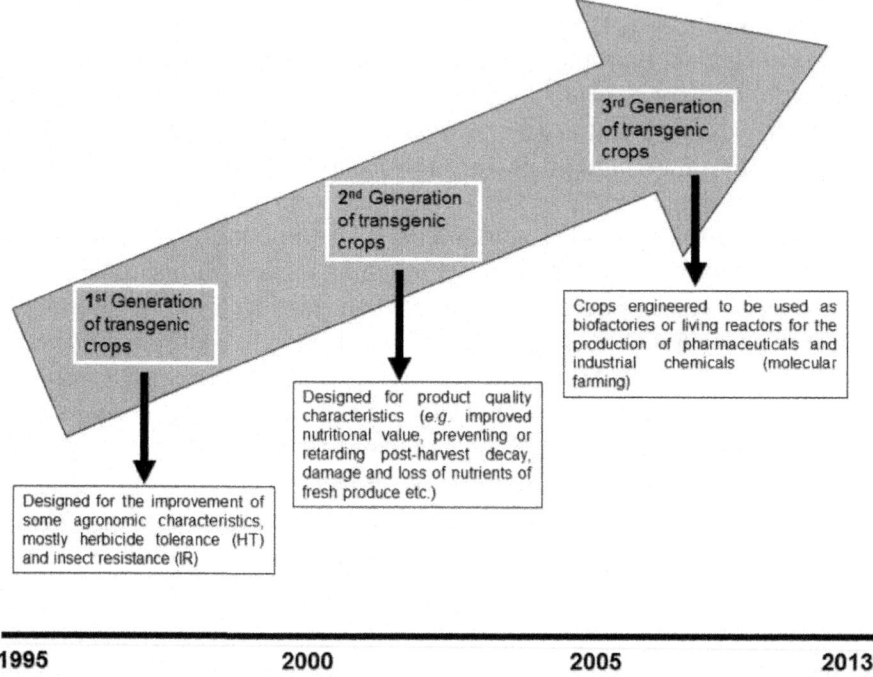

Figure 1. Development and characterization of genetically engineered crops.

Field Crops

Similar to the wide acceptance of hybrid corn due to its yield advantage more than sixty years ago, farmers in the United States now commonly plant transgenic cultivars of several field crops possessing traits that increase their value or yield potential. The typical transgenic traits approved for US crops are herbicide tolerance (bromoxynil, glufosinate, glyphosate, sulfonylurea), insect resistance (Bt kurstaki, Bt tenebrionis), virus resistance (Papaya ringspot virus, cucumber mosaic virus , zucchini yellow mosaic virus, watermelon mosaic virus, potato leaf roll virus, potato virus Y), male sterility (barnase/barstar), modified ripening (ACC synthase, ACC deaminase, SAM hydrolase, polygalacturonase), and modified oils (high lauric , myristic, oleic acids) [7] . Currently, transgenic acreage in several countries (4 - 17 different countries) for soybean, cotton, maize and canola account for 79%, 70%, 32% and 24% respectively of the total planting [1] . Table 1 lists different transgenic cultivars have been released or are at different stages of development.

The transgenes currently employed in weed management confer resistance to crops allowing use of effective broad-spectrum herbicides. The gene characterizing bacterial enzyme conferring tolerance to glyphosate has been most widely used, however, transgenes conferring tolerance to bromoxynil (Buctril), glufosinate (Liberty), and sulfonylurea (Glean) are also registered [7] . Two classes of Cry genes from Bacillus thuringiensis have been incorporated to produce Cry protein in a number of plants species to control a lipidopteran and co-lipidopteran insects [7] . The protein breaks down in the insect intestine releasing a lethal toxin called delta-endotoxin. Duke [85] enumerated three strategies for alleviating the need for herbicidal kill utilizing transgenes, namely, 1) transgenic alteration of biocontrol agents to make them more effective in managing weeds, 2) transgenes that produce a more competitive crop or provide the ability to produce natural phytotoxins (allelochemicals) and 3) cover crops that will self-destruct near the time of main crop planting alleviating the need for herbicidal kill. The disease resistances of commercial value have been incorporated into papaya (Carica papaya L.) to bestow resistance to papaya-ringspot-virus [86] and into rice (Oryza sativa L.) to give bacterial leaf blight resistance [87] .

Most of phosphorus (P) in the soil is present in insoluble form limiting its availability to the plant. However, López et al. [88] observed that engineering to overproduce citrate enhanced the ability of tobacco (Nicotiana tabacum L.) plants to use insoluble P in the soil with potential for commercial utilization. Researchers are attempting to improve photosynthesis in C_3 plants with the aim of boosting yield by achieving overexpression of C_4-cycle enzymes [89] -[91] but much work remains to accomplish this goal. In a recent review [92] , the potential of transgenic approach in improving rice production has been shown by identifying a range of traits that can be addressed to develop transgenic for commercial adoption and cultivation. Response of a plant to environmental stress can range from altered gene expression and cellular metabolism to changes in crop yields [93] . Genes have been introduced into cereals to fortify the grain nutritional value, e.g., cereals fortified with vitamin A to prevent partial or full blindness and with iron to relieve anemia due to iron deficiency in pregnant woman and children. Golden rice (*Oryza sativa* L.) is the result of the introduction of two genes from daffodils (*Narcissus pseudonarcissus* L.) and one from a microorganism resulting in increased production of vitamin A precursor beta-carotene [94] . Also in rice, genes involved in the production of an iron-binding protein and enzyme that facilitates its availability have been introduced to develop genotypes exhibiting elevated levels of iron [95] . Considerable amount of research has been devoted to enhance nutritive value of different crops with varied success. Researchers have also attempted to incorporate improved quality traits (delayed ripening, enhanced appearance, extended functionality, enhanced sweetness, and fats and oils) and nutritive value (fats and oils, protein, carbohydrate, caroteniod and vitamin E) into crops (Table 1).

Table 1. Selected examples of transgenic crops targeting a specific trait or quality for crop improvement

Scientific name	Common name	Trait/quality	Target gene product/ gene/source	Reference
Beta vulgaris	Sugar beet	Herbicide tolerance (glyphosate)	5-Enolypyruvylshikimate-3-phosphate synthase (EPSPS), CP4 strain of Agrobacterium tumefaciens	[8]
Glycine max L.	Soybean	Herbicide tolerance (glufosinate ammonium)	Phosphinothricin acetyltransferase (PAT), Streptomyces viridochromogenes	[9] [10]
		Carotenoids and vitamin E	Alpha-tocopherol	[11] [12]
		Fats and oil	Omega-3-fatty acid; increased oleic acid	[13]
		Protein	Methionine enriched glycinin	[14]
		Heat shock protein	P5CR-Increased proline accumulation	
Gossypium hirsutum L.	Cotton	Insect-resistance	cry1F gene, from Bacillus thuringiensis var Aizawa	[15]
		Increased ethanol production	ADH (Alcohol dehydrogenase)	[16]
Helianthus annuus	Sunflower	Herbicide tolerance (imidazolinone)	By selection of a naturally occurring mutant	[17]
		Fats and oil	Docosahexaenoic	[18]
Lens culinaris	Lentil	Chemically induced seed mutagenesis	Acetohydroxyacid synthase (AHAS)	[19]
Linum usitatissimum L.	Flax, Linseed	Herbicide tolerance (sulfonylurea)	Acetolactate synthase (ALS)	[20]
Brassica napus	Argentine Canola	Herbicide tolerance (glufosinate ammonium)	Phosphinothricin acetyltransferase (PAT), S. viridochromogenes	[21]

Medicago sativa	Alfalfa	Herbicide tolerance (glyphosate) (lucerne)	5-Enolypyruvylshikimate-3-phosphate synthase (EPSPS), CP4 strain of Agrobacterium tumefaciens	[22] [23]
		Drought and freezing		[24] [25]
			Sod, Mn-Sod, Mn superoxidase dismutase	
Oryza sativa	Rice	Herbicide tolerance (imidazolinone, imazethapyr)	Acetolactate synthase (ALS), ethyl methanesulfonate (EMS)	[26] [27]
		Protein	Beta-phaseolin	[28]
		Salinity and drought	HVA 1	[29] [30]
		Submergence tolerance	pdc1, Sub1	[31] [32] [33]
		Cold tolerance	Wx-control amylase synthesis, Gs2-chloroplastic glutamine synthetase	
		Metal toxicity tolerance	Parβ-Glutathione S-transferase	
Triticum aestivum	Wheat	Herbicide tolerance (imidazolinone)	Acetohydroxyacid synthase (AHAS), ALS	[34] [35]
		Glyphosate tolerance	5-Enolpyruvylshikimate-3-phosphate synthase (EPSPS), Agrobacterium tumefaciens, strain CP4	[36] [37] [38] [39] [40]
		Enhanced functionality	Protein modification of high molecular weight glutenins, flour functionality	[41] [42]
		Drought tolerance		
			DREB1A, HVA1	

Zea mays L.	Maize	Insect-resistance	cry1Ab, Bacillus thuringiensis subsp. kurstaki,	[43]
		Herbicide tolerance (glyphosate)		[44]
		Corn rootworm resistance	modified 5-enolpyruvyl shikimate-3-phosphate synthase (EPSPS)	[45]
				[46] [47]
		Drought tolerance	modified cry3A gene, E. coli	
			NF-YB, MAPK	
	Sugarcane	Insect resistance	Cry1a(b), synthetic cry1Ac, cry1Aa3, cry1Ab	[48] [49] [50] [51] [52] [53]
Lycopersicon esculentum	Tomato	Delayed ripening	1-Aminocyclopropane-1-carboxyllic acid (ACC) synthase	[54]
		Freezing tolerance		[55]
			afa3	
Cucurbita pepo	Squash	Cucumber mosiac virus (CMV), zucchini yellows mosaic (ZYMV) and watermelon mosaic virus (WMV)	Coat protein (CP)	[56]
Solanum tuberosum L.	Potato	Colorado potato beetle resistance	cry3A, Bacillus thuringiensis (subsp. Tenebrionis)	[57]
				[58]
		Carbohydrate	Amylose and amylopectin structure/ratio	[59]
		Enhanced sweetness		[60]
		Freezing	Monellin	[61]
		Heat shock protein	BetA	
			DcHSP17.7-improved cellular membrane stability and enhanced in-vitro tuberization	
	Cassava	Carbohydrate	Amylose and amylopectin structure/ratio	[62]
	Sweet potato	Protein	Essential amino acid rich protein	[63]
	Carrot	Salt tolerance (up to 400 mM)	Badh	[64]
	Cucumber	Enhanced sweetness	Thaumatin	[65]
Carica papaya	Papaya	Papaya ringspot virus (PRSV) resistant	Coat protein (CP)	[66]

Cucumis melo	Melon	Reduced accumulation of S-adenosylmethionine (SAM), and consequently reduced ethylene synthesis,	S-Adenosylmethionine hydrolase	[67]
Musa acuminata	Banana	Carbohydrate	Amylose and amylopectin structure/ratio	[68]
Prunus domestica	Plum	Plum pox virus (PPV) resistance	Coat protein (CP) gene	[69]
Nicotiana tabacum L.	Tobacco	Herbicide tolerance (bromoxynil, ioxynil) Reduced nicotine content Salinity Drought Chilling Waterlogging tolerance Metal toxicity tolerance	Nitrilase, Klebsiella pneumoniae Tobacco quinolinic acid phosphoribosyltransferase (QTPase) BetA, Mltd p5cs, TPS1, SacB ad7, Des9 ACC MsFer-Ferritin (ion storage); arsC	[70] [71] [72] [73] [55] [74] [75] [76] [77] [78] [79] [80]
Cichorium intybus	Chicory	Male sterility	Barnase ribonuclease, Bacillus amyloliquefaciens	[81]
Coffea arabica	Coffee	Resistance to pests (coffee leaf miner, coffee berry borer and nematodes)	Bt	[82] [83] [84]

Industrial Crops

The crop grown as a feedstock in the production of a commodity rather than for direct human consumption is referred to as industrial crop. The industrial crops contribute to the farm income and provide economic stimulus to rural areas. The products from industrial crops also provide a pool of substitutes for classic imports from other nations. Some examples of industrial crops are: oil palm, rape seed, soybean, safflower, peas, plantago, potato, fiber hemp, flax, guar, agave, cassava, jojoba, kenaf. Fiber crops are amongst the most common industrial crops. Some transgenic plant lines, which produce compounds for specific industrial applications, are already commercialized, and many more are at different stages of development by biotechnology companies, research institutes and universities. Progress has been slow in developing transgenic plants for important industrial uses. However, immense potential exists in employing GE for producing crops for biofuel, pulp and paper, plastics, oil and lubricants, and soil remediation.

Biofuel, Paper and Plastics

Lignocellulosic biomass constitutes an important category of feedstock whose conversion into biofuel is accretive to the environment due to its high carbon sequestration credential. However, there is a major impediment to the process of conversion of the biomass cellulose to ethanol as it is embedded within the lignin of the plant cell walls and not easily separable for direct chemical reaction. This recalcitrance adds extra and cumbersome steps to the conversion process and the release of all cell wall cellulose for utilization remains problematic. Researchers have genetically engineered the switchgrass (Panicum virgatum) to address the recalcitrance and reduced ethanol yield [96] . Overexpression of PvMYB4, a general transcriptional repressor of the phenylpropanoid/lig- nin biosynthesis pathway, has shown promising results by increasing cellulosic ethanol yield from switchgrass by 2.6-fold and a dramatic reduction of recalcitrance [96] . Genetically engineered PvMYB4-OX switchgrass can thus provide a novel system for further understanding cell wall recalcitrance. Similar studies were done on transgenic alfalfa lines where it was found that recalcitrance to both acid pretreatment and enzymatic digestion is directly proportional to lignin content. Modifying the lignin biosynthetic enzymes yielded nearly twice as much sugar from cell walls as wild-type plants [97] . Lignin modification, by genetic engineering, could bypass the need for acid pretreatment and thereby facilitate bioprocess consolidation. Moreover, induction of genes that accelerates lignin degradation can be valuable for pulp and paper production from lignocellulosic biomass as well. Laccase (benzenediol: oxygen oxidoreductase, EC 1.10.3.2) enzyme present in white rot fungi has attracted much attention because of its ability to degrade lignin [98] . Hood et al. [99] generated transgenic maize plants by employing an Agrobacterium-mediated system with fungal laccase gene (EMBL accession no. U44430) which showed highest expression in the maize embryo-preferred globulin 1 promoter and targeting of the protein to the cell wall. High oil germplasm was used to increase germination, as well as to assist in increasing expression 20-fold in five generations through breeding and selection. It was hypothesized that the high oil lines might provide substrate (i.e. oil) for the laccase-generated free radicals to act upon, thereby preventing the accumulation of free radicals that alter seed physiology, such as increased lignification. Downregulation of one of the major enzymes involved in lignin biosynthesis, 4-coumarate: coenzyme A ligase (Pt4CL1) in transgenic aspen (Populus tremuloldes), resulted in a 45% decrease in lignin with a compensation of 15% increase in cellulose, doubling the plant cellulose:lignin ratio without any change in lignin composition and without any apparent harm to plant growth, development or structural integrity [100] . Changing levels of cinnamyl alcohol dehydrogenase (CAD) have also been found to modify lignin synthesis [101] . Poplar (Populus tremula × Populus alba) with CAD antisense constructs grows similar to control trees but with an increase in the proportion of free phenolic groups in lignin facilitating solubilization and fragmentation [102] .

Poly 3-hydroxyalkanoates (PHAs) are a class of microbially produced polyesters comprising of at least 100 different PHA constituents and at least five different dedicated PHA biosynthetic pathways [103] with potential application as biodegradable plastics. Arai et al. [104] was able to transfer from Aeromonas caviae FA440 modified PHA synthase gene (phaCAc) into Arabidopsis thaliana that enabled the plant to accumulate PHA in its tissues.

Industrially Desirable Fatty Acids

Cahoon et al. [105] reported that expression of a gene from pot marigold encoding an enzyme that introduces conjugated double bonds into polyunsaturated fatty acids resulted in the accumulation of calendic acid, a novel conjugated polyunsaturated fatty acid, to amounts of 20% - 25% of the reported total soybean seed oil. Calendic acid is even more oxidatively unstable than linolenic acid, thus improving the drying properties of coating applications. However, the level of calendic acid concentration in soybean at 20% - 25% remains much lower than 55% concentration found in the marigold.

Castor (Ricinus communis L.) oil contains high levels (up to 90%) of ricinoleic acid needed for conversion to substitutes for petroleum derived lubricants, emulsifiers, inks, and nylons. Unfortunately, castor cultivation is prohibited in most countries as the seeds also contain toxin ricin. The level of ricinoleic acid achieved in tobacco and Arabidopsis have been only to the amount of 1% and 17%, respectively [97] [98] [106] [107] . Singh et al. [108] observed that although a single gene (FAH12) may regulate ricinoleic acid synthesis, its accumulation in triglycerol most likely required involvement of other genes. Napier [109] noted that, thus far, it has been difficult to attain levels of industrially desirable fatty acids in transgenic plants similar to found in the non-agrono- mic source plants.

McKeon [110] reported the ongoing efforts to enhance industrial chemical constituents in important crops. Canola with high laurate for detergent and soybean with high oleate for food and monomers have reached commercial stage, while canola with petroselenate for food and monomers, soybean with vernolate for plasticizers and coatings, cotton with low-saturates for food uses are in development.

Phytoremediation

There are many reports of transgenic plants exhibiting tolerance to varying levels of heavy metals, a trait useful for phytoremediation of contaminated soils [111] -[118] . Arabidopsis thaliana transformed by type 2 MT (metallothioneins) gene (tyMT) from cattail (Typha latifolia) exhibited an increased tolerance to Cu^{2+} and Cd^{2+} [119] . Indian mustard (Brassica juncea) plants overexpressing ATP sulphurylase were shown to have higher shoot Se concentrations and enhanced Se tolerance compared to wild type when grown in the presence of selenite [120] [121] . Family of sulfur rich peptides termed phytochelatins (PCs) are able to bind to Cd and some other heavy metals [122] and transgenic tobacco plants over expressing cysteine synthase in either the cytosol or chloroplasts were more tolerant to metals such as Cd, Se and Ni [123] . Transgenic plants have been developed with altered transporter genes with the aim to exclude a toxic metal ion, transporting the metal into the apoplastic space and vacuole where metal would be less likely to exert a toxic effect [124] . Phytoremediation uses different plant processes and mechanisms normally involved in the accumulation, complexation, volatilization, and degradation of organic and inorganic pollutants [125] . Table 2 shows categorization of different processes used by some of the model transgenic plants in phytoremediation. Most of the information available today is either from laboratory and or greenhouse experiments. Elaborated field testing is required to validate and establish the effectiveness of these transgenic plants for actual cleanup of contaminated metal sites.

Pharmaceutical Crops

Genetic transformation studies have shown the potential of producing recombinant proteins, including pharmaceuticals and industrial proteins, and other secondary metabolites in plants. Several substances have already been produced in transgenic plants and are in different stages of clinical trials (Table 3) but none of them were approved as pharmaceutical for humans until 2012. Recently, an enzyme, for treating the rare hereditary Gaucher disease (Elelyso, developed by Protalix) generated in carrot tissue became the first Plant Made Pharmaceutical (PMP) for human use to gain regulatory approval by the US FDA [140] . At present, many crops are being developed to produce drugs or biologics for the diagnosis, treatment, or prevention of diseases in human and animals. These include enzymes, hormones, anticoagulant factors, vaccines, and monoclonal antibodies targeted at a variety of disease, such as cystic fibrosis or non-Hodgkin's lymphoma [141] .

Current therapeutic proteins biosynthesis systems utilizing prokaryotes, yeasts and cultured mammalian cell mediums are handicapped in several ways [142] . Prokaryotes can only biosynthesize simple therapeutic proteins such as insulin, interferon or growth hormone as complex proteins produced in them are not always properly folded or processed for proper level of biological activity [143] . Complex proteins produced from yeasts have co- and post-translational modifications (PTMs) problems while cultured mammalian cells are difficult to scale up to large volume productions due to lack of bioreactor capacities, high operating cost, and are subject to virus or prion contamination [138] . Recent studies have suggested transgenic plants could be suitable alternatives for large scale, low cost, and safe production of complex therapeutic mammalian proteins [144] . A range of plant/ crop species have been used in production of commercial pharma/crop products. An extensive review by Saklani and Kutty [145] describes the status on plant derived compounds in clinical trials with a special emphasis on plant-based anticancer drugs.

Using plants as bioreactor can substantially alleviate capacity problem especially in cases of therapeutic antibodies production where rapid volume increases are of paramount concern. Plants also provide a wide array of species suited for use as bioreactors.

Table 2. Selected example of transgenic plants overexpressing genes for improved phytoextraction and phytovolatilization efficiency and phytodegradation potential

Target plant	Gene	Product/source	Performance	Reference
Phytoextraction efficiency				
Tobacco, rapeseed	MT2	Metallothionein—human	Enhanced Cd tolerance	[126]
Tobacco, cauliflower	MT1	Metallothionein—mouse	Tolerated 200 mM CdCl$_2$	[127]
			Tolerated 400 mM CdCl$_2$ in hydroponic medium	[128]
Arabidopsis	PsMTA	Metallothionein—pea	8× higher Cu accumulation	[129]
Indian mustard	gshl	γ-Glutamyl-cystein (E. coli)	3 - 5× higher γ -ECS and GSH levels and 90% higher shoot Cd concentrations	[130]
Poplar	gshl	γ-Glu-cys synthetase (E. coli)	Cd tolerance and increase of total sulfur in shoot	[131]
Phytovolatilization efficiency				

Arabidop-sis	merB	Organomecurial lyase-bacteria	Volatilization of up to 763 ng Hg (0) min^{-1}·g^{-1}	[132]
Yellow poplar	merApe9	Hg(II) reductase- muta-genized merA	Volatilization of 10× more mercury than WT plants	[133]
Indian mustard	SMT	Selenocysteine methyltrans-ferase A. bisulcatus	Volatilized 2.5× more Se than WT plants when sup-plied with selenate	[134]
Phytodegdration potential				
Tobacco	Onr	PETN reductase	Enhanced detoxification of nitroglycerin	[135]
Poplar	gshl	y-ECS	Elevated herbicide toler-ance and rapid herbicide degradation	[136]
Indian mustard	gshI and gshII	y-ECS and GS	Enhanced tolerance and 2 - 12× increase in nonprotein thiol level	[137]
Rice, Potato	P450C-YP	Mammalian cytochrome	Enhanced detoxification and cross tolerance toward several herbicides	[138] [139]

Table 3. Plant made pharmaceuticals for human use at different stages of regulatory approval

Product	Crop	Class	Applica-tion	Status	Organiza-tion	Refer-ence
Glucocer-ebrosidase ELELYSO/ UPLYSO	Carrot cell culture	Therapeu-tic enzyme	Gaucher's disease	US FDA approved, May, 2012	Protalix	[165]
Alpha-galactosidase (PRX-102)	Carrot cell culture	Therapeu-tic enzyme	Fabry's disease	Phase I/II	Protalix	[166]
Acetylcholes-terase (PRX-105)	Carrot cell culture	Therapeu-tic enzyme	Biodefense	Phase I	Protalix	[167]
Apo-A1Mi-lano	Safflower	Therapeu-tic protein	Cardio-vascular disease	Preclinical	SemiBio-Sys Genet-ics	[168]
Human serum albumin	Flax	Therapeu-tic protein	Main-tenance of blood plasma pressure	Preclinical	Agragen	[169]

A recombinant protein yield of up to 20 kg/ha have been reported using tobacco, corn, soybean or alfalfa [146] [147]. Newly developed plant expression systems in duckweed, moss, algae, and higher plant suspension-cultured cells offer the opportunity of fast turnover and high yield molecular farming in highly contained and completely controlled environment [142] . Unlike yeasts, transgenic plants possess the ability to carry out most PTMs needed to produce complex proteins such as plasma proteins, antigens, hormones, cytokines, enzymes and antibodies [144] [148] . In most cases, several PTMs including proteolytic cleavage(s), oligomerization and glycosylation are required to obtain therapeutic proteins of proper biological activity, pharmacokinetics, stability, and solubility [142] . Expression levels for therapeutic proteins in plants remain low and currently several approaches including codon optimization [149] -[151] , RNA silencing [152] [153] , targeted secretion of recombinant proteins by the roots [154] [155] , stored into seed endosperm [156] and seed oilbodies [157] , production in chloroplast [158] are being experimented to improve yields. Difference in the protein molecule N-glycans produced by plants from mammalian glycoproteins may also trigger immune responses in humans and induce their fast clearance from the bloodstream [159] [160] . Strategies to obtain plant-derived antibodies with human-compatible carbohydrate profiles include their retention in the endoplasmic reticulum [161] [162] or alternatively transformation of plants with mammalian glycosyltransferase [163] [164] . As a result of these advances, plants have emerged as a potential safe and cost-effective alternative to microbial or mammalian expression systems especially for large quantity multimeric recombinant proteins.

CURRENT STATUS OF GENETICALLY ENGINEERED CROPS

In the United States, the Animal and Plant Health Inspection Service (APHIS) regulates the development and release of transgenic crops. APHIS classifies phenotypic traits under AP (agronomic properties), BR (bacterial resistance), FR (fungal resistance), HT (herbicide tolerance), IR (insect resistance), MG (marker gene), NR (nematode resistance), OO (others), PQ (product quality), and VR (virus resistance) categories. APHIS issued a total of 2192 permits for different phenotypic traits in 2013. Other traits (OO), herbicide tolerance (HT) and agronomic properties (AP) topped the category by contributing 34.53, 21.94 and 20.12 percent [170] respectively (Figure 2). Permit for other traits also topped in year 2014 with more than 75% of total permit issued. One permit can contain multiple phenotypes of phenotype categories while each phenotype category may include one to several traits. The FAO data [171] on developing countries showed more than thousand different GMOs under various stages of commercialization (Table 4).

Since 1996 the increase in acreage of transgenic crops in industrial countries has moderated but the pace of planting these crops in developing countries has accelerated (Figure 3). The total area planted in genetically engineered crops has increased from 1.7 million hectares in 1996 to 175 million hectares in 2013 reflecting a remarkable increase global hectarage of genetically engineered crops by 100 folds. In 2013, transgenic crops were grown in a total of 27 countries where the top ten countries each grew more than 1 million hectares. A record 18 million farmers, in 27 countries, have planted 175 million hectares (432 million acres) in 2013 [1] . If a trend like this continues, it is expected that 40 or more countries will adopt genetically engineered crops by 2015, the final year of the second decade of commercialization.

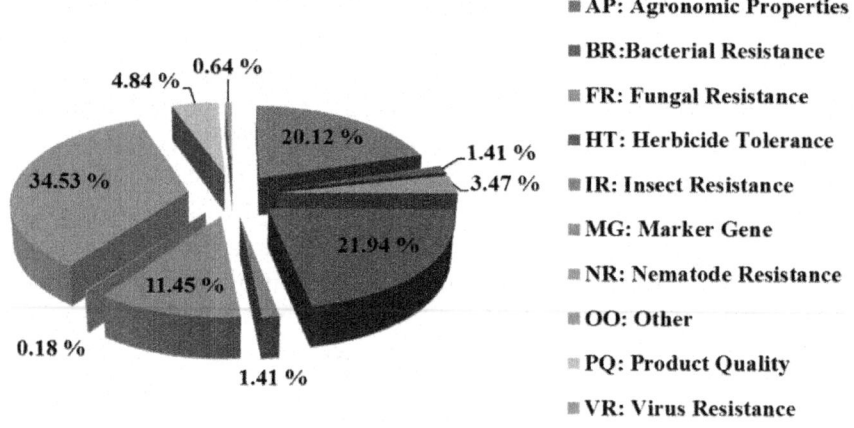

Figure 2. Total permits issued by phenotype for the transgenic crops in the United States during year 2013.

Table 4. Number of genetically modified organisms (GMOs) in developing countries under different stages of development

Region/(GMOs)		Experimental phase	Field trial	Commercial phase	Not specified
Asia (679)		453	119	33	74
Africa (85)		39	36	10	-
Latin America and Caribbean (306)		99	185	15	7
Europe (28)		18	6	2	2
Near East (51)		31	16	2	2
Grand total	1149	640	362	62	85

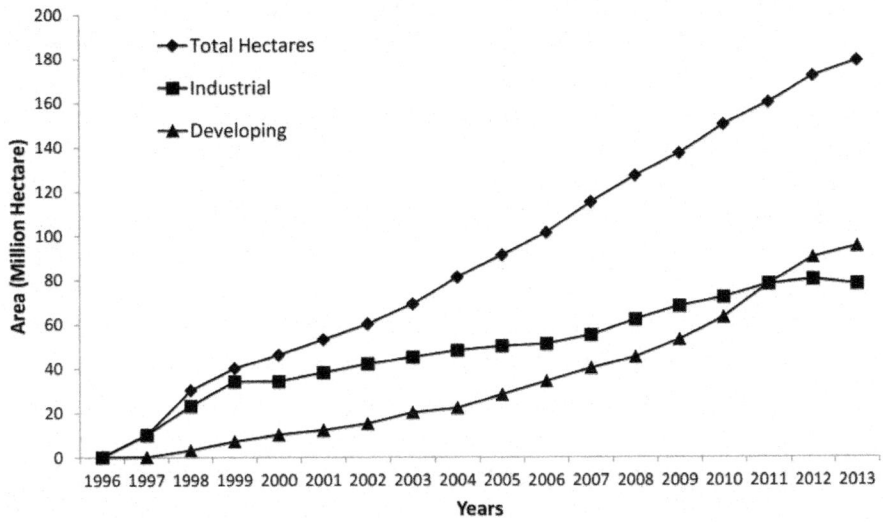

Figure 3. Trend of global growth in acreage of genetically engineered crop production between developing and industrial countries (adapted from Reference [1]).

HURDLES TO GENETIC ENGINEERING AND OVERCOMING STRATEGIES

The rapid commercialization of transgenic technology is not without controversy. In addition to public suspicion or outright opposition, the main scientific concerns regarding deployment of transgenic crops include potential harmful effects to non-target organisms, gene flow into related wild species and persistence of gene products in the environment.

Transgenic technology is still in inception and faces a number of challenges. To start with, current methods of gene insertion are not precise to control the location and number of copies of the gene inserted. As the location has considerable influence on gene expression, if not placed suitably, "silencing" of the inserted gene can occur or inserted gene can silence the native genes. Transgenic traits are not always stable over generation and such instability could be deleterious to fertility, yield or other parameters of plant fitness or usefulness. Oftentimes, the inserted DNAs contain multiple stacked genes with the potential to cause undesirable interactions in the host plant. Pleiotropy is a common form of gene expression. Thus a gene known for a trait deemed desirable may produce unintended harmful side effects. King et al. [172] noted that glyphosate applied for weed control in soybean decreased biomass and seed yields when soil moisture was limiting.

There is fear that year-after year production of Bt crops will produce pest population resistant to cry proteins. This apprehension is appropriate as it has happened in the past after an insecticide was applied repetitively. The only way to overcome the threat from the acquired resistance by insects is to infuse into the plant new defense mechanism or build multiple barriers of defense. The incorporation of two Bt genes, cry1Ac and cry2Ab2 in cotton variety Bollgard II is an example of complex defense barrier through stacking of two resistance genes with different modes of action making it difficult for the pests to develop resistance to two proteins simultaneously. Eventually though, invasion of the barrier by the pests should be expected, and thus as with conventional breeding, transgenic breeding will also have to remain a continuous process.

The public and scientific debate on the long-term consequences of unnecessary DNA-sequences introduced into the plant during the transformation process continues. It is feared that widespread use of antibiotic-resistant genes for markers could lead to the evolution of bacterial varieties resistant to antibiotics. Hohn et al. [173] has reviewed strategies employed to eliminate selectable marker recombinant DNA-sequences from the transgenic plants. Zuo et al. [174] effectively removed marker genes from transgenic plants using the Cre/loxP site-specific recombinase. Some transformation systems, now, can be performed with any selectable marker systems [175] .

Consumers rightfully should have a choice to consume or not to consume food grown out of transgenic crops. The separation of a crop produced by both transgenic and non-transgenic seeds on the same farm can be costly and bear chance of mix-up and resulting costly food recall similar to Starlink corn episode. Keenan and Stemmer [176] suggest transgene deletion from the food portion (grain, tuber) of the transgenic plants with appropriate recombinase to eliminate the need for identity preservation of food produced from non-transgenic crop varieties. The recombinase protein itself should also be rapidly degradable. Regulatory guidelines as to the acceptable level of transgenic DNA "residue" in food should assure the public safeguard.

There is also concern that pollen from a transgenic variety will drift and cross-fertilize the nontransgenic variety of the same crop in the vicinity or cross between herbicide resistant transgenics and related wild species will create "superweed". Ways to prevent transgene outflow could consist of measures to ensure its inheritance only through maternally derived genome by engineered male sterility or introduction of transgene into the chloroplast rather than nuclear genome [168] [169] [177] [178] , but these techniques have potential pitfalls. Male sterility will cause problems for farmers intending to save seeds for the next crop. The transfer of 0.1% - 0.5% of chloroplast traits via pollen [179] in tobacco as well as parental inheritance of chloroplast DNA in other higher plants has been observed [180] [181] . In addition, Huang et al. [182] observed that transgenes can migrate from the chloroplast to the nuclear genome. Singh et al. [183] described a strategy based on epigenetic inheritance (imprinting) and post-transcriptional gene silencing (PTGS)/RNA interference (RNAi) that would allow all seeds from self-pollinated transgenic plants to be harvested and re-sown, without the need for specific treatments, while retaining all of the transgenes present in the parent while preventing outcrossing via either male or female gametophyte.

FUTURE SCOPE

The overall prospects of acceptance of transgenic crops and products will depend on positive public perception of this technology, especially as the agrobiotech industry complies with regulations, conducts rigorous research on biosafety and successfully completes field trials of these crops. Like any other newly developed technology, transgenic crop agrobiotechnology industry faces its own unique challenges and hurdles, especially relating to consumer concerns on health risks and environmental safety and barriers to world-wide trade. Long term effects of GE foods must be rigorously studied but guidelines and regulations for field-testing and marketing of GM products likewise must be clearly defined to remove ambiguity and potential law suits.

There are numerous technological challenges that must be overcome while attempting to introduce specific traits into a crop. Furthermore, the high cost (approximately US$100 - 136 million [184] [185]) of developing and obtaining authorization for the commercialization of a transgenic event limits the development of transgenic crops only for selecting traits of wide interest. Agro-based companies like Monsanto, Dupont, Syngenta, Dow Agro, and Bayer Crop Science have daunting task of generating profit for share holders while they are also sensitive to the farming community and have a humanistic, consumer-oriented approach.

On positive note, numerous successes have been documented especially with regard to agronomic traits like herbicide resistance, pest resistance and drought tolerance. The high adoption rate of genetically engineered crops and controlled field trials and research in many developing countries and some EU countries are optimistic developments. The need to feed a growing world population, enhancing ability of crops to withstand climate change, and public preference for plant based industrial and pharmaceutical products should drive further research in GM plants and their production worldwide. Successful commercialization and marketing of transgenic crops and products would require mutual understanding and implementation of international standards and trade policies among nations.

REFERENCES

1. James, C. (2013) Global Status of Commercialized Biotech/GM Crops: 2013. ISAAA Brief No. 46, ISAAA, Ithaca, NY.

2. Carozzi, N. and Koziel, M.G. (1997) Transgenic Maize Expressing a Bacillus thuringiensis Insecticidal Protein for Control of European Corn Borer. In: Carozzi, N. and Koziel, M.G., Eds., Advances in Insect Control: The Role of Transgenic Plants, Taylor and Francis, London, 63-74.

3. Perlak, F.J., Deaton, R.W., Armstrong, T.A., Fuchs, R.L., Sims, S.S., Greenplate, J.T. and Fischhoff, D.A. (1990) Insect Resistant Cotton Plants. Nature Biotechnology, 8, 939-943. http://dx.doi.org/10.1038/nbt1090-939

4. Perlak, F.J., Stone, T.B., Muskopf, Y.M., Peterson, L.J., Parker, G.B., McPherson, S.A., et al. (1993) Genetically Improved Potatoes: Protection from Damage by Colorado Potato Beetles. Plant Molecular Biology, 22, 313-321. http://dx.doi.org/10.1007/BF00014938

5. Padgette, S.R., Kolacz, K.H., Delannay, X., Re, D.B., LaVallee, B.J., Tinius, C.N., et al. (1995) Development, Identification, and Characterization of Glyphosate-Tolerant Soybean Line. Crop Science, 35, 1451-1461.http://dx.doi.org/10.2135/cropsci1995.0011183X003500 050032x

6. Naranjo, M.A. and Vicente, O. (2008) Transgenic Plants for the Third Millennium Agriculture. Bulletin of UASVM, Horticulture, 65, 38-43.

7. Clark, D., Klee, H. and Dandekar, A. (2004) Despite Benefits, Commercialization of Transgenic Horticultural Crops Lags. California Agriculture, 58, 89-98.http://dx.doi.org/10.3733/ca.v058n02p89

8. Monsanto (2004) http://ceragmc.org/index.php?action=gm_crop_databa se&mode=ShowProd&data=GTSB77&frmat=LONG

9. Bayer Crop Science (Aventis CropScience(AgrEvo) (1996)http://ceramc. org/index.php/GmCropDatabaseEvent/event/60

10. Romer, S., Fraser, J., Kiano, C., Shipton, C., Misawa, W., Schuch, W. and Bramley, P.M. (2000) Elevation of the Pro- vitamin A Content of Transgenic Tomato Plants. Nature Biotechnology, 18, 666-669. http://dx.doi.org/10.1038/76523

11. Kinney, A.J. (1996) Designer Oils for Better Nutrition. Nature Biotechnology, 14, 946.http://dx.doi.org/10.1038/nbt0896-946

12. Lui, K.S. and Brown, E.A. (1996) Enhancing Vegetable Oil Quality through Plant Breeding and Genetic Engineering. Food Technology, 50, 67-71.

13. Kim, C.S., Kamiya, S., Sato, T., Utsumi, S. and Kito, M. (1990) Improvement of Nutritional Value and Functional Properties of Soybean Glycinin by Protein Engineering. Protein Engineering, Design and Selection, 3, 725-731.http://dx.doi.org/10.1093/protein/3.8.725

14. Malik, M.K., Solvin, J.P., Hwang, C.H. and Zimmerman, J.L. (1999) Modified Expression of a Carrot Small Heat Shock Protein Genes, Hsp17.7, Results in Increased or Decreased Thermo Tolerance. Plant Journal, 20, 89-99. http://dx.doi.org/10.1046/j.1365-313X.1999.00581.x

15. Dow Agro Sciences LLC (2004) http://ceragmc.org/index.php?action=gm_crop_database&mode=ShowProd&data=281-24-236

16. Dennis, E.S., Dolferus, R., Ellis, M., Rahman, M., Wu, Y., Hoeren, F.U., et al. (2000) Molecular Strategies for Improving Waterlogging Tolerance in Plants. Journal of Experimental Botany, 51, 89-97. http://dx.doi.org/10.1093/jexbot/51.342.89

17. BASF (2005) http://ceragmc.org/index.php?hstIDXCode%5B%5D=30&auDate1=&auDate2=&action=gm_crop_database&mode=Submit

18. Knutzon, D. (1999) Polyunsaturated Fatty Acids in Plants (International Patent Publication No. WO 99/64614, Date Filed: 6/10/1999). World Intellectual Property Organization, Geneva.

19. BASF (2004) http://ceragmc.org/index.php?hstIDXCode%5B%5D=49&auDate1=&auDate2=&action=gm_crop_database&mode=Submit

20. University of Saskatchewan, Crop Dev. Centre (1999) http://ceragmc.org/index.php?hstIDXCode%5B%5D=4&gType%5B%5D=&auDate1=&auDate2=&action=gm_crop_database&mode=Submit

21. Bayer CropScience (Aventis CropScience(AgrEvo) (1998) http://ceragmc.org/index.php?action=gm_crop_database&mode=ShowProd&data=T45+%28HCN28%29

22. Monsanto (2004) http://ceragmc.org/index.php?hstIDXCode%5B%5D=18&auDate1=&auDate2=&action=gm_crop_database&mode=Submit

23. Hightower, R., Baden, C., Penzes, E., Lund, P. and Dunsmuir, P. (1991) Expression of Antifreeze Proteins in Transgenic Plants. Plant Molecular Biology, 17, 1013-1021.http://dx.doi.org/10.1007/BF00037141

24. McKersie, B.D., Bowley, S.R. and Jones, K.S. (1999) Winter Survival of Transgenic Alfalfa Overexpressing Superoxide Dismutase. Plant Physiology, 119, 839-848.http://dx.doi.org/10.1104/pp.119.3.839

25. McKersie, B.D., Chen, Y., de Beus, M., Bowley, S.R., Bowler, C., Inze, D., D'Halluin, K. and Botterman, J. (1993) Superoxide Dismutase Enhances Tolerance for Freezing Stress in Transgenic Alfalfa (Medicago

sativa L.). Plant Physiology, 103, 1155-1163.http://dx.doi.org/10.1104/pp.103.4.1155

26. BASF (2002) http://ceragmc.org/index.php?action=gm_crop_database&mode=ShowProd&data=CL121%2C+CL141%2C+CFX51

27. Zheng, A., Sumi, K., Tanaka, K. and Murai, N. (1995) The Bean Seed Storage Protein β-Phaseolin Is Synthesized, Processed and Accumulated in the Vacuolar Type-II Protein Bodies of Transgenic Rice Endosperm. Plant Physiology, 109, 777-786.

28. Xu, D., Duan, X., Wang, B., Hong, B., Ho, T.H.D. and Wu, R. (1996) Expression of a Late Embryogenesis Abundant Protein Gene, HVA1, from Barley Confers Tolerance to Water Deficits and Salt Stress in Transgenic Rice. Plant Physiology, 110, 249-257.

29. Qunimio, C.A., Torrizo, L.B., Setter, T.L., Ellis, M., Grover, A., Abrigo, E.M., et al. (2000) Enhancement of Submergence Tolerance in Transgenic Rice Overproducing Pyruvate Decarboxylase. Journal of Plant Physiology, 156, 516- 521.http://dx.doi.org/10.1016/S0176-1617(00)80167-4

30. Xu, K., Xu, X., Fukao, T., Canlas, P., Maghirang-Rodriguez, R., Heuer, S., et al. (2006) Sub1A Is an Ethylene-Response-Factor-Like Gene that Confers Submergence Tolerance to Rice. Nature, 442, 705-708. http://dx.doi.org/10.1038/nature04920

31. Hirano, H.Y. and Sano, Y. (1998) Enhancement of Wx Gene Expression and the Accumulation of Amylose in Response to Cool Temperatures during Seed Development in Rice. Plant & Cell Physiology, 39, 807-812. http://dx.doi.org/10.1093/oxfordjournals.pcp.a029438

32. Hoshida, H., Tanaka, Y., Hibino, T., Hayashi, Y., Tanaka, A. and Takabe, T. (2000) Enhance Tolerance to Salt Stress in Transgenic Rice over Expression Chloroplast Glutamine Synthetase. Plant Molecular Biology, 43, 103-111.http://dx.doi.org/10.1023/A:1006408712416

33. Conway, G. and Toenniessen, G. (1999) Feeding the World in the Twenty-First Century. Nature, 402, C55-C58. http://dx.doi.org/10.1038/35011545

34. BASF (2007) http://ceragmc.org/index.php?action=gm_crop_database&mode=ShowProd&data=BW7

35. Monsanto (2004) http://ceragmc.org/index.php?action=gm_crop_database&mode=ShowProd&data=MON71800

36. Shewry, P., Tatham, A., Barro, F., Barcelo, P. and Lazzeri, P. (1995) Biotechnology of Breadmaking: Unraveling and Manipulating the Multi-Protein Gluten Complex. Nature Biotechnology, 13, 1185-1190. http://

dx.doi.org/10.1038/nbt1195-1185

37.   Blechl, A. and Anderson, O. (1996) Expression of a Novel High-Molecular-Weight Glutenin Subunit Gene in Transgenic Wheat. Nature Biotechnology, 14, 875-879.http://dx.doi.org/10.1038/nbt0796-875

38.   Barro, F., Rooke, L., Bekes, F., Gras, P., Tatham, F.R., Lazzeri, P., et al. (1997) Transformation of Wheat with High Molecular Weight Subunit Genes Results in Improved Functional Properties. Nature Biotechnology, 15, 1295-1299.http://dx.doi.org/10.1038/nbt1197-1295

39.   Shewry, P., Tatham, A. and Lazzeri, P. (1997) Biotechnology of Wheat Quality. Journal of the Science of Food and Agriculture, 73, 397-406. http://dx.doi.org/10.1002/(SICI)1097-0010(199704)73:4<397::AID-JSFA758>3.0.CO;2-Q

40.   Vasil, I. and Anderson, O. (1997) Genetic Engineering of Wheat Gluten. Trends in Plant Science, 2, 292-297. http://dx.doi.org/10.1016/S1360-1385(97)89950-5

41.   Pellegrineschi, A., Reynolds, M., Pacheco, M., Bitro, R.M., Almeraya, R., Yamaguchi-Shinazaki, K. and Hoisington, D. (2004) Stress Induced Expression in Wheat of the Arabidopsis thaliana DREB1A Gene Delays Water Stress Symptoms under Greenhouse Conditions. Genome, 47, 493-500. http://dx.doi.org/10.1139/g03-140

42.   Sivamani, E., Bahieldin, A., Wraith, J.M., Al-Niemi, T., Dyer, W.E., Ho, T.H.D. and Qu, R.D. (2000) Improved Biomass Productivity and Water Use Efficiency under Deficit Conditions in Transgenic Wheat Constitutively Expressing the Barley HVA1 Gene. Plant Science, 155, 1-9. http://dx.doi.org/10.1016/S0168-9452(99)00247-2

43.   Syngenta Seeds, Inc. (1995) http://ceragmc.org/index.php?action=gm_crop_database&mode=ShowProd&data=176

44.   Syngenta Seeds, Inc. (1996) http://ceragmc.org/index.php?action=gm_crop_database&mode=ShowProd&data=GA21

45.   Syngenta Seeds, Inc. (2007) http://ceragmc.org/index.php?action=gm_crop_database&mode=ShowProd&data=MIR604

46.   Shou, H., Bordallo, P. and Wang, K. (2004) Expression of the Nicotiana Protein Kinase (NPK1) Enhanced Drought Tolerance in Transgenic Maize. Journal of Experimental Botany, 55, 1013-1019. http://dx.doi.org/10.1093/jxb/erh129

47.   Xiao, B.Z., Chen, X., Xiang, C.B., Tang, N., Zhang, Q.F. and Xiong, L.Z. (2009) Evaluation of Seven Function- Known Candidate Genes for Their Effects on Improving Drought Resistance of Transgenic Rice under

Field Conditions. Molecular Plant, 2, 73-83.http://dx.doi.org/10.1093/mp/ssn068

48. Arencibia, A., Vázquez, R.I., Prieto, D., Téllez, P., Carmona, E.R., Coego, A.H.L., et al. (1997) Transgenic Sugarcane Plants Resistant to Stem Borer Attack. Molecular Breeding, 3, 247-255. http://dx.doi.org/10.1023/A:1009616318854

49. Braga, D.P.V., Arrigoni, E.D.B., Burnquist, W.L., Silva Filho, M.C., Ulian, E.C. and Hogarth, D.M. (2001) A New Approach for Control of Diatraea saccharalis (Lepidoptera: Crambidae) through the Expression of an Insecticidal CryIa(b) Protein in Transgenic Sugarcane. Proceedings of International Society of Sugarcane Technologists, 24, 331-336.

50. Braga, D.P.V., Arrigoni, E.D.B., Silva Filho, M.C. and Ulian, E.C. (2003) Expression of the Cry1Ab Protein in Genetically Modified Sugarcane for the Control of Diatraea saccharalis (Lepidoptera: Crambidae). Journal of New Seeds, 5, 209-221.http://dx.doi.org/10.1300/J153v05n02_07

51. Weng, L.X., Deng, H.H., Xu, J.L., Li, Q., Wang, L.H., Jiang, Z.D., et al. (2006) Regeneration of Sugarcane Elite Breeding Lines and Engineering of Strong Stem Borer Resistance. Pest Management Science, 62, 178-187. http://dx.doi.org/10.1002/ps.1144

52. Kalunke, R.M., Kolge, A.M., Babu, K.H. and Prasad, D.T. (2009) Agrobacterium Mediated Transformation of Sugarcane for Borer Resistance Using Cry 1Aa3 Gene and One-Step Regeneration of Transgenic Plants. Sugar Tech, 11, 355-359.http://dx.doi.org/10.1007/s12355-009-0061-1

53. Arvinth, S., Arun, S., Selvakesavan, R.K., Srikanth, J., Mukunthan, N., Kumar, P.A., et al. (2010) Genetic Transformation and Pyramiding of Aprotinin-Expressing Sugarcane with cry1Ab for Shoot Borer (Chilo infuscatellus) Resistance. Plant Cell Reports, 29, 383-395.http://dx.doi.org/10.1007/s00299-010-0829-5

54. DNA Plant Technology Corporation (1994) http://ceragmc.org/index.php?action=gm_crop_database&mode=ShowProd&data=1345-4

55. Kishor, P.B.K., Hong, Z., Miao, G.H., Hu, C.A. and Verma, D.P.S. (1995) Overexpression of δ-Pyrroline-5-Carboxylate Synthetase Increases Proline Production and Confers Osmotolerance in Transgenic Plants. Plant Physiology, 108, 1387-1394.

56. Asgrow (USA) Seminis Vegetable Inc. (1994) http://ceragmc.org/index.php?action=gm_crop_database&mode=ShowProd&data=CZW-3

57. Monsanto (1996) http://cera-gmc.org/index.php?action=gm_crop_database&mode=ShowProd&data=ATBT04-6%2C+ATBT04-

27%2C+ATBT04-30%2C+ATBT04-31%2C+ATBT04-36%2C+SPBT02-5%2C+SPBT02-7

58. Visser, R., Somhorst, I., Kuipers, G., Ruys, N., Feenstra, W. and Jacobsen, E. (1991) Inhibition of the Expression of the Gene for Granule-Bound Starch Synthase in Potato by Antisense Constructs. Molecular and General Genetics, 225, 289-296.http://dx.doi.org/10.1007/BF00269861

59. Pennarubia, L., Kim, R., Giovannoni, J., Kim, S. and Fischer, R. (1992) Production of the Sweet Protein Monellin in Transgenic Plants. Nature Biotechnology, 10, 561-564.http://dx.doi.org/10.1038/nbt0592-561

60. Holmberg, N. and Bulow, L. (1998) Improving Stress Tolerance in Plants by Gene Transfer. Trends in Plant Science, 3, 61-66. http://dx.doi.org/10.1016/S1360-1385(97)01163-1

61. Ahn, Y.J. and Zimmerman, L. (2006) Introduction of Carrot HSP17.7 into Potato (Solanum tuberosum L.) Enhances Cellular Membrane Stability and Tuberization in Vitro. Plant, Cell and Environment, 29, 95-104. http://dx.doi.org/10.1111/j.1365-3040.2005.01403.x

62. Visser, R., Suurs, L., Steeneken, P. and Jacobsen, E. (1994) Some Physicochemical Properties of Amylose-Free Potato Starch. Starch/Stärke, 49, 443-448.http://dx.doi.org/10.1002/star.19970491104

63. Katsube, T., Kurisaka, N., Ogawa, M., Maruyama, N., Ohtsuka, R., Utsumi, S. and Takaiwa, F. (1999) Accumulation of Soybean Glycinin and Its Assembly with Glutelins in Rice. Plant Physiology, 120, 1063-1074. http://dx.doi.org/10.1104/pp.120.4.1063

64. Kumar, S., Dhingra, A. and Daniell, H. (2004) Plastid Expressed Betaine Aldehyde Dehydrogenase Gene in Carrot Cultured Cells, Roots and Leaves Confers Enhanced Salt Tolerance. Plant Physiology, 136, 2843-2854. http://dx.doi.org/10.1104/pp.104.045187

65. Scwacka, M., Krzymowskab, M., Kowalczyk, M.E. and Osuch, A. (1999) Transgenic Cucumber Plants Expressing the Thaumatin Gene. In: Bielecki, S., Tramper, J. and Polak, J., Eds., Food Biotechnology. Proceedings of an International Symposium, Elsevier, Amsterdam, 43-48.

66. Cornell University (1997) http://ceragmc.org/index.php?action=gm_crop_database&mode=ShowProd&data=55-1%2F63-1

67. Agritope Inc. (1999) http://ceragmc.org/index.php?hstIDXCode%5B%5D=16&auDate1=&auDate2=&action=gm_crop_database&mode=Submit

68. Schwall, G., Safford, R., Westcott, R., Jeffcoat, A., Tayal, A., Shi, Y.C.,

et al. (2000) Production of Very-High-Amy- lose Potato Starch by Inhibition of SBE A and B. Nature Biotechnology, 18, 551-554. http://dx.doi.org/10.1038/75427

69.   USDA (2009) http://ceragmc.org/index.php?hstIDXCode%5B%5D=59 &auDate1=&auDate2=&action=gm_crop_database&mode=Submit

70.   National Operating Company Tobacco and Matches (1999)http://ceragmc.org/index.php?action=gm_crop_database&mode=ShowProd& data=C%2FF%2F93%2F08-02

71.   Vector Tobacco Inc. (2002) http://ceragmc.org/index.php?action=gm_ crop_database&mode=ShowProd&data=Vector+21-41

72.   Lilius, G., Holmberg, N. and Bulow, L. (1996) Enhanced NaCl Stress Tolerance in Transgenic Tobacco Expressing Bacterial Choline Dehydrogenase. Nature Biotechnology, 14, 177-180. http://dx.doi.org/10.1038/nbt0296-177

73.   Tarczynski, M., Bohnert, H. and Jense, R.G. (1993) Stress Protection of Transgenic Tobacco by Production of the Osmolyte Mannitol. Science, 259, 508-510.http://dx.doi.org/10.1126/science.259.5094.508

74.   Holmstrom, K.O., Mantyla, E., Welin, B., Mandal, A., Palva, E.T., Tunnela, O.E. and Londesborough, J. (1996) Drought Tolerance in Tobacco. Nature, 379, 683-684.http://dx.doi.org/10.1038/379683a0

75.   Pilon-Smits, E.A.H., Ebskamp, M.J.M., Paul, M.J., Jeuken, M.J.W., Weisbeek, P.J. and Smeekens, S.C.M. (1995) Improved Performance of Transgenic Fructan-Accumulating Tobacco under Drought Stress. Plant Physiology, 107, 125-130.

76.   Murata, N., Sato, N., Takahashi, N. and Hamazaki, Y. (1982) Composition and Positional Distribution of Fatty Acids in Phospholipids from Leaves of Chilling Sensitive and Chilling Resistant Plants. Plant and Cell Physiology, 23, 1071- 1079.

77.   Kodama, H., Hamada, T., Horiguchi, G., Nishimura, M. and Iba, K. (1994) Genetic Enhancement of Cold Tolerance by Expression of a Gene for Chloroplast-Fatty Acid Desaturase in Transgenic Tobacco. Plant Physiology, 105, 601-605.

78.   Grichko, V.P. and Glick, B.R. (2001) Flooding Tolerance of Transgenic Tomato Plants Expressing the Bacterial Enzyme ACC Deaminase Controlled by the 35S rolD or PRB-1b Promoter. Plant and Cell Physiology, 42, 245-249.

79.   Deak, M., Horvarth, G.V., Davletova, S., Torok, K., Sass, L., Vass, I., et al. (1999) Plants Ectopically Expressing the Iron-Binding Protein Ferratin

Are Tolerant to Oxidative Damage and Pathogen. Nature Biotechnology, 17, 192-196.http://dx.doi.org/10.1038/6198

80. Dhankher, O.P., Shast, N.A., Rosen, B.P., Fuhrmann, M. and Meagher, R.B. (2003) Increased Cadmium Tolerance and Accumulation by Plants Expressing Bacterial Arsenate Reductase. New Phytologist, 159, 431-441. http://dx.doi.org/10.1046/j.1469-8137.2003.00827.x

81. Bejo Zaden, B.V. (1997) http://ceragmc.org/index.php?hstIDX Code%5B%5D=13&auDate1=&auDate2=&action=gm_crop_ database&mode=Submit

82. Carneiro, M.F. (1997) Coffee Biotechnology and Its Application in Genetic Transformation. Euphytica, 96, 167-172. http://dx.doi. org/10.1023/A:1002969429010

83. Santana-Buzzy, N., Rojas-Herrara, R., Galaz-Avalos, R.M., Ku-Cauich, J.R., Mijangos-Cortes, J., Gutierrez-Pancheco, L.C., et al. (2007) Advances in Coffee Tissue Culture and Its Practical Applications. In Vitro Cellular & Developmental Biology-Plant, 43, 507-520.http:// dx.doi.org/10.1007/s11627-007-9074-1

84. Vega, F.E., Ebert, A.W. and Ming, R. (2008) Coffee Germplasm Resources, Genomics, and Breeding. Plant Breeding Reviews, 30, 415-447.

85. Duke, S.O. (2003) Weeding with Transgenes. Trends in Biotechnology, 21, 192-195.http://dx.doi.org/10.1016/S0167-7799(03)00056-8

86. Gonsalves, D. (1998) Control of Papaya Ringspot Virus in Papaya: A Case Study. Annual Review of Phytopathology, 36, 415-437.http:// dx.doi.org/10.1146/annurev.phyto.36.1.415

87. Zhai, W., Li, L.X., Tian, W., Zhou, Y., Pan, X., Cao, S., et al. (2000) Introduction of a Rice Blight Resistance Gene, Xa21, into Five Chinese Rice Varieties through an Agrobacterium-Mediated System. Science in China Series C, 43, 361-368.http://dx.doi.org/10.1007/BF02879300

88. López-Bucio, J., de la Vega, O.M., Guevara-García, A. and Herrera-Estrella, L. (2000) Enhanced Phosphorous Uptake in Transgenic Tobacco Plants that Overproduce Citrate. Nature Biotechnology, 18, 450-453. http://dx.doi.org/10.1038/74531

89. Häusler, R.E., Hirsch, H.J., Kreuzaler, F. and Peterhänsel, C. (2002) Overexpression of C_4-Cycle Enzymes in Transgenic C_3 Plants: A Biotechnological Approach to Improve C_3-Photosynthesis. Journal of Experimental Botany, 53, 591-607.http://dx.doi.org/10.1093/ jexbot/53.369.591

90. Leegood, R.C. (2002) C_4 Photosynthesis: Principles of CO_2 Concentration and Prospects for Its Introduction into C_3 Plants. Journal of Experimental Botany, 53, 581-590.http://dx.doi.org/10.1093/jexbot/53.369.581

91. Taniguchi, Y., Ohkawa, H., Masumoto, C., Fukuda, T., Tamai, T., Lee, K., et al. (2008) Overproduction of C_4 Photosynthetic Enzymes in Transgenic Rice Plants: An Approach to Introduce the C_4-Like Photosynthetic Pathway into Rice. Journal of Experimental Botany, 59, 1799-1809. http://dx.doi.org/10.1093/jxb/ern016

92. Kathuria, H., Giri, J., Tyagi, H. and Tyagi, A.K. (2007) Advances in Transgenic Rice Biotechnology. Critical Reviews in Plant Sciences, 26, 65-103.http://dx.doi.org/10.1080/07352680701252809

93. Bajaj, S. and Mohanty, A. (2005) Recent Advances in Rice Biotechnology—Towards Genetically Superior Transgenic Rice. Plant Biotechnology Journal, 3, 275-307.http://dx.doi.org/10.1111/j.1467-7652.2005.00130.x

94. Ye, X., Al-Babili, S., Kloti, A., Zhang, J., Lucca, P. and Potrykus, I. (2000) Engineering the Provitamin A (β-Carotene) Biosynthetic Pathway into (Carotinoid-Free) Rice Endosperm. Science, 287, 303-305. http://dx.doi.org/10.1126/science.287.5451.303

95. Goto, F., Yoshihara, T., Shigemoto, N., Toki, S. and Takaiwa, F. (1999) Iron Fortification of Rice Seed by the Soybean Ferritin Gene. Nature Biotechnology, 17, 282-286.http://dx.doi.org/10.1038/7029

96. Shen, H., Poovaiah, C.R., Ziebell, A., Tschaplinski, T.J., Pattathil, S., Gjersing, E., et al. (2013) Enhanced Characteristics of Genetically Modified Switchgrass (Panicum virgatum L.) for High Biofuel Production. Biotechnology for Biofuels, 6, 71.http://dx.doi.org/10.1186/1754-6834-6-71

97. Chen, F. and Dixon, R.A. (2007) Lignin Modification Improves Fermentable Sugar Yields for Biofuel Production. Nature Biotechnology, 25, 759-761.http://dx.doi.org/10.1038/nbt1316

98. Thurston, C.F. (1994) The Structure and Function of Fungal Laccases. Microbiology, 140, 19-26. http://dx.doi.org/10.1099/13500872-140-1-19

99. Hood, E.E., Bailey, M.R., Beifuss, K., Magallanes-Lundback, M., Horn, M.E., Callaway, E., et al. (2003) Criteria for High Level Expression of a Fungal Laccase Gene in Transgenic Maize. Plant Biotechnology Journal, 1, 129-140. http://dx.doi.org/10.1046/j.1467-7652.2003.00014.x

100. Dean, J.F.D. (2005) Synthesis of Lignin in Transgenic and Mutant Plants. In: Steinbuchel, A., Doi, Y. and Weinheim, D.E., Eds., Biotechnology of Biopolymers. From Synthesis to Patents, Wiley-VCH Verlag, Weinheim,

4-26.

101. Baucher, M., Monties, B., Van Montagu, M. and Boerjan, W. (1998) Biosynthesis and Genetic Engineering of Lignin. Critical Reviews in Plant Sciences, 17, 125-197.http://dx.doi.org/10.1016/S0735-2689(98)00360-8

102. Lapierre, C., Pollet, B., Petit-Conil, M., Toval, G., Romero, J., Pilate, G., et al. (1999) Structural Alterations of Lignins in Transgenic Poplars with Depressed Cinnamyl Alcohol Dehydrogenase or Caffeic Acid O-Methyltransferase Activity Have an Opposite Impact on the Efficiency of Industrial Kraft Pulping. Plant Physiology, 119, 153-164.http://dx.doi.org/10.1104/pp.119.1.153

103. Madison, L.L. and Huisman, G.W. (1999) Metabolic Engineering of Poly(3-Hydroxyalkanoates): From DNA to Plastic. Microbiology and Molecular Biology Reviews, 63, 21-53.

104. Arai, Y., Nakashita, H., Suzuki, Y., Kobayashi, Y., Shimizu, T., Yasuda, M., et al. (2002) Synthesis of a Novel Class of Polyhydroxyalkanoates in Arabidopsis Peroxisomes, and Their Use in Monitoring Short-Chain-Length Intermediates of β-Oxidation. Plant & Cell Physiology, 43, 555-562. http://dx.doi.org/10.1093/pcp/pcf068

105. Cahoon, E.B., Ripp, K.G., Hall, S.E. and Kinney, A.J. (2001) Formation of Conjugated Δ^8,Δ^{10}-Double Bonds by Δ^{12}-Oleic Acid Desaturase-Related Enzymes. Biosynthetic Origin of Calendic Acid. The Journal of Biological Chemistry, 276, 2637-2643.http://dx.doi.org/10.1074/jbc.M009188200

106. Broun, P. and Somerville, C. (1997) Accumulation of Ricinoleic, Lesquerolic, and Densipolic Acids in Seeds of Transgenic Arabidopsis Plants that Express a Fatty Acyl Hydroxylase cDNA from Castor Bean. Plant Physiology, 113, 933-942.http://dx.doi.org/10.1104/pp.113.3.933

107. van de Loo, F.J., Broun, P., Turner, S. and Somerville, C. (1995) An Oleate 12-Hydroxylase from Ricinus communis L. Is a Fatty Acyl Desaturase Homolog. Proceedings of the National Academy of Sciences of the United States of America, 92, 6743-6747.http://dx.doi.org/10.1073/pnas.92.15.6743

108. Singh, S.P., Zhou, X.R., Liu, Q., Stymne, S. and Green, A.G. (2005) Metabolic Engineering of New Fatty Acids in Plants. Current Opinion in Plant Biology, 8, 197-203.http://dx.doi.org/10.1016/j.pbi.2005.01.012

109. Napier, J.A. (2007) The Production of Unusual Fatty Acids in Transgenic Plants. Annual Review of Plant Biology, 58, 295-319.http://dx.doi.org/10.1146/annurev.arplant.58.032806.103811

110. McKeon, T.A. (2003) Genetically Modified Crops for Industrial Products and Processes and Their Affects on Human Health. Trends in Food Science and Technology, 14, 229-241. http://dx.doi.org/10.1016/S0924-2244(03)00071-2

111. Berken, A., Mulholland, M.M., LeDuc, D.L. and Terry, N. (2002) Genetic Engineering of Plants to Enhance Selenium Phytoremediation. Critical Reviews in Plant Sciences, 21, 567-582. http://dx.doi.org/10.1080/0735-260291044368

112. Eapen, S. and D'Souza, S.F. (2005) Prospects of Genetic Engineering of Plants for Phytoremediation of Toxic Metals. Biotechnology Advances, 23, 97-114.http://dx.doi.org/10.1016/j.biotechadv.2004.10.001

113. Eapen, S., Singh, S. and D'Souza, S. (2007) Advances in Development of Transgenic Plants for Remediation of Xenobiotic Pollutants. Biotechnology Advances, 25, 442-451.http://dx.doi.org/10.1016/j.biotechadv.2007.05.001

114. Macek, T., Kotrba, P., Svatos, A., Novakova, M., Demnerova, K. and Mackova, M. (2008) Novel Roles for Genetically Modified Plants in Environmental Protection. Trends in Biotechnology, 26, 146-152. http://dx.doi.org/10.1016/j.tibtech.2007.11.009

115. Doty, S.L. (2008) Enhancing Phytoremediation through the Use of Transgenics and Endophytes. New Phytologist, 179, 318-333. http://dx.doi.org/10.1111/j.1469-8137.2008.02446.x

116. Vangronsveld, J., Herzig, R., Weyens, N., Boulet, J., Adriaensen, K., Ruttens, A., et al. (2009) Phytoremediation of Contaminated Soils and Groundwater: Lessons from the Field. Environmental Science and Pollution Research, 16, 765-794.http://dx.doi.org/10.1007/s11356-009-0213-6

117. Kotrba, P., Najmanova, J., Macek, T., Ruml, T. and Mackova, M. (2009) Genetically Modified Plants in Phytoremediation of Heavy Metal and Metalloid Soil and Sediment Pollution. Biotechnology Advances, 27, 799-810.http://dx.doi.org/10.1016/j.biotechadv.2009.06.003

118. Van Aken, B., Correa, P.A. and Schnoor, J.L. (2010) Phytoremediation of Polychlorinated Biphenyls: New Trends and Promises. Environmental Science and Technology, 44, 2767-2776. http://dx.doi.org/10.1021/es902514d

119. Zhang, Y.W., Tam, N.F.Y. and Wong, Y.S. (2004) Cloning and Characterization of Type 2 Metallothionein-Like Gene from a Wetland Plant, Typha latifolia. Plant Science, 167, 869-877. http://dx.doi.org/10.1016/j.plantsci.2004.05.040

120. Pilon-Smits, E.A.H., Hwang, S., Lytle, C.M., Zhu, Y., Tai, J.C., Bravo, R.C., et al. (1999) Overexpression of ATP Sulfurylase in Indian Mustard Leads to Increased Selenate Uptake, Reduction and Tolerance. Plant Physiology, 119, 123-132.http://dx.doi.org/10.1104/pp.119.1.123

121. Van Huysen, T., Terry, N. and Pilon-Smits, E.A.H. (2004) Exploring the Selenium Phytoremediation Potential of Transgenic Indian Mustard Overexpressing ATP Sulfurylase or Cystathionine-γ-Synthase. International Journal of Phytoremediation, 6, 111-118.http://dx.doi.org/10.1080/16226510490454786

122. Cobbett, C. and Goldsbrough, P. (2002) Phytochelatins and Metallothioneins: Roles in Heavy Metal Detoxification and Homeostasis. Annual Review of Plant Biology, 53, 159-182. http://dx.doi.org/10.1146/annurev.arplant.53.100301.135154

123. Kawashima, C.G., Noji, M., Nakamura, M., Ogra, Y., Suzuki, K.T. and Saito, K. (2004) Heavy Metal Tolerance of Transgenic Tobacco Plants Over-Expressing Cysteine Synthase. Biotechnology Letters, 26, 153-157.http://dx.doi.org/10.1023/B:BILE.0000012895.60773.ff

124. Tong, Y.P., Kneer, R. and Zhu, Y.G. (2004) Vacuolar Compartmentalization: A Second-Generation Approach to Engineering Plants for Phytoremediation. Trends in Plant Science, 9, 7-9. http://dx.doi.org/10.1016/j.tplants.2003.11.009

125. Cherian, S. and Oliveira, M.M. (2005) Transgenic Plants in Phytoremediation: Recent Advances in New Possibilities. Environmental Science and Technology, 39, 9377-9390.http://dx.doi.org/10.1021/es0511341

126. Misra, S. and Gedamu, L. (1989) Heavy Metal Tolerant Transgenic Brassica napus L. and Nicotiana tabacum L. Plants. Theoretical and Applied Genetics, 78, 161-168.http://dx.doi.org/10.1007/BF00288793

127. Pan, A., Yang, M., Tie, F., Li, L., Chen, Z. and Ru, B. (1994) Expression of Mouse Metallothionein-I Gene Confers Cadmium Resistance in Transgenic Tobacco Plants. Plant Molecular Biology, 24, 341-351. http://dx.doi.org/10.1007/BF00020172

128. Hasegawa, I., Terada, E., Sunairi, M., Wakita, H., Shinmachi, F., Noguchi, A., et al. (1997) Genetic Improvement of Heavy Metal Tolerance in Plants by Transfer of the Yeast Metallothionein Gene (CUP1). Plant and Soil, 196, 277-281.http://dx.doi.org/10.1023/A:1004222612602

129. Evans, K.M., Gatehouse, J.A., Lindsay, W.P., Shi, J., Tommey, A.M. and Robinson, N.J. (1992) Expression of the Pea Metallothionein-Like Gene PsMT$_A$ in Escherichia coli and Arabidopsis thaliana and Analysis

of Trace Metal Ion Accumulation: Implications for PsMT$_A$Function. Plant Molecular Biology, 20, 1019-1028.http://dx.doi.org/10.1007/BF00028889

130. Zhu, Y., Pilon-Smits, E.A.H., Tarun, A., Weber, S.U., Jouanin, L. and Terry, N. (1999) Cadmium Tolerance and Accumulation in Indian Mustard Is Enhanced by Overexpressing γ-Glutamylcysteine Synthetase. Plant Physiology, 121, 1169-1177.http://dx.doi.org/10.1104/pp.121.4.1169

131. Arisi, A.C.M., Noctor, G., Foyer, C.H. and Jouanin, L. (1997) Modification of Thiol Contents in Poplars (Populus tremula × P. alba) Overexpressing Enzymes Involved in Glutathione Synthesis. Planta, 203, 362-373. http://dx.doi.org/10.1007/s004250050202

132. Bizily, S.P., Rugh, C.L., Summers, A.O. and Meagher, R.B. (1999) Phytoremediation of Methylmercury Pollution: merB Expression in Arabidopsis thaliana Confers Resistance to Organomercurials. Proceedings of the National Academy of Sciences of the United States of America, 96, 6808-6813. http://dx.doi.org/10.1073/pnas.96.12.6808

133. Rugh, C.L., Wilde, H.D., Stack, N.M., Thompson, D.M., Summers, A.O. and Meagher, R.B. (1996) Mercuric Ion Reduction and Resistance in Transgenic Arabidopsis thaliana Plants Expressing a Modified Bacterial merA Gene. Proceedings of the National Academy of Sciences of the United States of America, 93, 3182-3187.http://dx.doi.org/10.1073/pnas.93.8.3182

134. LeDuc, D.L., Tarun, A.S., Montes-Bayon, M., Meija, J., Malit, M.F., Wu, C.P., et al. (2004) Overexpression of Selenocysteine Methyltransferase in Arabidopsis and Indian Mustard Increases Selenium Tolerance and Accumulation. Plant Physiology, 135, 377-383. http://dx.doi.org/10.1104/pp.103.026989

135. French, C.E., Rosser, S.J., Davies, G.J., Nicklin, S. and Bruce, N.C. (1999) Biodegradation of Explosives by Transgenic Plants Expressing Pentaerythritol Tetranitrate Reductase. Nature Biotechnology, 17, 491-494. http://dx.doi.org/10.1038/8673

136. Gullner, G., Komives, T. and Rennenberg, H. (2001) Enhanced Tolerance of Transgenic Poplar Plants Overexpressing γ-Glutamylcysteine Synthetase towards Chloroacetanilide Herbicides. Journal of Experimental Botany, 52, 971-979.http://dx.doi.org/10.1093/jexbot/52.358.971

137. Flocco, C.G., Lindblom, S.D., Elizabetha, A.H. and Smits, P. (2004) Overexpression of Enzymes Involved in Glutathione Synthesis Enhances Tolerance to Organic Pollutants in Brassica juncea. International Journal of Phytoremediation, 6, 289-304.http://dx.doi.

org/10.1080/16226510490888811

138. Ohkawa, H., Imaishi, H., Shiota, N., Yamada, T. and Inui, H. (1999) Cytochrome P450s and Other Xenobiotic Metabolizing Enzymes in Plants. In: Brooks, J.T. and Roberts, T.R., Eds., Pesticide Chemistry and Biosciences: The Food-Environment Challenge, Royal Society of Chemistry, Cambridge, 259-264.

139. Inui, H., Kodama, T., Ohkawa, Y. and Ohkawa, H. (2000) Herbicide Metabolism and Cross-Tolerance in Transgenic Potato Plants Co-Expressing Human CYP1A1, CYP2B6 and CYP2C19. Pesticide Biochemistry and Physiology, 66, 116-129.http://dx.doi.org/10.1006/pest.1999.2454

140. Protalix Biotherapeutics (2014) http://www.genengnews.com/gen-news-highlights/protalix-pfizer-report-fda-approval-of-plant-derived-b-gaucher-b/81246710/

141. Union of Concerned Scientists (2006) Position Paper: Pharmaceutical and Industrial Crops. http://www.ucsusa.org

142. Liénard, D., Sourrouille, C., Gomord, V. and Faye, L. (2007) Pharming and Transgenic Plants. Biotechnology Annual Review, 13, 115-147. http://dx.doi.org/10.1016/S1387-2656(07)13006-4

143. Walsh, G. and Jefferis, R. (2006) Post-Translational Modifications in the Context of Therapeutic Proteins. Nature Biotechnology, 24, 1241-1252. http://dx.doi.org/10.1038/nbt1252

144. Gomord, V. and Faye, L. (2004) Posttranslational Modification of Therapeutic Proteins in Plants. Current Opinion in Plant Biology, 7, 171-181.http://dx.doi.org/10.1016/j.pbi.2004.01.015

145. Saklani, A. and Kutty, S.K. (2008) Plant-Derived Compounds in Clinical Trials. Drug Discovery Today, 13, 161-171. http://dx.doi.org/10.1016/j.drudis.2007.10.010

146. Ma, J.K., Hiatt, A., Hein, M., Vine, N.D., Wang, F., Stabila, P., et al. (1995) Generation and Assembly of Secretory Antibodies in Plants. Science, 268, 716-719.http://dx.doi.org/10.1126/science.7732380

147. Khoudi, H., Laberge, S., Ferullo, J.M., Bazin, R., Darveau, A., Castonguay, Y., et al. (1999) Production of a Diagnostic Monoclonal Antibody in Perennial Alfalfa Plants. Biotechnology and Bioengineering, 64, 135-143. http://dx.doi.org/10.1002/(SICI)1097-0290(19990720)64:2<135::AID-BIT2>3.3.CO;2-H

148. Gomord, V., Chamberlin, P., Jefferis, R. and Faye, L. (2005) Biopharmaceutical Production in Plants: Problems, Solutions, and

Opportunities. Trends in Biotechnology, 23, 559-565. http://dx.doi. org/10.1016/j.tibtech.2005.09.003

149. Perlak, F.J., Fuchs, R.L., Dean, D.A., McPherson, S.L. and Fischhoff, D.A. (1991) Modification of the Coding Sequence Enhances Plant Expression of Insect Control Protein Genes. Proceedings of the National Academy of Sciences of the United States of America, 88, 3324-3328. http://dx.doi.org/10.1073/pnas.88.8.3324

150. Batard, Y., Hehn, A., Nedelkina, S., Schalk, M., Pallet, K., Schaller, H., et al. (2000) Increasing Expression of P450 and P450-Reductase Proteins from Monocots in Heterologous Systems. Archives of Biochemistry and Biophysics, 379, 161-169.http://dx.doi.org/10.1006/abbi.2000.1867

151. Hamada, A., Yamaguchi, K.I., Ohnishi, N., Harada, M., Nikumaru, S. and Honda, H. (2005) High Level Production of Yeast (Schwanniomyces occidentalis) Phytase in Transgenic Rice Plants by a Combination of Signal Sequence and Codon Modification of the Phytase Gene. Plant Biotechnology Journal, 3, 43-55.

152. Scholthof, H.B., Scholthof, K.B. and Jackson, A.O. (1995) Identification of Tomato Bushy Stunt Virus Host-Specific Symptom Determinants by Expression of Individual Genes from a Potato Virus X Vector. The Plant Cell, 7, 1157- 1172.http://dx.doi.org/10.1105/tpc.7.8.1157

153. Voinnet, O., Rivas, S., Mestre, P. and Baulcombe, D. (2003) An Enhanced Transient Expression System in Plants Based on Suppression of Gene Silencing by p19 Protein of Tomato Bushy Stunt Virus. The Plant Journal, 33, 949-956.http://dx.doi.org/10.1046/j.1365-313X.2003.01676.x

154. Komarnytsky, S., Borisjuk, N., Yakoby, N., Garvey, A. and Raskin, I. (2006) Cosecretion of Protease Inhibitor Stabilizes Antibodies Produced by Plant Roots. Plant Physiology, 141, 1185-1193. http://dx.doi. org/10.1104/pp.105.074419

155. Drake, P.M., Chargelegue, D.M., Vine, N.D., van Dolleweerd, C.J., Obregon, P. and Ma, J.K. (2003) Rhizosecretion of a Monoclonal Antibody Protein Complex from Transgenic Tobacco Roots. Plant Molecular Biology, 52, 233-241.http://dx.doi.org/10.1023/A:1023909331482

156. Arcalis, E., Marcel, S., Altmann, F., Kolarich, D., Drakakaki, G., Fischer, R., et al. (2004) Unexpected Deposition Patterns of Recombinant Proteins in Post-Endoplasmic Reticulum Compartments of Wheat Endosperm. Plant Physiology, 136, 3457-3466.http://dx.doi.org/10.1104/ pp.104.050153

157. Nykiforuk, C.L., Booth, J.G., Murray, E.W., Keon, R.G., Goren, H.J., Markley, N.A. and Moloney, M.M. (2006) Transgenic Expression and

Recovery of Biologically Active Recombinant Human Insulin from Arabidopsis thaliana Seeds. Plant Biotechnology Journal, 4, 77-85. http://dx.doi.org/10.1111/j.1467-7652.2005.00159.x

158. Faye, L. and Daniell, H. (2006) Novel Pathway for Glycoprotein Import into Chloroplasts. Plant Biotechnology Journal, 19, 71-74.

159. Bardor, M., Faveeuw, C., Gilbert, A.C., Gilbert, D., Galas, L., Trottein, F., et al. (2003) Immunoreactivity in Mammals of Two Typical Plant Glyco-Epitopes, Cor-α(1,3)-Fucose and Core Xylose. Glycobiology, 13, 427-434. http://dx.doi.org/10.1093/glycob/cwg024

160. Gomord, V., Sourouille, C., Fitchette, A.C., Bador, M., Pagny, S., Lerouge, P. and Faye, L. (2004) Production and Glycosylation of Plant-Made Pharmaceuticals: The Antibody as a Challenge. Plant Biotechnology Journal, 2, 83-100. http://dx.doi.org/10.1111/j.1467-7652.2004.00062.x

161. Sriraman, R., Bardor, M., Sack, M., Vaquero, C., Faye, L., Fischer, R., et al. (2004) Recombinant Anti-hCG Antibodies Retained in the Endoplasmic Reticulum of Transformed Plants Lack Core Xylose and Core-α(1,3)-fucose Residues. Plant Biotechnology Journal, 2, 279-287. http://dx.doi.org/10.1111/j.1467-7652.2004.00078.x

162. Triguero, A., Cabrera, G., Cremata, J., Yuen, C.T., Wheeler, J. and Ramirez, N.I. (2005) Plant-Derived Mouse IgG Monoclonal Antibody Fused to KDEL Endoplasmic Reticulum-Retention Signal Is N-glycosylated Homogenously throughout the Plant and Mostly High-Mannose Type N-glycans. Plant Biotechnology Journal, 3, 449-457. http://dx.doi.org/10.1111/j.1467-7652.2005.00137.x

163. Palacpac, N.Q., Yoshida, S., Sakai, H., Kimura, Y., Fujiyama, K., Yoshida, T. and Seki, T. (1999) Stable Expression of Human β1,4-galactosyltransfrase in Plant Cells Modifies N-Linked Glycosylation Patterns. Proceedings of the National Academy of Sciences of the United States of America, 96, 4692-4697. http://dx.doi.org/10.1073/pnas.96.8.4692

164. Bakker, H., Bardor, M., Molhoff, J., Gomord, V., Elbers, I., Stevens, L., et al. (2001) Humanized Glycans on Antibodies Produced by Transgenic Plants. Proceedings of the National Academy of Sciences of the United States of America, 98, 2899-2904. http://dx.doi.org/10.1073/pnas.031419998

165. Protalix Biotherapeutics (2014) http://www.protalix.com/products/elelyso-taliglucerase-alfa.asp

166. Protalix Biotherapeutics (2014) http://www.protalix.com/development-pipeline/prx-102-fabry-disease.asp

167. Protalix   Biotherapeutics   (2014)   http://www.thestreet.com/story/11306621/1/protalixs-acetylcholinesterase-demonstrates-potential-role-in-the-treatment-of-parkinsons-disease.html

168. SemiBioSys Genetics (2014) http://www.semibiosys.com

169. Agragen http://www.plantpharma.org

170. USDA-APHIS (2014) http://www.aphis.usda.gov/biotechnology/about.shtml

171. FAO (Food and Agriculture Organization) (2009) http://www.fao.org

172. King, C., Purcell, L. and Vories, E. (2001) Plant Growth and Nitrogenase Activity of Glyphosate-Tolerant Soybean in Response to Foliar Glyphosate Applications. Agronomy Journal, 93, 179-186. http://dx.doi.org/10.2134/agronj2001.931179x

173. Hohn, B., Levy, A.A. and Puchta, H. (2001) Elimination of Selection Markers from Transgenic Plants. Current Opinion in Biotechnology, 12, 139-143.http://dx.doi.org/10.1016/S0958-1669(00)00188-9

174. Zuo, J.R., Niu, Q.W., Moller, S.G. and Chua, N.H. (2001) Chemical-Regulated, Site-Specific DNA Excision in Transgenic Plants. Nature Biotechnology, 19, 157-161.http://dx.doi.org/10.1038/84428

175. Goedeke, S., Hensel, G., Kapusi, E., Gahrtz, M. and Kumlehn, J. (2007) Transgenic Barley in Fundamental Research and Biotechnology. Transgenic Plant Journal, 1, 104-117.

176. Keenan, R.J. and Stemmer, W.P.C. (2002) Nontransgenic Crops from Transgenic Plants. Nature Biotechnology, 20, 215-216. http://dx.doi.org/10.1038/nbt0302-215

177. Daniell, H. (2002) Molecular Strategies for Gene Containment in Transgenic Crops. Nature Biotechnology, 20, 581-586. http://dx.doi.org/10.1038/nbt0602-581

178. Kuvshinov, V., Koivu, K., Kanerva, A. and Pehu, E. (2001) Molecular Control of Transgene Escape from Genetically Modified Plants. Plant Science, 160, 517-522.http://dx.doi.org/10.1016/S0168-9452(00)00414-3

179. Avni, A. and Edelman, M. (1991) Direct Selection for Parental Inheritance of Chloroplasts in Sexual Progeny of Nicotiana. Molecular and General Genetics, 225, 273-277.http://dx.doi.org/10.1007/BF00269859

180. Corriveau, J.P. and Coleman, A.W. (1988) Rapid Screening Method to Detect Potential Biparental Inheritance of Plastid DNA and Results for over 200 Angiosperm Species. American Journal of Botany, 75, 1443-1458. http://dx.doi.org/10.2307/2444695

181. Wang, T., Li, Y., Shi, Y., Reboud, X., Darmency, H. and Gressel, J. (2004) Low Frequency Transmission of a Plastid- Encoded Trait in Setaria italica. Theoretical and Applied Genetics, 108, 315-320. http://dx.doi.org/10.1007/s00122-003-1424-8

182. Huang, C.Y., Ayliffe, M.A. and Timmis, J.N. (2003) Direct Measurement of the Transfer Rate of Chloroplast DNA into the Nucleus. Nature, 408, 796-815.

183. Singh, D.P., Jermakkow, A.M. and Swain, S.M. (2007) Preliminary Development of a Genetic Strategy to Prevent Transgene Escape by Blocking Effective Pollen Flow from Transgenic Plants. Functional Plant Biology, 34, 1055-1060.http://dx.doi.org/10.1071/FP06323

184. Monsanto (2005) Annual Report for Fiscal Year Ended August 31, Form 10-K. U.S. Securities and Exchange Commission, Washington DC.http://www.monsanto.com/investors/documents/pubs/2005/mon_2005_10-k.pdf

185. McDougall, P. (2011) The Cost and Time Involved in the Discovery, Development and Authorisation of a New Plant Biotechnology Derived Trait. Consultancy Study for Crop Life International by P McDougall, Midlothian, 1-24.

Chapter 4

RECENT ADVANCES IN THE GENETIC TRANSFORMATION OF COFFEE

M. K. Mishra[1] and A. Slater[2]

[1]Central Coffee Research Institute, Coffee Research Station, Chikmagalur, Karnataka 577117, India

[2]The Biomolecular Technology Group, Faculty of Health and Life Sciences, De Montfort University, Gateway, Leicester LE1 9BH, UK

ABSTRACT

Coffee is one of the most important plantation crops, grown in about 80 countries across the world. The genus Coffea comprises approximately 100 species of which only two species, that is, Coffea arabica (commonly known as arabica coffee) and Coffea canephora (known as robusta coffee), are commercially cultivated. Genetic improvement of coffee through traditional breeding is slow due to the perennial nature of the plant. Genetic transformation has tremendous potential in developing improved coffee varieties with desired agronomic traits, which are otherwise difficult to achieve through traditional breeding. During the last twenty years, significant progress has been made in coffee biotechnology, particularly in the area of transgenic technology. This paper provides a detailed account of the advances made in the genetic transformation of coffee and their potential applications.

INTRODUCTION

Coffee is one of the most important agricultural commodities, ranking second in international trade after crude oil. The total global production of green coffee is above 134.16 million bags (60 kg capacity) with a retail sales value in excess of $22.7 billion during 2010-11 in the world market [1]. Coffee is grown in about 10.2 million hectares land spanning over 80 countries in the tropical and subtropical regions of the world especially in Africa, Asia, and Latin America. The economics of many coffee growing countries depends heavily on the earnings from this crop. More than 100 million people in the coffee growing areas worldwide derive their income directly or indirectly from the produce of this crop.

Coffee trees belong to the genus Coffea in the family Rubiaceae. The genus Coffea L. comprises more than 100 species [2], of which only two species, that is, C. arabica (arabica coffee) and C. canephora (robusta coffee), are commercially cultivated. Another coffee species, Coffea liberica is also cultivated in a small scale to satisfy local consumption. Almost all the coffee species are diploid (2 n = 2 x = 2 2) and generally self-incompatible except C. arabica which is a natural allotetraploid (2 n = 4 x = 4 4) self-fertile species [3]. In the consumer market, C. arabica is preferred for its beverage quality, aromatic characteristics, and low-caffeine content compared to robusta, which is characterized by a stronger bitterness, and higher-caffeine content. Arabica contributes towards 65% of global coffee production [4].

C. arabica is mainly native to the highlands of Southwestern Ethiopia with additional populations in South Sudan (Boma Plateau) and North Kenya (Mount Marsabit) [5–8]. The C. arabica varieties grown all over the world are derived from either the "Typica" or "Bourbon" genetic base, which has resulted in low-genetic diversity among cultivated arabicas. In contrast, C. canephora has a wide geographic distribution, extending from the western to central tropical and subtropical regions of the African continent, from Guinea and Liberia to Sudan and Uganda with high genetic diversity in the Democratic Republic of Congo [9]. C. canephoramaintains heterozygosity due to its cross-pollinating nature.

COFFEE BREEDING AND ITS LIMITATIONS

Coffee breeding is largely restricted to the two species, C. arabica and C. canephora, that dominate world coffee production. However, C. liberica and C. congensis have contributed useful characters to the gene pool of C. arabica and C. canephora, respectively, through natural and artificial interspecific hybridisation. In C. arabica, initial breeding objectives were to increase productivity and adaptability to local conditions. To achieve these objectives, breeding strategies were directed towards identification of superior plants in the population and their propagation and crossing with existing cultivars. These early breeding efforts, which were carried out from 1920 to 1940, had considerable success in identifying and developing vigorous and productive cultivars. Several of these varieties such as Kents and S.288 from India, Mundo Novo, Caturra and Catuai from Brazil, and Blue Mountain from Jamaica, are still under commercial cultivation. These cultivars are suggested to have a larger degree of genetic variability than the base population [10]. The appearance of coffee leaf rust (Hemileia vastatrix Berk and Br) in epidemic scale in Southeast Asia between 1870 and 1900 had a devastating effect on arabica coffee cultivation in several coffee growing countries. This has changed the breeding focus worldwide with emphasis now given to disease resistance. This has resulted in the introduction of other tolerant species, especially C. canephora, in many countries. Until now, C. canephora has provided the major source of disease and pest resistance traits such as coffee leaf rust (H. vastatrix), coffee berry disease (Colletotrichum kahawae), and root-knot nematode (Meloidogyne spp.) not available in C. arabica. Besides, C. canephora, other diploid species such as C. liberica has been used as source of resistance to leaf rust [11] and C. racemosa for imparting resistance to coffee leaf minor [12]. Further, the cultivation of C. arabica with other diploid species such as C. canephora and C. liberica in close proximity has resulted in spontaneous hybrids in many countries. Natural interspecific hybrids such as Hybrido-de Timor (a hybrid between C. arabica and C. canephora [13] from Timor island), Devamachy (a hybrid between C. arabica and C. canephora),and S.26 (a hybrid between C. arabica and C. liberica, which both originated in India [14]) are the main source of resistance to pest and disease and extensively used in C. arabica breeding programmes.

Like C. arabica, improvement of C. canephora was originally aimed at increasing productivity, and improving bean size and liquor quality. The breeding methods adopted for C. canephora involvedmass selection and intra- as well as interspecific hybridization. Varieties like Apoata of Brazil, S.274 of India, and Nemaya of Central America were derived through mass selection. The spontaneous diploid interspecific hybrid between C. canephora var. ugandae and the C. congensis (called Congusta in Indonesia), and the C × R hybrid variety developed through artificial hybridization between C. congensis and C. canephora in India are examples of improved robusta cultivars developed through interspecific hybridization.

Although conventional breeding is mainly used for coffee improvement, it is a long process involving several different techniques, namely selection, hybridization, and progeny evaluation. A minimum of 30 years is required to develop a new cultivar using any of these methods. Further, the long generation time of the coffee tree, the high cost of field trials, the lack of accuracy of the breeding process, the differences in ploidy level between C. arabica and other diploid species, and the incompatibility are all major limitations associated with conventional coffee breeding. In addition to these, genetic resistance to coffee white stem borer (Xylotrechus quadripes) and coffee berry borer (Hypothenamus hampei), drought and cold tolerance, and herbicide resistance are some of the features that are not easily available in the coffee gene pool or are difficult to incorporate using conventional breeding. Another constraint that hinders the arabica coffee improvement programme is the selection of genetically diverse parental lines for hybridization and the identification of hybrids at an early stage of plant growth based on morphological traits. This is because most of the commercial arabica cultivars are morphologically identical and not easily distinguishable from one another. Uniformity of morphological traits in C. arabica could be attributed to the origin of the species, its narrow genetic base and self-fertile nature. In coffee, identification of cultivars is mainly based upon phenotypic features, but this approach is not reliable and is subject to environmental influences, mainly because of the long generation time of the coffee trees. In some countries of Asia, Latin America, and Africa, coffee is cultivated under shade in varied agroclimatic conditions and displays remarkably different morphologies in various microclimatic zones. In view of the above, it becomes imperative to develop alternative techniques that are reliable, quick, and efficient for discriminating between coffee cultivars. Among the various markers available for genetic analysis in coffee, molecular markers are more efficient, precise, and reliable than other markers for discriminating closely related species and cultivars. The DNA-based markers have the potential of complimenting coffee breeding and improvement program in form of marker-assisted selection (MAS).

MOLECULAR MARKERS AND COFFEE GENETIC IMPROVEMENT

Various molecular markers, such as restriction fragment length polymorphism (RFLP), random amplified polymorphic DNA (RAPD), amplified fragment length polymorphism (AFLP), intersimple sequence repeat (ISSR), simple sequence repeats (SSR), and expressed sequence tag derived simple sequence repeats (EST-SSR) have been used in coffee genetic diversity studies [46–51]. In addition to the above, a large number of commercial coffee samples of American, Indian, and African origin were also analyzed using highly polymorphic SSR markers which revealed that Indian cultivars were genetically diverse from the American and African cultivars [52]. More recently, a new type of molecular marker known as a sequence related amplified polymorphism (SRAP) was used in genetic diversity analysis of coffee cultivars and species [53, 54]. SRAP markers were also successfully used to discriminate between parents in hybrid identification [55, 56] and therefore has great potential in coffee breeding programmes. In addition to the above, single-nucleotide polymorphisms (SNPs) and PCR-RFLP markers were used in coffee genome analysis, which revealed that in C. arabica, polymorphisms are created by paralogous chromosomes, whereas homozygosity of many genes is maintained by the self-fertile nature of the species [57]. These results further demonstrated that in allopolyploid C. arabica, the two parental genomes remain separated and exhibit multiple allelic inheritance patterns, and these findings will be very important for designing strategies and decisions in breeding programmes as well as in sequencing projects. In recent years, concerted efforts have been made by several laboratories across the world, under the International Coffee Genome Network (ICGN) programme, to sequence the coffee genome by using high-throughput sequencing technology which will unravel several key aspects of the coffee genome that may be useful for coffee genetic improvement. In addition to molecular markers, a two-dimensional protein mapping technique was also used to differentiate green coffee samples [58]. A detailed review of the role of various molecular markers in coffee is already available [59] and therefore beyond the scope of the present review.

THE NEED FOR GENETIC TRANSFORMATION OF COFFEE

Since its initial application to plants more than 25 years ago, genetic transformation has become an indispensable tool in plant molecular biology and functional genomics research [60]. Genetic transformation technology is considered as an extension of conventional plant breeding technologies [61]. It offers unique breeding opportunities by introducing novel genetic material irrespective of the species barrier and creating phenotypes with desired traits that are not available in the germplasm pool of crop plants. The major objectives for using genetic engineering technique in coffee are to introduce new traits in to elite coffee genotypes, develop new cultivars with desirable traits such as pest and disease resistance, herbicide resistance, drought and frost tolerance, and improved cup quality, which are not possible to incorporate using traditional breeding techniques. The recent developments in coffee transcriptomics and the availability of large amounts of expressed sequence tag (EST) data from both C. canephora and C. arabica [62–64], as well as the development of coffee bacterial artificial chromosome (BAC) genomic libraries [65, 66], have opened up new possibilities in the area of coffee functional genomics. A key component of most functional genomics approaches is the availability of a highly efficient transformation system useful for designing strategies for gene identification, elucidation of gene functions, regulation and interaction of genes and gene expression analysis to understand the involvement of genes, in coffee biological processes. This will help in precisely targeting the trait of interest using various transformation tools (genes and promoters) with increase probability of success in reducing economic costs.

Genetically modified coffee plants have been produced by different research groups in the world [21, 25, 26,29, 34, 37, 67]. Despite significant advances over the last 20 years, coffee transformation is far from a routine procedure in many laboratories [35]. The objective of this paper is to provide an update on coffee genetic transformation over the last decade, including the in vitro methods used for plant generation.

IN VITRO PLANT REGENERATION

The establishment of an efficient regeneration system is important for genetic transformation of coffee. Various in vitro multiplication methods such as somatic embryogenesis, meristem and axillary bud culture, and induction of adventitious buds have been reported using different types of tissue in various coffee species [68, 69].

Somatic Embryogenesis

The initiation and development of embryos from somatic tissues without the involvement of sexual fusion are known as somatic embryogenesis. In coffee, induction of somatic embryogenesis and plant regeneration was first reported from the internodal explants of C. canephora [70]. In C. arabica, calluses were successfully induced from seeds, leaves, and anthers of two different cultivars, that is, Mundo Novo and Bourbon Amarelo [71]. During the last 35 years, a number of protocols for somatic embryogenesis have been developed for various genotypes of coffee [68]. The first protocol to obtain calli with high embryogenic potential from the leaf explants of C. arabica cv. Bourbon used two different culture media compositions: a first "conditioning" medium and a second "induction" medium [72, 73]. The availability of auxins is critical for the induction of embryogenic calli [72]. In coffee, both high-frequency somatic embryogenesis (HFSE) and low-frequency somatic embryogenesis (LFSE) were established. 2,4-D strongly increases HFSE in combination in primary cultures where as IBA and NAA combined with K increase LFSE. During somatic embryo induction in C. arabica cv. Caturra Rojo, two types of cell clusters, embryogenic and nonembryogenic were observed [74]. The differences in gene expression at both RNA and protein levels were observed between the embryogenic and nonembryogenic cell clusters. Further, it was observed that the number of genes turned off in somatic cells to allow for the change from somatic to embryogenic state is higher than those genes that are turned on [74].

In coffee, somatic embryogenesis follows two distinct developmental patterns: (1) direct somatic embryogenesis, where embryos originate directly from the explants and (2) indirect somatic embryogenesis, where embryos are derived from an embryogenic dedifferentiated tissue (callus). However, both direct and indirect somatic embryos of coffee formed from leaf segments and callus, respectively, have a unicellular origin [75]. Various attempts were made to reduce the time needed for embryogenesis and increase the embryogenesis frequency in coffee. Triacontanol, silver nitrate ($AgNO_3$), salicylic acid, thidiazuron, and 6-(3-methyl-3-butenylamino) purine (2ip) are the widely used growth regulators in coffee embryogenesis. Interestingly, picomolar concentrations of salicylates reported to induce cellular growth and enhance somatic embryogenesis in C. arabica tissue culture [76]. Similarly, triacontanol, as well as silver nitrate, at low concentration in combination with indole-3-acetic acid (IAA) and benzyladenine (BA) induced direct somatic embryogenesis in both species of C. arabica and C. canephora [77, 78]. Additionally, thidiazuron (TDZ) also induced direct somatic embryos from the cultured leaf explants of C. canephora cv. C × R [79]. In C. canephora, the embryogenic response of the explants has been shown to increase by the addition of polyamines, either alone or in combination with silver nitrate. It has been observed that incorporation in the in vitro culture medium of inhibitors of the polyamine biosynthetic pathway such as D,L-. alpha-difloromethylornithine (DFMO) and D,L-.alpha-difloromethylarginine (DFMA) significantly reduced the embryogenic response of the explants in C. canephora, indicating the pivotal role played by polyamines in coffee somatic embryogenesis [80]. Besides the polyamines, indoleamines (melatonin and serotonin) as well as calcium and calcium ionophores (A23187) have also been shown to be beneficial in inducing somatic embryogenesis [81]. Apart from exogenous growth hormones, ethylene and dissolved oxygen concentration play a crucial role in coffee somatic embryogenesis [82, 83].

The use of somatic embryos on an industrial scale was achieved by inducing somatic embryos of C. arabica in liquid medium using bioreactors [84, 85]. The yield of embryos achieved was about 46,000 embryos/3L Erlenmeyer flask (after 7 weeks of culture). Various other workers also reported the production of somatic embryos for industrial use [86, 87]. Extensive studies were carried out in the use of conventional and temporary immersion system for coffee somatic embryo production [68, 88, 89]. However, to date the major obstacle associated with production of somatic embryos on a commercial scale is synchronisation of embryogenesis and conversion of plantlets.

Micropropagation

The coffee plant has a single apical meristem with each axil leaf having 4-5 dormant orthotropic buds and two plagiotropic buds. The plagiotropic buds only start development from the 10th to 11th node. For apical meristem culture and the culture of dormant buds, both orthotropic and plagiotropic buds were cultured for obtaining plantlets. Microcuttings or nodal culture comprise a tissue culture approach which involves culturing nodal stem segments carrying dormant auxiliary buds and stimulating them to develop. Each single segment can provide 7–9 microcuttings every eighty days. Most of these studies were carried out during the 1980s, and these topics have been reported in an earlier review [68].

Several studies have been carried out with a view to micropropagating superior coffee genotypes using apical or axillary meristem culture and nodal culture [68]. A maximum of nine shoots was obtained per one shoot explant [90]. Culture of microcuttings in a temporary immersion system resulted in a 6-fold increase in the multiplication rate, in comparison with microcuttings multiplied on solid medium [91, 92]. The field performance of embryo-generated plants was reported and showed a normal response in terms of physiology and yield. The genetic fidelity of micropropagated plants of C. canephora obtained through somatic embryogenesis was assessed in a large-scale field trial [93]. A total number of 5067 trees regenerated from five to 7-month-old embryogenic cell suspension cultures were planted in the Philippines and in Thailand for comparison with control plants derived from auxiliary budding in vitro. No significant differences in yield and morphological features were observed between the somatic seedlings and microcutting derived trees [93]. However, in contrast to the above, several studies have clearly demonstrated culture-induced variation and regeneration of somaclonal variants in coffee obtained through direct and indirect somatic embryogenesis [94–96]. Detailed molecular analysis of the plantlets of C. arabica derived from high-frequency somatic embryogenesis revealed alterations in both the nuclear and mitochondrial genomes [97]. These reports therefore proposed a critical evaluation of tissue culture-derived plants both at phenotypic and molecular level.

Adventitious shoot development is an alternative method of coffee micropropagation. Shoots originating in tissues located in areas other than leaf axils or shoot tips are subjected to one phase of dedifferentiation followed by differentiation and morphogenesis [68, 98]. Rooting is the most difficult and expensive phase of the micropropagation process, and the success of newly formed plantlets is closely linked to the ability of the root system to adapt to the autotrophic conditions. Several studies have been carried out to improve the rate of rooting of micropropagated plants [68].

DEVELOPMENT OF TRANSGENIC COFFEE

Genetic engineering research on coffee has been pursued for the past fifteen years with two major objectives:

- to elucidate the function, regulation, and interaction of agronomically important genes through a functional genomics approach and
- to improve coffee genotypes with desirable traits through the introduction of targeted genes.

Candidate Genes

In recent years, the development in high-throughput sequencing (HTS) technologies has allowed the rapid acquisition of significant amounts of sequence data, and this has increased our understanding of the genomics of a particular species. During the last decade, significant progress has been made in developing an EST database for coffee. Initial efforts in developing ESTs in Coffea arabica were initiated by the University of Trieste, Italy, and the EST sequences have been placed in the public domain (http://www.coffee.dna.net/). In a private/public collaboration between Nestle and Cornell University, 47000 ESTs from C. canephora were established comprising 13175 unigenes [63]. Subsequently the Brazilian government funded an ambitious coffee genome program, and this has resulted in the establishment of 200 000 ESTs which led to the identification of 30000 genes [64]. Very recently, the Italian group has generated an additional 161 660 ESTs which will be publicly available at the website (http://www.coffeedna.net/ [99]). In parallel with the development of EST database, BAC libraries of coffee species, C. arabica and C. canephora were established [65, 66]. Such maps are of central strategic importance for marker-assisted breeding, positional cloning of agronomical important genes, and analysis of gene structure and function.

Due to the concerted efforts on coffee genomics, many candidate genes from coffee have been identified and some of them have been cloned and are currently being characterized. These include a caffeine biosynthesis gene [28, 100], a sucrose synthase gene [66], osmotic stress response genes [101], genes for seeds oil content [102] and several pathogen resistance genes such as Mex-1 gene [103], SH$_3$ gene [104], and Ck-1 gene to CBD [105]. An efficient genetic transformation protocol is necessary in order to validate the structural and functional aspects of these intrinsic genes. In coffee transformation experiments, genes isolated from coffee as well as from heterologous sources have been used. Some of the genes isolated from coffee used in transformation experiments include a theobromine synthase gene (CaMXMTI) for suppressing caffeine biosynthesis [28, 106] and an ACC oxidase gene involved in ethylene biosynthesis [34]. The genes introduced to coffee from heterologous sources include a cry1Ac gene from Bacillus thuringiensis targeted against leaf miner [26], the α-AII gene from common bean for imparting resistance to coffee berry borer [30], the BARgene for herbicide tolerance [21, 35], and the homeobox gene WUSCHEL (WUS) from Arabidopsis responsible for stem cell identity [36]. The functional significance and expression of these transgenes in coffee plants are described subsequently in this paper.

Promoters

In most of the coffee transformation experiments reported so far, with few exceptions, the CaMV35S promoter derived from cauliflower mosaic virus is extensively used in transgenic constructs (See Table 1). In a comparative study made earlier, the efficacy of different promoters driving uidA transient expression in endosperm, somatic embryos, and leaf explants of C. arabica was analyzed using microprojectile bombardment [18, 19]. It was observed that the EF-1α promoter (from Arabidopsis thaliana EF-1α translation elongation factor) directed maximum transient expression of the uidA gene compared to other promoters [18]. Therefore, this promoter was used subsequently by the same group in Agrobacterium gene constructs, driving the cry1Ac gene for the control of leaf miner in coffee [26, 33]. The efficacy of the CaMV35S viral promoter was also compared with two coffee promoters (α-tubulin and α-arabicin) which have revealed a similar level of transient uidA gene expression [19]. These findings have opened up the possibility of using coffee specific promoter in transformation experiments.

Table 1: Summary of transformation studies in Coffea sp[*]

DNA delivery method	Coffea species	Explants used	Binary vector	Agrobacterium strain	Promoter	Selection marker	Target gene	Results	Reference	Country
Direct delivery electroporation	C. arabica	Protoplast	NA	—	pGA472	nptII	uidA	TE	[15]	USA
	C. arabica	SE	pCAMBIA 3201	—	CaMV35S	bar	uidA	TGI	[16]	Venezuela
	C. canephora	Endosperm	pCAMBIA 1301	—	CaMV35S	hpt	uidA	TE	[17]	India
Microprojectile bombardment	C. arabica	Leaf	pPIGK	—	EF-1a	bar	uidA	TE	[18]	France
	C. arabica	ET	pCAMBIA 2301	—	CaMV35S	nptII	uidA	TGI	[19]	Colombia
	C. arabica	ET	pBI-426	—	CaMV35S	nptII	uidA	TGI, PR	[20]	Brazil
	C. canephora	ET	pCAMBIA 3301	—	CaMV35S	bar	uidA	TGI, PR	[21]	Brazil
	C. arabica	SE	pCAMBIA 2301	—	CaMV35S	nptII	uidA	TE	[22]	Costa Rica
	C. arabica	ET	pBI-426	—	CaMV35S	nptII	uidA	TGI, PR	[23]	Brazil

Method	Species	Explant	Vector	Strain	Promoter	Selection	Reporter	Result	Ref	Country
Indirect delivery A. tumefaciens	C. arabica	Protoplast	pGV2260	NA	CaMV35S	hpt	uidA	TE	[24]	France
	C. canephora	ET	pIGI121-Hm	EHA101		hpt	uidA	TGI, PR	[25]	Japan
	C. arabica and C. canephora	SE	pBIN19	LBA4404	EF-1α	csr-1-1	cry1Ac	TGI, PR	[26]	France
	C. canephora	ZE, SE, H	pBECKS 400	EHA101	CaMV35S	hpt	uidA	TGI,	[27]	India
	C. arabica and C. canephora	ET, SE	pHIBI –IG	EHA101	CaMV35S	hpt	gfp & CaMX-MTI	RNAi, PR	[28]	Japan
	C. canephora	H	pBECKS 400	EHA 101	CaMV35S	hpt	uidA	TGI, PR	[29]	India
	C. canephora	ET	pCAMBIA 3301	EHA105	CaMV35S	ppt	uidA	TGI	[30]	Brazil
	C. canephora	ET, Leaf	pER10W-35SRed	C58	CaMV35S	nptII	DsRFP	TGI, PR	[31]	Mexico
	C. canephora	SE	pCAMBIA 1381	A4, EHA 101	CaMV35S	hpt	NMT	RNAi	[32]	India
	C. canephora	SE	pBIN19	LBA 4404	EF-1	csr-1-1	cry1Ac	Field test	[33]	France
	C. arabica	ET	pCAMBIA 3300	EHA 105	CaMV35S	bar	ACC-oxidase	AE	[34]	Brazil
	C. canephora	ET	pCAMBIA 3301	EHA 105	CaMV35S	bar	uidA	TGI, PR	[35]	Brazil
	C. canephora	Embryos	pER10W-35SRed	C58C1	CaMV35S	WUS-CHEL	DsRFP	TGI,PR	[36]	Mexico
	C. canephora and C. arabica	ET	pBECKS 2000	EHA 101, EHA 105, LBA 4404, AGL 1	CaMV35S	hpt, visual	uidA, sgfp	TGI, PR	[37]	India
	C. arabica	ET	pMDC32	LBA1119	CaMV35S	hpt	sgfp	PR	[38]	France
A. rhizozenes	C. arabica and C. canephora	SE	pBIN 19	A4	CaMV35S	hpt	uidA	TGI, PR	[39]	France
	C. arabica	Leaf	NA	IFO 14554		NA	NA	TGI, PR	[40]	Japan
	C. canephora	SE	pBIN 19	A4	NA	csr-1-1	cry1Ac	TGI, PR	[41]	France
	C. canephora	SE	pCAMBIA 1301	A4	CaMV35S	hpt	uidA	TGI, PR	[42]	India
	C. arabica	H	pBIN 19	A4	CaMV35S	visual	uidA	TGI, PR	[43]	France
	C. arabica	H	pCAMBIA 2300	A4	CaMV35S	visual	gfp	TGI, PR	[44]	France
					CaMV35S					

*Updated from Etienne et al. [45]; ET: embryogenic tissue; SE: somatic embryos; H: hypocotyls; TE: transient gene expression; TGI: target gene integration; PR: plant rregeneration; RNAi: RNA interference, AE: antisense expression.

Reporter Genes

Reporter genes are used in gene constructs to optimize the transformation procedure. In the majority of coffee transformation experiments, the uidA gene is used as a reporter gene (See Table 1). Only recently have the sgfp(synthetic green fluorescence protein) and DsRFP (Red fluorescent protein) been used in coffee transformation [28, 31, 37]. However, most transient expression studies have been carried out using the uidA reporter gene. In an effort to optimize the Agrobacterium transformation protocol in coffee, the expression of the uidA gene driven by a CaMV35S promoter was compared in various tissues under different cocultivation conditions [27]. It was observed that the endogenous GUS activity was reduced substantially when 20% methanol was added to the GUS staining solution. Marked differences in GUS activity were observed between the endogenous, and transformed tissue as the transformed plants exhibited a deep blue colour in reaction with X-gluc, while nontransformed plants only exhibited a pale blue coloration. Several factors such as of cocultivation period, preculture of explants, and acetosyringone-influenced GUS activity. Furthermore, in a comparative study, it was observed that GUS expression in leaf explants was more pronounced in the cut ends of the veins, whereas zygotic and somatic embryos, hypocotyls, and the adjacent region are the main target sites [27]. Recently, in another experiment, the expression of p35S. GUS expression was found to be stronger in the root tip and central vascular system compared to other regions in the root system [43].

Recently, the expression of both uidA and gfp genes driven by the CaMV35S promoter was monitored using various Agrobacterium strains and culture conditions [37]. Expression of the sgfp-S65T gene driven by the CaMV35S promoter (signified by green fluorescence) was observed in cocultivated calli just 2 days after cocultivation. Initially, green fluorescence appeared as discrete spots but subsequently, calli that showed green fluorescence increased in size, producing a bright fluorescence mass after 15 days cocultivation. The expression of both uidA and sgfp was intense in globular- and torpedo-shaped embryos until the development of cotyledonary leaves. In older leaves, green florescence was weak due to the interference of chlorophyll, which emits red fluorescence at the same activating wavelength. The pattern of expression of GUS and gfp genes driven by the CaMV35S promoter was similar, being much more pronounced in the leaf veins and root tips and in vascular zones. Using the CaMV35S promoter, similar expression patterns were obtained for DsRFP in somatic embryos [31] and for gfp in roots [44].

TRANSFORMATION SYSTEMS

Both direct and indirect DNA delivery systems have been employed to transform coffee by various workers and the details are described below.

Direct DNA Delivery

Electroporation

Electroporation is a process through which permeability of the cell plasma membrane is significantly increased by the external application of electrical field. It is usually used in molecular biology as a way of introducing some substance into a cell, such as a drug that can change the cell's function, or a piece of coding DNA. Electroporation was used to integrate foreign DNA into protoplasts of C. arabica [15]. Regeneration of transformed embryos and plantlets resistant to kanamycin was obtained but the plantlets did not survive due to a weak root system. In another experiment, various parameters influencing transformation of coffee somatic embryos using electroporation of pCAMBIA 3201 plasmid carrying the uidA gene were described [16]. The results showed that the electroporation of somatic embryos at torpedo stage can be a promising approach to coffee transformation since transformed torpedo-shaped embryos produced significantly higher numbers of gus positive secondary embryos in the culture medium. Recently, the expression of uidA gene driven by the N-methyltransferase (NMT) promoter was studied by electroporation of coffee endosperm [17]. The results indicated that uidA gene expression driven by the NMT promoter is targeted to the external surface of the vacuoles.

Microprojectile Bombardment

Genetic transformation of coffee via microprojectile bombardment was described for the first time, using a gunpowder driven device and several target explants [18]. The study compared different promoters and demonstrated that the EF-1α promoter from Arabidopsis thaliana is more effective than the CaMV35Spromoter in driving transient GUS expression in leaves of microcuttings. The interaction between osmotic preconditioning and physical parameters of helium gun device was studied in C. arabica suspension cells. It was observed that four hours of pretreatment of the target tissue with mannitol and sorbitol before bombardment increased the number of cells expressing GUS gene without causing cell necrosis [19].

Successful regeneration of transgenic coffee plants (C. canephora) by microprojectile bombardment was achieved by using the pCambia3301 plasmid containing the uidA and bar genes [21]. The study demonstrated the effectiveness of the bar gene for selection of transformants in vitro and in vivo identification of transgenic coffee plants. In C. arabica, the plasmid pBI-426 carrying the nptII and uidA genes was employed in particle bombardment of embryogenic calli, and transformants were selected using kanamycin [20]. The transgenic status of the regenerated plants was confirmed by PCR analysis. Recently, a C. arabica suspension culture with a high regenerative capacity for secondary somatic embryogenesis was used for transformation using microprojectile bombardment [22]. However, no transformants could be regenerated due to damage to bombarded tissue. Very recently, successful regeneration of transgenic C. arabica was reported using bombardment of embryogenic calli followed by kanamycin selection [23]. The authors reported the normal growth of the transgenic plants and obtained T_1 progeny presenting 3 : 1 segregation of the uidA transgene.

Indirect DNA Delivery

Agrobacterium tumefaciens. The Agrobacterium tumefaciens mediated transformation technique has been extensively used for genetic transformation of coffee and is the method of choice for many workers (see Table1). An initial report of A. tumefaciens-mediated transformation of coffee involved the cocultivation of protoplast with different Agrobacterium strains carrying nptII and uidA genes [24]. Transient GUS expression was demonstrated in the callus tissue derived from protoplasts but plant regeneration was not obtained. The first successful A. tumefaciens transformation and transgenic plant regeneration were achieved in C. canephora[25] and subsequently in C. arabica [26]. In C. canephora, various parameters that influence T-DNA delivery to coffee tissue were studied using transient GUS gene expression [27]. It was reported that preculture of explants prior to Agrobacterium cocultivation, addition of acetosyringone to cocultivation medium and the duration of the cocultivation period significantly influenced T-DNA delivery to coffee tissue. In addition to the culture conditions, the A. tumefaciens strains also influence the transformation efficiency in coffee [37]. A. tumefaciens strains and their plasmids are classified by the function of the opine genes they carry. These opines, which are synthesized in the infected plant cells, are mainly agropine, nopaline, and octopine. For coffee transformation, various A. tumefaciens strains such as LBA 4404 pAL4404 (octopine), C58CI pMP90 (nopaline), EHA 101 pEHA101 (agropine), EHA 105 pEHA101 (agropine), and AGL1 pTiBo542 (agropine) have been used by various workers (see Table 1). Recently, a study was carried out to compare the efficiency of four different Agrobacterium strains (LBA 4404, EHA101, EHA105, AGL1) in coffee transformation using pBECKS 2000 vector constructs carrying uidA and gfp reporter genes [37]. It was observed that EHA 105 and EHA101 were more efficient compared to LBA4404 in T-DNA delivery and transgenic plant regeneration. Based on this improved protocol, mass production of transgenic coffee plants of both C. arabica and C. canephora was achieved [37].

In order to improve the efficiency of A. tumefaciens mediated transformation, sonication and vacuum infiltration methods were incorporated during cocultivation [31, 35]. In most of the A. tumefaciens mediated transformation, embryogenic tissues and/or somatic embryos were used as the target material for cocultivation (see Table 1). In C. canephora, a highly efficient A. tumefaciens transformation and regeneration protocol were established using hypocotyl explants as the target material (Figures 1(a)–1(l)) [29]. In C. canephora, the collar region of the hypocotyls was found to be more suitable for A. tumefaciens transformation [107]. However, inC. arabica, embryogenic calli were found to be more suitable for A. tumefaciens transformation [38]. The methodology for genetic transformation of Coffea using Agrobacterium was described in detail [108]. They were able to transform 20 different genotypes either belonging to C. arabica or C. canephora by cocultivation of embryogenic calli with A. tumefaciens.

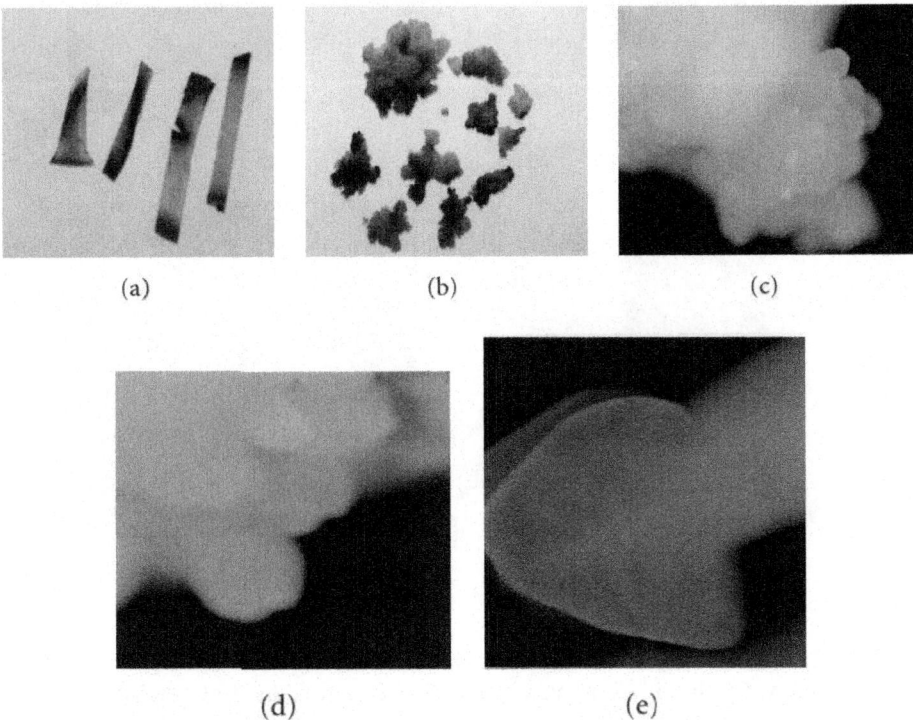

(a)   (b)   (c)

(d)   (e)

(f)   (g)   (h)

(i)   (j)

(k)   (l)

Figure 1: Agrobacterium tumefaciens mediated transformation and regeneration of C. canephora cv. C × R and C. arabica genotypes. (b, c, d, f, and k) C. arabica (a, e, g, h, i, j, and l) C. canephora. (a) Cocultivated hypocotyls of in vitro seedlings expressing transient GUS expression; (b) embryogenic calli showing transient GUS expression following cocultivation; (c) initiation of somatic embryos from the transformed calli showing green fluorescence; (d) mass of heart shaped somatic embryo showing green fluorescence; (e) germinating somatic embryo with well-developed cotyledon leaves with bright green fluorescence; (f) Gus expression in the root tips of a germinated transformed plant; (g) strong GFP expression in transgenic root; (h) in vitro plant regeneration; (i) Gus staining of the leaf of a transgenic plant; (j) GFP expression in the developing leaf; (k) GUS staining of the regenerated transformed plant; (l) transgenic plant in the soil.

A. tumefaciens mediated transformation has also been used for gene silencing using RNAi technology and several genes such as theobromine synthase (CaMXMTI) and N-methyltransferase (NMT) (both involved in caffeine biosynthesis) and the ACC oxidase gene (involved in ethylene biosynthesis) were targeted to coffee tissue in reverse orientation for obtaining stable silencing [28, 32, 34]. Recently, transgenic C. canephora plants incorporating a homeobox gene WUSCHEL (WUS) responsible for stem cell identity were regenerated [36].

Agrobacterium rhizogenes. Agrobacterium rhizogenes mediated transformation of both C. canephora and C. arabica species was reported as early as 1993 [39]. Subsequently, many workers have reported successful A. rhizogenes mediated transformation and plantlet regeneration in both C. arabica and C. canephora using different explants [40, 42, 44, 109]. In almost all cases, Agrobacterium strain A4 was used except in one case where bacterial strain IFO 14554 has been used (Table 1). Most of the binary constructs used in A. rhizogenes transformation are either based on a pBIN19 or pCambia backbone. A. rhizogenes transformation is very useful for functional analysis of genes involved in resistance to root knot nematode in coffee.

SELECTION OF TRANSFORMANTS

The successful recovery of stable transformants depends upon the choice of a suitable selective agent, its optimal concentration, timing, and frequency of selection. An effective selective agent must allow the growth and development of transformed tissue but simultaneously restrict the proliferation of nontransformed cells in the same culture medium. Therefore, incorporation of the right selectable marker gene in gene constructs is critical to the success of transformation. In coffee transformation, various selection marker genes (hpthygromycin-R, nptII kanamycin-R, csr1-I chlor and sulfuron-R, ppt phosphinothricin-R) were used (Table 1), and their efficacy has been evaluated [18, 27]. The first work in coffee transformation was carried out using kanamycin as the selective agent [39]. However, contradictory results were obtained with regard to the efficacy of kanamycin for selection of coffee transformants. Many workers [15, 39, 41, 110] have reported the development of nontransformed somatic embryos at higher kanamycin concentrations and have attributed this to a poor capacity for transformant selection. However, there have also been several reports of successful selection of transformed coffee plants using kanamycin as the sole selective agent [20, 23, 31]. In many transformation experiments, hygromycin at concentrations of 20–100 mg/L was used successfully for selection of transformants [25, 27–29]. The efficiency of hygromycin as a selective agent was tested in different transformed and nontransformed tissue of C. canephora and it was observed that 25 mg/L hygromycin severely checked nontransformed somatic embryo growth and proliferation [27]. Similar results were also obtained by using hygromycin in A. rhizogenes mediated coffee transformation [42].

In addition to antibiotics, several other types of selection markers such as herbicide selection and positive selection have also been used in coffee transformation. The reliability of chlorsulfuron (csr1-I), phosphinothricin (ppt), and ammonium glufosinate (bar) as selection markers to regenerate transformed tissue were already confirmed in both C. arabica and C. canephora using various transformation methods [21, 26, 30]. As an alternative to negative markers such as antibiotics and herbicides, positive selection marker genes such as phosphomannose isomerase (pmi) and xylose isomerase (xylA) have also been used for coffee transformation, producing transformants able to grow in the presence of mannose and xylose, respectively, without an additional carbohydrate source. The study indicated that compared to mannose, xylose is an effective selective agent for coffee transformation [111].

In recent times, transgenic crops regenerated carrying antibiotic and herbicide resistance genes have generated public disquiet about food safety and environmental impact. This has stimulated research into utilizing visual selection markers instead of using antibiotic and/or herbicide selection markers. In coffee, green fluorescent protein, (gfp), and red fluorescent protein (DsRFP) were used for visual selection of transformed tissue following A. tumefaciens mediated transformation [28, 31, 37]. The regeneration of transformed roots of C. arabica using visual selection of green epifluorescence without using any selective agent was achieved throughA. rhizogenes mediated transformation [44]. Recently transgenic plants of both C. canephora and C. arabicawere regenerated by employing visual selection of green fluorescent protein as the sole screen following A. tumefaciens mediated transformation [112].

TRANSGENE EXPRESSION

Expression of Reporter Genes

Studies pertaining to expression of transgenes in a perennial crop like coffee are very important. However, such reports are very limited. The expression of uidA and gfp transgenes driven by a CaMV35S viral promoter was monitored at different stages of plant growth of C. canephora following A. tumefaciens mediated transformation [27, 37]. It was observed that maximum expression of both uidA and gfp genes were obtained at the globular- and torpedo-shaped somatic embryos. Following the development of cotyledon leaves, GUS expression was scattered, with pronounced expression shifted to vascular regions in the well-developed leaves. In roots, maximum expression was obtained in the root tips and root hairs compared to the main roots. Recently, the expression of uidA gene driven by the double CaMV35S promoter was monitored in the flowers and fruits of C. arabica transgenic plants obtained through microprojectile bombardment [23].

Expression of Insect Resistance Genes

Transgenic coffee (C. canephora) plants incorporating synthetic cry genes (cry1Ac) from Bacillus thuringiensiswere regenerated [26]. Field assessment demonstrated the resistance of transgenic plants to coffee leaf miner (Perileucoptera coffeella) indicating the functional stability of the transgenes [33]. In another study, transgenic plants of C. canephora incorporating α-A11 from common bean were produced via A. tumefaciens mediated transformation, and bioassays with the insect are underway to confirm functional validation of its protein in coffee [30].

Expression of Herbicide Resistance Genes

Transgenic coffee plants incorporating the csr-1-1 gene were produced using A. tumefaciens mediated transformation [26]. In this study, plants were selected using chlorsulfuron but several nontransformed escapes were obtained, which suggests that the herbicide is not a very tight selective agent. In another study, transgenic C. canephora plants were produced using the bar gene [35]. Regenerated plants were sprayed with the herbicide ammonium glufosinate under green house conditions and showed no phytotoxicity effects.

Modification of Expression of Genes Controlling Biochemical and Physiological Traits

The expression of genes, involved in the caffeine biosynthesis pathway, was modified using RNAi technology to reduce the level of CaMXMT1 (theobromine synthase) [28]. Transgenic plants expressing antisense ACC oxidase (involved in fruit maturation and ethylene production) have also been produced and stable expression of the transgene observed [34].

APPLICATIONS OF TRANSGENIC TECHNOLOGY

Genetic transformation technology has potential applications in coffee agriculture by incorporating desirable traits such as disease and insect resistance, drought and frost, tolerance and herbicide resistance. Transgenic technology can also be used to increase nutritional value and improve cup quality, produce varieties with caffeine-free beans, and for production of hybrid crops for molecular farming. Identification of target specific genes is one of the pre-requisites for developing transgenic crops. The availability of a large number of EST sequences in coffee and initiation of coffee genome sequencing may speed up the gene discovery and accelerate transgenic research efforts in coffee.

Insect Resistance

Production of insect resistant coffee plants is one of the major objectives of the breeding programs. The major pests attacking coffee include coffee berry borer (CBB, Hypothenemus hampei), white stem borer (WSB,Xylotrechus quadripes), leaf miner (Perileucoptera coffeela), and root nematodes (Meloidogyne spp. andPratylenchus spp.). The CBB is present in almost all the coffee growing countries and considered to the most devastating pest in coffee. To date, there is no reported source of resistance to CBB in the coffee gene pool. Like CBB, WSB is another serious pest in arabica coffee in India and several other East Asian countries. Both CBB and WSB belong to the order Coleoptera. For India, controlling WSB is the biggest challenge and has therefore become the highest research priority. Robusta is generally resistant to WSB but the interspecific robusta arabica hybrids are susceptible to WSB. Although leaf miner is not yet a serious pest in India and other East Asian countries, it is an economically important pest in East Africa and Brazil.

Effective chemical control of CBB and WSB is difficult due to the nature of their life cycle inside the berry and stem, respectively, as well as environmental concern regarding the use of pesticides. Biological control measures are adapted to combat these insect pests with varying degrees of success. Developing coffee plants resistant to these pests using genetic transformation technology could be one of the alternative strategies to counter pest damage.

For insect resistance, several different classes of proteins from bacterial, plant, and animal sources have been isolated and their insecticidal properties tested against many important pests. Amongst these proteins, Bt toxins are most important and several transgenic crops expressing Bt genes have been commercialized. Coffee transgenic plants carrying a synthetic version of the cry1Ac gene have been produced [26, 67]. Indeed, this was the first report that an important agronomic trait has been introduced into a coffee plant. The transgenic plants presented similar features in growth and development compared to normal plants. Transgenic plants highly resistant to leaf minor under greenhouse conditions were tested under field conditions in French Guyana for 4 years for field resistance [33, 113]. From a total of 54 independent transformation events, 70% of the events were resistant to leaf minor. Unfortunately, the field trial was vandalized which led to the termination of the experiment [114].

The effectiveness of Bt genes in controlling coleopteran pests is well documented in corn and potato, which indicates that Bt genes might be effective against CBB. The high toxicity of B. thuringiensis serovar israelensisagainst first instar larvae of CBB has been demonstrated [115]. In another experiment, an α-amylase inhibitor from Phaseolus vulgaris was tested against CBB and found to have an inhibitory effect on its growth and development [116].

In addition to the CBB and WSB, arabica coffee varieties are also susceptible to endoparasitic root-knot nematode (Meloidogyne spp.) [117]. So far, 15 species have been reported to be parasites of coffee. Controlling nematodes is extremely difficult and currently seedling grafting with robusta rootstock is followed as one of the control measure. Sources of resistance specific to root-knot nematodes have been identified in coffee trees [118] and the Mex-1 gene conferring resistance to M. exigua in C. arabica is in the process of isolation [103]. The functional analysis of nematode resistant genes could be carried out by using the A. rhizogenes mediated transformation protocol already developed in coffee (See Table 1). Root-specific promoters could also be used in the vector constructs to drive transgene expression in the root.

Cultivation of insect resistant transgenic coffee should address ecological concerns related to insects and soil microorganisms. Transgene stability also needs to be studied since a coffee plantation can remain for several years without replantation. The constant selection pressure of the transgenic plants on the targeted insect population should be monitored, as there may be a chance of emergence of insect resistant populations.

Tolerance to Abiotic Stress

In many coffee growing countries, abiotic stress such as drought and frost are the major climatic factors that limit coffee production. Changes in climatic patterns due to global climate change is considered increasingly important for coffee cultivation. Drought induces water stress in plants, which affects vegetative growth and vigour, and triggers floral abnormalities and poor fruit set. It also indirectly increases the incidence of pests and diseases in the plants. Arabica is generally more tolerant to water stress than robusta, partly due to its extensive deep root system. C. racemosa is known to be a good source of drought tolerance. In India, hybrids between C. canephora and C. racemosa have been obtained and are currently under evaluation for drought tolerance.

In many coffee growing countries, coffee is propagated in marginal areas where the annual rainfall is below 1000 mm, with prolonged dry spells of over 4-5 months. In those areas, water shortage and unfavourable temperatures constitute major constraints, and the growth and productivity of robusta coffee are badly affected.

As with drought, periodic frost also affects coffee production in parts of Brazil. The introduction of drought and frost tolerant genes through genetic transformation would be of great importance for alleviating these problems. Research is now being carried out by several groups to identify genes involved in biotic as well as abiotic stress. The most promising approach of genetic engineering for drought tolerance includes the use of functional or regulatory genes as well as the transfer of transcription factors. In recent years, plants tolerant to high temperature and water stress have been the subject of intense research [119–121]. For achieving drought tolerance, genes that have been targeted include those encoding enzymes involved in detoxification or osmotic response metabolism, enzymes active in signalling, proteins involved in the transport of metabolites, and regulating the plant energy status [119–121]. The dissection of molecular mechanisms related to signal transduction and transcriptional regulation might help in engineering drought tolerance in coffee.

Disease Resistance

Coffee leaf rust (CLR) caused by the fungus Hemileia vastatrix is the most important disease in coffee with substantial loss to coffee production and productivity in all the coffee growing countries. In addition to leaf rust, coffee berry disease (CBD) caused by fungus Colletotrichum kahawae can be a devastating anthracnose causing substantial crop loss in Africa. Several other fungal and bacterial diseases may also affect coffee, causing economic damage to a small extent. Arabica is more susceptible to many diseases than robusta coffee. Though most of the disease control measures rely upon chemical control, they are more expensive and labour intensive. The long-term solution is the breeding of resistant varieties, which is the focus of many breeding programmes. However, breeding for disease resistant varieties is time consuming due to the perennial nature of coffee, with its long gestation period. India has a long history of arabica coffee breeding especially for leaf rust resistance and is the first country to demonstrate the existence of multiple races of leaf rust. Resistance to CLR is conditioned primarily by a number of major (*SH*) genes, and coffee genotypes are classified into different resistance groups based on their interaction with different rust races pathogen [122]. Currently C. canephora provides the main source of resistance to pests and diseases including CLR (H. vastatrix) and CBD (C. kahawae) and is therefore used in breeding programs. Other diploid species like C. liberica and C. racemosahave been used as a source of resistance to coffee leaf rust and coffee leaf miner, respectively [12, 123].

The development of coffee varieties resistant to major fungal diseases such as CLR and CBD using transgenic technology will benefit the coffee industry immensely. During the last 15 years, significant progress has been made in the area of host-pathogen interactions [124, 125] and many resistance genes involved in recognizing invading pathogens have been identified and cloned [126]. A number of signalling pathways, which are induced following pathogen infection, have been dissected [127]. Many antifungal compounds that are synthesized by plants to combat fungal infection have been identified [128]. Understanding the specific induction of targeted pathways and identification of specific pathways responsible for particular fungal resistance is important in order to employ this strategy in transgenic technology. The recent investigation of gene expression during coffee leaf rust infection could give an insight into the defence pathways operating in coffee [129, 130]. Efforts have been made to identify and clone resistance genes from coffee for achieving durable resistance. Recently, the genetic and physical map of two resistance genes, that is, the *SH* 3 gene conferring resistance to rust [104] and the Ck1 gene conferring resistance to C. kahawae CBD [105] have been established. These genes could be used for molecular marker assisted breeding programmes.

Production of Low-Caffeine Coffee

Low-caffeine and decaffeinated coffee represent around 10% of the coffee sales around the world [35]. The industrial process for coffee decaffeination can be expensive and affects the original flavour and aroma in coffee [131].

Transgenic coffee plants with suppressed caffeine synthesis using RNA interference (RNAi) technology have been obtained [28, 106]. Specific sequences in the 3' untranslated region of the theobromine synthase gene (CaMXMT1) were selected for construction of RNAi short and long fragments.

The caffeine and theobromine content of the transgenic plants reduced by up to 70% compared to the untransformed plants. In C. canephora, RNAi technology has also been employed to silence the N-methyl transferase gene involved in caffeine biosynthesis [17]. Recently the promoter of an N-methyltransferase (NMT) gene involved in caffeine biosynthesis was cloned [100], which will be very useful for studying the regulation of caffeine biosynthesis.

Improvement in Cup Quality

Improvement in coffee cup quality requires elaborate knowledge of the chemical constituents as well as the metabolic pathways involved in the elaboration of quality. The constituents of coffee beans include minerals, proteins, carbohydrates caffeine, chlorogenic acids (CGA), glycosides, lipids, and many volatile compounds that give flavour to coffee by roasting. Among these, the role of three major constituents: sucrose, CGA, and trigonelline have been studied in coffee. The sucrose content of coffee bean is associated with coffee flavour; the higher the sucrose content in green beans, the more intense will be the cup flavour [132, 133]. The sucrose content of C. arabica (8.2-8.3%) is higher than C. canephora (3.3–4.0%). The sucrose:amino acids ratio in green beans determines the profile of volatile compounds. Manipulating sucrose content in coffee bean is therefore important in improving cup quality. Recently, the sucrose synthase gene (CcSUS2) from C. canephora has been cloned and sequenced [66]. This provides an opportunity to manipulate the sucrose content in coffee.

Chlorogenic acids are products of phenylpropanoid metabolism. Chlorogenic acids are a group of hydroxycinnamoyl quinic acids (HQA) formed by esterification between caffeic acids, coumaric acids, and quinic acids [134]. They are present in relatively large quantities in the coffee bean and are the precursor of phenolic compounds in roasted beans. Robusta beans contain higher CGA (10%) compared to arabica beans (6-7%). CGA are known to have antioxidant properties as well as being associated with disease resistance [135]. Genetic manipulation of genes involved in CGA synthesis can serve either of these purposes by up- or downregulating the pathway.

Phenylalanine ammonia-lyase (PAL) catalyzes the first step of the phenylpropanoid pathway leading to the synthesis of a wide range of chemical compounds including flavonoids, coumarins, hydroxycinnamoyl esters, and lignins [136]. Recently, the full length cDNA and corresponding genomic sequences of PAL from C. canephora was isolated, characterized, and functionally validated [137, 138]. This has opened up new possibilities for manipulating the level of the PAL enzyme in coffee which in turn will be useful for cup quality improvement and manipulating antioxidant properties in coffee.

Fruit Ripening

Uniformity during fruit ripening is decisively related to cup quality in coffee, and consequently to the value of the product. Fruits at the ideal ripening stage produce the best organoleptic characteristics for coffee. The presence of overripened or green fruits changes the acidity, the bitterness and consequently the cup quality. In order to maximise uniform ripening of coffee fruits, it is essential to control the action of genes involved in the last step of maturation process. Ethylene is known to trigger ripening, and increasing ethylene biosynthesis is associated with various stages of ripening process [139]. To control coffee fruit maturation, two of the major genes involved in ethylene biosynthesis, namely, ACC synthase and ACC oxidase, have been cloned [139, 140]. Introduction of the ACC oxidase gene in antisense orientation has been achieved in both C. arabica and C. canephora [34]. The effect of the transgene on ethylene production and fruit maturation has yet to be reported. The inhibition of genes downstream to the initial ethylene burst is also an option to control coffee fruit maturation [141].

GM COFFEE AND CONSUMER APPROVAL

Considering that the technology for coffee transformation is available, and given the rapid progress in gene discovery, it may not be very long before transgenic coffee hits the market. Since transgenic coffee is at the initial stages of commercial development and needs to be integrated into the main breeding programmes for evaluation, it will take at least 15–20 years for field release of GM coffee. However, the main obstacle will be consumer approval and acceptance of genetically modified coffee. With a rapid increase in the cultivation of various transgenic crops around the globe, consumer perception towards transgenic coffee may be more positive than it is today. Undoubtedly, GM coffee must undergo rigorous testing on both health and environmental effects before it is released for commercial cultivation.

CONCLUSION AND FUTURE PERSPECTIVES

During the last 15 years, transgenic crops became an integral part of the agricultural landscape. The number of transgenic crops and the area under cultivation is rapidly increasing in many parts of the world. This has been made possible by the application of genetic transformation technology and its integration with plant breeding programmes. Despite significant advances made over the last 15 years, coffee transformation is still time consuming and laborious. In addition, a genotype independent transformation protocol is not yet achieved in coffee. Genetic transformation of coffee has two major applications: (1) a tool for the validation of gene function and (2) production of transgenic crops with agronomically important traits. In order to achieve these goals, a simple, efficient, genotype-independent routine transformation protocol needs to be developed for coffee. Public concern regarding the use of antibiotic marker genes in transgenic technology need to be addressed. In this regard, coffee transformation should be based on several strategies such as use of positive selection markers and GFP for transformant selection, and on the cre/lox system for elimination of selectable marker genes. Development of a clean gene transgenic technology for coffee based on the gfp gene for visual selection and the cre/lox vector system is currently in progress. All coffee transformation programs should address the expression of transgenes in appropriate tissues, for which tissue specific promoters need to be used. The stable expression of transgenes should be monitored at every stage of plant growth and development. In addition, genomic technologies such as transgenics, molecular marker assisted breeding, genomics, proteomics, and metabolomics should complement traditional breeding efforts for hastening the genetic improvement of coffee.

Further coffee transformation programmes and investments should involve public and private companies. At the same time, researchers must make an effort to educate the public and help them understand the real advantages and risks associated with the use of genetically modified coffee. This is the only way to address irrational fears about the transgenic crops, and this will pave the way to the use of transgenic technology for coffee improvement.

ACKNOWLEDGMENTS

M. K. Mishra thanks the Department of Biotechnology, Government of India for an overseas Associateship to De Montfort University, UK and Coffee Board, India for deputation.

REFERENCES

1.   ICO. International Coffee Organization, "ICO Annual Review 2010," 2010, http://www.ico.org/.

2.   A. P. Davis, R. Govaerts, D. M. Bridson, and P. Stoffelen, "An annotated taxonomic conspectus of the genus Coffea (Rubiaceae)," Botanical Journal of the Linnean Society, vol. 152, no. 4, pp. 465–512, 2006. ·

3.   A. Charrier and J. Berthaud, "Botanical classification of coffee," in Coffee, Botany, Biochemistry and Production of Beans and Beverage,

M. N. Clifford and K. C. Wilson, Eds., pp. 13–47, Croom Helm, London, UK, 1985.

4.   A. Lécolier, P. Besse, A. Charrier, T. N. Tchakaloff, and M. Noirot, "Unraveling the origin of Coffea arabica "Bourbon pointu" from la Réunion: a historical and scientific perspective," Euphytica, vol. 168, no. 1, pp. 1–10, 2009.

5.   T. W. M. Gole, T. Demel, M. Denich, and T. Bosch, "Diversity of traditional coffee production systems in Ethiopia and their contribution to conservation of genetic diversity," in Proceedings of the International Agricultural Research for Development, Deutscher Tropentag, Bonn, Germany, 2001.

6.   A. S. Thomas, "The wild arabica coffee on the Boma plateau of Anglo-Egyptian Sudan," Empirical Journal of Experimental Agriculture, vol. 10, pp. 207–212, 1942.

7.   F. Anthony, J. Berthaud, J. L. Guillaumet, and M. Lourd, "Collecting wild coffee species in Kenya and Tanzania," Plant Genetic Resources Newsletter, vol. 69, pp. 23–29, 1987.

8.   E. W. Githae, M. Chuah-Petiot, J. K. Mworia, and D. W. Odee, "A botanical inventory and diversity assessment of Mt. Marsabit forest, a sub-humid montane forest in the arid lands of Northern Kenya,"African Journal of Ecology, vol. 46, no. 1, pp. 39–45, 2008. ·

9.   M. F. Carneiro, "Coffee biotechnology and its application in genetic transformation," Euphytica, vol. 96, no. 1, pp. 167–172, 1997.

10. H. A. M. van der Vossen, "Coffee selection and breeding," in Coffee: Botany, Biochemistry and Production of Beans and Beverage, M. N. Clifford and K. C. Wilson, Eds., pp. 48–96, Croom Helm, London, UK, 1985.

11. K. H. Srinivasan and R. L. Narasimhaswamy, "A review of coffee breeding work done at the Government coffee experiment station, Balehonnur," Indian Coffee, vol. 34, pp. 311–321, 1975.

12. O. Guerreiro Filho, M. B. Silvarolla, and A. B. Eskes, "Expression and mode of inheritance of resistance in coffee to leaf miner Perileucoptera coffeella," Euphytica, vol. 105, no. 1, pp. 7–15, 1999.

13. A. Bettencourt, Considerações gerais sobre o 'Híbrido de Timor', Instituto Agronomico de Campinas. Circular, no. 31, Instituto Agronomico de Campinas, 1973.

14. "Coffee Guide," Central Coffee Research Institute, Coffee Board, Bangalore, India, 2000.

15. C. R. Barton, T. L. Adams, and M. A. Zarowitz, "Stable transformation of foreign DNA into Coffea arabica plants," in Proceedings of the 14th International Conference on Coffee Science (ASIC '91), pp. 460–464, San Francisco, Calif, USA, 1991.

16. R. Fernandez-Da Silva and A. Menéndez-Yuffá, "Transient gene expression in secondary somatic embryos from coffee tissues electroporated with the genes GUS and BAR," Electronic Journal of Biotechnology, vol. 6, no. 1, pp. 29–35, 2003.

17. V. Kumar, K. V. Satyanarayana, A. Ramakrishna, A. Chandrashekar, and G. A. Ravishankar, "Evidence for localization of N-methyltransferase (MMT) of caffeine biosynthetic pathway in vacuolar surface ofCoffea canephora endosperm elucidated through localization of GUS reporter gene driven by NMT promoter," Current Science, vol. 93, no. 3, pp. 383–386, 2007.

18. J. van Boxtel, M. Berthouly, C. Carasco, M. Dufour, and A. Eskes, "Transient expression of beta-glucuronidase following biolistic delivery of foreign DNA into coffee tissues," Plant Cell Reports, vol. 14, no. 12, pp. 748–752, 1995.

19. A. G. Rosillo, J. R. Acuna, A. L. Gaitan, and M. de Pena, "Optimised DNA delivery into Coffea arabicasuspension culture cells by particle bombardment," Plant Cell, Tissue and Organ Culture, vol. 74, no. 1, pp. 45–49, 2003.

20. W. G. Cunha, F. R. B. Machado, G. R. Vianna, J. B. Teixeira, and E. V. S. Albuquerque, Obtencao De Coffea Arabica Geneticamente Modificadas Por Bombardmento De Calos Embriogenecos, vol. 73 ofBoletim de Pesquisa e desenvolvimento, Embrapa, Brasilia, Brazil, 2004.

21. A. F. Ribas, A. K. Kobayashi, L. F. P. Pereira, and L. G. E. Vieira, "Genetic transformation of Coffea canephora by particle bombardment," Biologia Plantarum, vol. 49, no. 4, pp. 493–497, 2005.

22. A. M. Gatica-Arias, G. Arrieta-Espinoza, and A. M. Espinoza Esquivel, "Plant regeneration via indirect somatic embryogenesis and optimisation of genetic transformation in Coffea arabica L. cvs. Caturra and Catuaí," Electronic Journal of Biotechnology, vol. 11, no. 1, pp. 1–11, 2008.

23. E. V. S. Albuquerque, W. G. Cunha, A. E. A. D. Barbosa et al., "Transgenic coffee fruits from Coffea arabica genetically modified by bombardment," In Vitro Cellular & Developmental Biology—Plant, vol. 45, no. 5, pp. 532–539, 2009.

24. J. Spiral and V. Petiard, "Protoplast culture and regeneration in coffee species," in Proceedings of the 14th International Conference on Coffee

Science (ASIC '91), pp. 383–391, San Francisco, Calif, USA, 1991.

25. T. Hatanaka, Y. E. Choi, T. Kusano, and H. Sano, "Transgenic plants of coffee Coffea canephora from embryogenic callus via Agrobacterium tumefaciens-mediated transformation," Plant Cell Reports, vol. 19, no. 2, pp. 106–110, 1999.

26. T. Leroy, A. M. Henry, M. Royer et al., "Genetically modified coffee plants expressing the Bacillus thuringiensiscry1Ac gene for resistance to leaf miner," Plant Cell Reports, vol. 19, no. 4, pp. 382–389, 2000.

27. M. K. Mishra, H. L. Sreenath, and C. S. Srinivasan, "Agrobacterium-mediated transformation of coffee: an assessment of factors affecting gene transfer efficiency," in Proceedings of the 15th Plantation Crops Symposium, K. Sreedharan, P. K. V. Kumar, Jayarama, and B. M. Chulaki, Eds., pp. 251–255, Mysore, India, December 2002.

28. S. Ogita, H. Uefuji, M. Morimoto, and H. Sano, "Application of RNAi to confirm theobromine as the major intermediate for caffeine biosynthesis in coffee plants with potential for construction of decaffeinated varieties," Plant Molecular Biology, vol. 54, no. 6, pp. 931–941, 2004.·

29. M. K. Mishra and H. L. Sreenath, "High-efficiency Agrobacterium-mediated transformation of coffee (Coffea canephora Pierre ex. Frohner) using hypocotyl explants," in Proceedings of the 20th International Conference on Coffee Science (ASIC '04), pp. 792–796, Bangalore, India, October 2004.

30. A. R. R. Cruz, A. L. D. Paixao, F. R. Machado, et al., Obtencao de plantas transformadas de Coffea canephora por co-cultivo de calos embriogenicos com A. Tumefaciens, vol. 73 of Boletim de pesquiza e Desenvolvimento, Embrapa, Brasilia, Brazil, 2004.

31. R. L. R. Canche-Moo, A. Ku-Gonzalez, C. Burgeff, V. M. Loyola-Vargas, L. C. Rodríguez-Zapata, and E. Castaño, "Genetic transformation of Coffea canephora by vacuum infiltration," Plant Cell, Tissue and Organ Culture, vol. 84, no. 3, pp. 373–377, 2006.

32. V. Kumar, K. V. Sathyanarayana, S. Saarala Itty, P. Giridhar, A. Chandrasekhar, and G. A. Ravishankar, "Post transcriptional gene silencing for down regulating caffeine biosynthesis in Coffea canephora P. ex Fr," in Proceedings of the 20th International Conference on Coffee Science (ASIC '04), pp. 769–774, Bangalore, India, 2004.

33. B. Perthuis, J. L. Pradon, C. Montagnon, M. Dufour, and T. Leroy, "Stable resistance against the leaf miner Leucoptera coffeella expressed by genetically transformed Coffea canephora in a pluriannual field experiment in French Guiana," Euphytica, vol. 144, no. 3, pp. 321–329,

2005.

34. A. F. Ribas, R. M. Galvao, L. F. P. Pereira, and L. G. E. Vieira, "Transformacao de Coffea arabica com o gene da ACC oxidase em orientacao antinsenso," in Proceedings of the 50th Congreso Brasileiro de Genetica, p. 492, Sao Pulo, Brazil, September 2005.

35. A. F. Ribas, A. K. Kobayashi, L. F. P. Pereira, and L. G. E. Vieira, "Production of herbicide-resistant coffee plants (Coffea canephora P.) via Agrobacterium tumefaciens-mediated transformation," Brazilian Archives of Biology and Technology, vol. 49, no. 1, pp. 11–19, 2006.

36. A. Arroyo-Herrera, A. Ku Gonzalez, R. Canche Moo et al., "Expression of WUSCHEL in Coffea canephora causes ectopic morphogenesis and increases somatic embryogenesis," Plant Cell, Tissue and Organ Culture, vol. 94, no. 2, pp. 171–180, 2008.

37. M. K. Mishra, H. L. Sreenath, Jayarama et al., "Two critical factors: Agrobacterium strain and antibiotics selection regime improve the production of transgenic coffee plants," in Proceedings of the 22nd International Association for Coffee Science (ASIC '08), pp. 843–850, Campinas, Brazil, 2008.

38. A. F. Ribas, E. Dechamp, A. Champion et al., "Agrobacterium-mediated genetic transformation of Coffea arabica (L.) is greatly enhanced by using established embryogenic callus cultures," BMC Plant Biology, vol. 11, article 92, 2011.

39. J. Spiral, C. Thierry, M. Paillard, and V. Petiard, "Obtention de plantules de Coffea canephora Pierre (Robusta) transformées par Agrobacterium rhizogenes," Comptes Rendus de l' Academie des Sciences de Paris, vol. 316, no. 1, pp. 1–6, 1993.

40. M. Sugiyama, C. Matsuoka, and T. Takagi, "Transformation of Coffea with Agrobacterium rhizogenes," in Proceedings of the 16th International Conference on Coffee Science (ASIC '95), pp. 853–859, Kyoto, Japan, 1995.

41. C. A. Giménez, A. Menéndez-Yuffá, and E. de García, "Efecto del antibiótico kanamicina sobre diferentes explantes del híbrido de café (Coffea sp.) Catimor," Phyton, vol. 59, pp. 39–46, 1996.

42. V. Kumar, K. V. Satyanarayana, S. Sarala Itty et al., "Stable transformation and direct regeneration inCoffea canephora P ex. Fr. by Agrobacterium rhizogenes mediated transformation without hairy-root phenotype," Plant Cell Reports, vol. 25, no. 3, pp. 214–222, 2006.

43. E. Alpizar, E. Dechamp, B. Bertrand, P. Lashermes, and H. Etienne, "Transgenic roots for functional genomics of coffee resistance genes

to root-knot nematodes," in Proceedings of the 21st International Conference on Coffee Science (ASIC '06), pp. 653–659, Montpellier, France, September 2006.

44. E. Alpizar, E. Dechamp, S. Espeout et al., "Efficient production of Agrobacterium rhizogenes-transformed roots and composite plants for studying gene expression in coffee roots," Plant Cell Reports, vol. 25, no. 9, pp. 959–967, 2006.

45. H. Etienne, P. Lashermes, A. Menendez-Yuffa, Z. de Guglielmo-Croquer, E. Alpizar, and H. L. Sreenath, "Coffee," in Compendium of Transgenic Crop Plants, Transgenic Plantation Crops, Ornamentals and Turf Grasses, C. Kole and T. Hall, Eds., pp. 57–84, Wiley Blackwell Publishers, London, UK, 2008.

46. M. C. Combes, S. Andrzejewski, F. Anthony et al., "Characterization of microsatellite loci in Coffea arabica and related coffee species," Molecular Ecology, vol. 9, no. 8, pp. 1178–1180, 2000.

47. D. L. Steiger, C. Nagai, P. H. Moore, C. W. Morden, R. V. Osgood, and R. Ming, "AFLP analysis of genetic diversity within and among Coffea arabica cultivars," Theoretical and Applied Genetics, vol. 105, no. 2-3, pp. 209–215, 2002.

48. L. E. C. Diniz, N. S. Sakiyama, P. Lashermes, E. T. Caixeta, A. C. B. Oliveira, Zambolim, et al., "Analysis of AFLP marker associated to the Mex-1 resistance locus in Icatu progenies," Crop Breeding and Applied Biotechnology, vol. 5, pp. 387–393, 2005.

49. M. P. Maluf, M. Silvestrini, L. M. C. Ruggiero, O. Guerreiro Filho, and C. A. Colombo, "Genetic diversity of Coffea arabica inbreed lines assessed by RAPD, AFLP and SSR marker system," Scientia Agricola, vol. 62, no. 4, pp. 366–373, 2005.

50. P. S. Hendre, R. Phanindranath, V. Annapurna, A. Lalremruata, and R. K. Aggarwal, "Development of new genomic microsatellite markers from robusta coffee (Coffea canephora Pierre ex A. Froehner) showing broad cross-species transferability and utility in genetic studies," BMC Plant Biology, vol. 8, article 51, 2008.

51. R. F. Missio, E. T. Caixeta, E. M. Zambolim, L. Zambolim, C. D. Cruz, and N. S. Sakiyama, "Polymorphic information content of SSR markers for Coffea spp," Crop Breeding and Applied Biotechnology, vol. 10, no. 1, pp. 89–94, 2010.

52. P. Tornincasa, R. Dreos, B. de Nardi, E. Asquini, J. Devasia, M. K. Mishra, et al., "Genetic diversity of commercial coffee (C. arabica L) from America, India and Africa assessed by simple sequence repeats

(SSRs)," in Proceedings of the 21st International Association for Coffee Science (ASIC '06), pp. 778–785, Montpellier, France, 2006.

53. M. K. Mishra, S. Nishani, and Jayarama, "Genetic relationship among indigenous coffee species from India using RAPD, ISSR and SRAP markers," Biharean Biologists, vol. 5, no. 1, pp. 17–24, 2011.

54. M. K. Mishra, S. Nishani, and Jayarama, "Molecular identification and genetic relationships among coffee species (Coffea L.) inferred from ISSR and SRAP marker analyses," Archives of Biological Sciences, vol. 63, no. 3, pp. 667–679, 2011.

55. M. K. Mishra, N. Suresh, A. M. Bhat et al., "Genetic molecular analysis of Coffea arabica (Rubiaceae) hybrids using SRAP markers," Revista de Biologia Tropical, vol. 59, no. 2, pp. 607–617, 2011.

56. M. K. Mishra, A. M. Bhat, N. Suresh et al., "Molecular genetic analysis of arabica coffee hybrids using SRAP marker approach," Journal of Plantation Crops, vol. 39, no. 1, pp. 41–47, 2011.

57. M. K. Mishra, P. Tornincasa, B. de Nardi, E. Asquini, R. Dreos, L. Del Terra, et al., "Genome organization in coffee as revealed by EST PCRRFLP, SNPs and SSR analysis," Journal of Crop Science and Biotechnology, vol. 14, no. 1, pp. 25–37, 2011.

58. M. T. Gil-Agusti, N. Campostrini, L. Zolla, C. Ciambella, C. Invernizzi, and P. G. Righetti, "Two-dimensional mapping as a tool for classification of green coffee bean species," Proteomics, vol. 5, no. 3, pp. 710–718, 2005.

59. P. S. Hendre, R. K. Aggarwal, and D. N. A. Markers, "Development and applications for genetic improvement of coffee," in Genomics Assisted Crop Improvement, R. K. Varshney and R. Tuberosa, Eds., vol. 2 of Genomics Application in Crops, pp. 399–434, Springer.

60. M. de Block, L. Herrera-Estrella, M. van Montagu, J. Schell, and P. Zambryski, "Expression of foreign genes in regenerated plants and in their progeny," The EMBO Journal, vol. 3, no. 8, pp. 1681–1689, 1984.

61. G. Y. Zhong, "Genetic issues and pitfalls in transgenic plant breeding," Euphytica, vol. 118, no. 2, pp. 137–144, 2001.

62. A. Pallavicini, L. Del Terra, M. R. Sondahl et al., "Transcriptomics of resistance response in Coffea arabica L," in Proceedings of the 20th International conference on coffee science (ASIC '04), pp. 66–67, Bangalore, India, October 2004.

63. C. Lin, L. A. Mueller, J. M. Carthy, D. Crouzillat, V. Pétiard, and S. D. Tanksley, "Coffee and tomato share common gene repertoires as revealed

by deep sequencing of seed and cherry transcripts,"Theoretical and Applied Genetics, vol. 112, no. 1, pp. 114–130, 2005.

64. L. G. E. Vieira, A. C. Andrade, C. Colombo, et al., "Brazillian coffee genome project: an EST-based genomic resource," Brazillian Journal of Plant Physiology, vol. 18, no. 1, pp. 95–108, 2006.

65. S. Noir, S. Patheyron, M. C. Combes, P. Lashermes, and B. Chalhoub, "Construction and characterisation of a BAC library for genome analysis of the allotetraploid coffee species (Coffea arabicaL.)," Theoretical and Applied Genetics, vol. 109, no. 1, pp. 225–230, 2004.

66. T. Leroy, P. Marraccini, M. Dufour et al., "Construction and characterization of a Coffea canephora BAC library to study the organization of sucrose biosynthesis genes," Theoretical and Applied Genetics, vol. 111, no. 6, pp. 1032–1041, 2005.

67. J. Spiral, T. Leroy, M. Paillard, and V. Petiard, "Transgenic coffee (Coffea sp.)," in Biotechnology in Agricultulture and Forestry, Y. P. S. Bajaj, Ed., pp. 55–76, Springer, Heidelberg, Germany, 1999.

68. M. F. Carneiro, "Advances in coffee biotechnology," AgBiotechNet, vol. 1, pp. 1–14, 1999.

69. V. Kumar, M. M. Naidu, and G. A. Ravishankar, "Developments in coffee biotechnology—in vitro plant propagation and crop improvement," Plant Cell, Tissue and Organ Culture, vol. 87, no. 1, pp. 49–65, 2006.

70. G. Staritsky, "Embryoid formation in callus tissues of coffee," Acta Botanica Neerlandica, vol. 19, pp. 509–514, 1970.

71. W. R. Sharp, L. S. Caldas, O. J. Crocomo, L. C. Monaco, and A. Carvalho, "Production of Coffea arabicacallus of three ploidy levels and subsequent morphogenesis," Phyton, vol. 31, pp. 67–74, 1973.

72. M. R. Sondahl and W. Sharp, "High frequency induction of somatic embryos in cultured leaf explants ofCoffea arabica L," Zeitschrift für Pflanzenphysiologie, vol. 81, pp. 395–408, 1977.

73. M. R. Sondahl, D. Spahlinger, and W. R. Sharp, "A histological study of high frequency and low frequency induction of somatic embryos in cultured leaf explants of Coffea arabica L," Zeitschrift für Pflanzenphysiologie, vol. 94, no. 2, pp. 101–108, 1979.

74. F. R. Quiroz-Figueroa, C. F. J. Fuentes-Cerda, R. Rojas-Herrera, and V. M. Loyola-Vargas, "Histological studies on the developmental stages and differentiation of two different somatic embryogenesis systems of Coffea arabica," Plant Cell Reports, vol. 20, no. 12, pp. 1141–1149, 2002.

75. F. R. Quiroz-Figueroa, M. Méndez-Zeel, F. Sánchez-Teyer, R. Rojas-

Herrera, and V. M. Loyola-Vargas, "Differential gene expression in embryogenic and non-embryogenic clusters from cell suspension cultures of Coffea arabica," Journal of Plant Physiology, vol. 159, no. 11, pp. 1267–1270, 2002.

76. F. R. Quiroz-Figueroa, M. Méndez-Zeel, A. Larqué-Saavedra, and V. M. Loyola-Vargas, "Picomolar concentrations of salicylates induce cellular growth and enhance somatic embryogenesis in Coffea arabica tissue culture," Plant Cell Reports, vol. 20, no. 8, pp. 679–684, 2001.

77. P. Giridhar, E. P. Indu, G. A. Ravishankar, and A. Chandrasekar, "Influence of triacontanol on somatic embryogenesis in Coffea arabica L. and Coffea canephora P. ex Fr," In Vitro Cellular & Developmental Biology—Plant, vol. 40, no. 2, pp. 200–203, 2004.

78. P. Giridhar, E. P. Indu, K. Vinod, A. Chandrashekar, and G. A. Ravishankar, "Direct somatic embryogenesis from Coffea arabica L. and Coffea canephora P ex Fr. under the influence of ethylene action inhibitor-silver nitrate," Acta Physiologiae Plantarum, vol. 26, no. 3, pp. 299–305, 2004.

79. P. Giridhar, V. Kumar, E. P. Indu, A. Chandrasekar, and G. A. Ravishankar, "Thidiazuron induced somatic embryogenesis in Coffea arabica L. and Coffea canephora P ex Fr," Acta Botanica Croatica, vol. 63, no. 1, pp. 25–33, 2004.

80. V. Kumar, P. Giridhar, A. Chandrashekar, and G. A. Ravishankar, "Polyamines influence morphogenesis and caffeine biosynthesis in in vitro cultures of Coffea canephora P. ex Fr," Acta Physiologiae Plantarum, vol. 30, no. 2, pp. 217–223, 2008. ·

81. A. Ramakrishna, P. Giridhar, M. Jobin, C. S. Paulose, and G. A. Ravishankar, "Indoleamines and calcium enhance somatic embryogenesis in Coffea canephora P ex Fr," Plant Cell, Tissue and Organ Culture, vol. 108, no. 2, pp. 267–278, 2012.

82. T. Hatanaka, E. Sawabe, T. Azuma, N. Uchida, and T. Yasuda, "The role of ethylene in somatic embryogenesis from leaf discs of Coffea canephora," Plant Science, vol. 107, no. 2, pp. 199–204, 1995. ·

83. M. de Feria, E. Jiménez, R. Barbón, A. Capote, M. Chávez, and E. Quiala, "Effect of dissolved oxygen concentration on differentiation of somatic embryos of Coffea arabica cv. Catimor 9722," Plant Cell, Tissue and Organ Culture, vol. 72, no. 1, pp. 1–6, 2003. ·

84. A. Zamarripa, J. P. Ducos, H. Tessereau, H. Bollon, A. B. Eskes, and V. Pétiard, "Développement d'un procédé demultiplication en masse du caféier par embryogenèse somatique en milieu liquide," inProceedings of the 14th International Scientific Colloquium on Coffee (ASIC '91), pp.

392–402, San Francisco, Calif, USA, 1991.

85. A. Zamarripa, Etude et development de l'embryogenese en millieu liquid du cafeier (Coffea canephora P., Coffea arabica L. Et Hybrid Arabusta) [Ph.D. thesis], Ecolo National Superieure Agronomique, Rennes, France, 1993.

86. J. P. Ducos, A. Zamarripa, A. B. Eskes, and V. Petiard, "Production of somatic embryos of coffee in a bioreactor," in Proceedings of the 15th International Conference on Coffee Science (ASIC '93), pp. 89–96, Montpellier, France, 1993.

87. C. Noega and M. R. Sondahl, "Arabica coffee micropropagation through somatic embryogenesis via bioreactors," in Proceedings of the 15th International Conference on Coffee Science (ASIC '93), pp. 73–81, Montpellier, France, 1993.

88. H. Etienne and M. Berthouly, "Temporary immersion systems in plant micropropagation," Plant Cell, Tissue and Organ Culture, vol. 69, no. 3, pp. 215–231, 2002. ·

89. J. Albarrán, B. Bertrand, M. Lartaud, and H. Etienne, "Cycle characteristics in a temporary immersion bioreactor affect regeneration, morphology, water and mineral status of coffee (Coffea arabica) somatic embryos," Plant Cell, Tissue and Organ Culture, vol. 81, no. 1, pp. 27–36, 2005.

90. M. F. Carneiro and T. M. O. Ribeiro, "In vitro meristem culture and plant regeneration in some genotypes of Coffea arabica," Broteria Genetica, vol. 85, pp. 127–138, 1989.

91. M. Berthouly, D. Alvarad, C. Carrasco, and C. Teisson, "In vitro micropropagation of coffee sp. By temporary immersion," in Abstracts Eighth International Congress of Plant Tissue and Cell Culture, p. 162, Florence, Italy, 1994.

92. C. Teisson, D. Alvard, M. Berthouly, F. Cote, J. V. Escalant, and H. Etienne, "Culture in vitro par immersion temporaire un nouveau récipient," Plantations, Recherche, Développement, vol. 2, no. 5, pp. 29–31, 1995.

93. J. P. Ducos, R. Alenton, J. F. Reano, C. Kanchanomai, A. Deshayes, and V. Pétiard, "Agronomic performance of Coffea canephora P. trees derived from large-scale somatic embryo production in liquid medium," Euphytica, vol. 131, no. 2, pp. 215–223, 2003. ·

94. M. R. Sondahl, W. R. Romig, and A. Bragin, "Induction and selection of somaclonal variation in coffee," US patent 5436395, 1995.

95. V. M. Loyola-Vargas, C. Fuentes, M. Monforte-Gonzalez, M. Mendez-

Zeel, R. Rojas, and J. Mijangos-Cortes, "Coffee tissue culture as a new modelfor the study of somaclonal variation," in Proceedings of the 18th International Conference on Coffee Science (ASIC '99), pp. 302–307, Paris, France, 1999.

96. L. F. Sanchez-Teyer, F. R. Quiroz-Figueroa, V. M. Loyola-Vargas, and D. Infante-Herrera, "Culture-induced variation in plants of Coffea arabica cv. Caturra rojo, regenerated by direct and indirect somatic embryogenesis," Molecular Biotechnology, vol. 23, no. 2, pp. 107–115, 2003.

97. V. Rani, K. P. Singh, B. Shiran et al., "Evidence for new nuclear and mitochondrial genome organizations among high-frequency somatic embryogenesis-derived plants of allotetraploid Coffea arabica L. (Rubiaceae)," Plant Cell Reports, vol. 19, no. 10, pp. 1013–1020, 2000.·

98. D. Ganesh and H. L. Sreenath, "Embryo culture in coffee: technique and applications," Indian Coffee, vol. 4, pp. 7–9, 1999.

99. G. de Moro, M. Modonut, E. Asquini, P. Tornincasa, A. Pallavicini, and G. Graziosi, "Development and analysis of an EST databank of Coffea arabica," in Proceedings of the 6th Solanaceae Genome Workshop, p. 127, New Delhi, India, 2009.

100. K. V. Satyanarayana, V. Kumar, A. Chandrashekar, and G. A. Ravishankar, "Isolation of promoter forN-methyltransferase gene associated with caffeine biosynthesis in Coffea canephora," Journal of Biotechnology, vol. 119, no. 1, pp. 20–25, 2005.

101. C. Hinniger, V. Caillet, F. Michoux et al., "Isolation and characterization of cDNA encoding three dehydrins expressed during Coffea canephora (Robusta) grain development," Annals of Botany, vol. 97, no. 5, pp. 755–765, 2006.

102. A. J. Simkin, T. Qian, V. Caillet et al., "Oleosin gene family of Coffea canephora: quantitative expression analysis of five oleosin genes in developing and germinating coffee grain," Journal of Plant Physiology, vol. 163, no. 7, pp. 691–708, 2006.

103. S. Noir, F. Anthony, B. Bertrand, M. C. Combes, and P. Lashermes, "Identification of a major gene (Mex-1) from Coffea canephora conferring resistance to Meloidogyne exigua in Coffea arabica," Plant Pathology, vol. 52, no. 1, pp. 97–103, 2003.

104. N. S. Prakash, D. V. Marques, V. M. P. Varzea, M. C. Silva, M. C. Combes, and P. Lashermes, "Introgression molecular analysis of a leaf rust resistance gene from Coffea liberica into C. arabica L.,"Theoretical and Applied Genetics, vol. 109, no. 6, pp. 1311–1317, 2004.

105. E. K. Gichuru, M. C. Combes, E. W. Mutitu, et al., "Characterization

and genetic mapping of a gene conferring resistance to coffee berry disease (Colletotrichum kahawae) in Arabica coffee (Coffea arabica)," in Proceedings of the 21st International Conference on Coffee Science (ASIC '06), pp. 786–793, Montpellier, France, 2006.

106. S. Ogita, H. Uefuji, Y. Yamaguchi, N. Koizumi, and H. Sano, "RNA interference: producing decaffeinated coffee plants," Nature, vol. 423, no. 6942, p. 823, 2003.

107. V. Sridevi, P. Giridhar, P. S. Simmi, and G. A. Ravishankar, "Direct shoot organogenesis on hypocotyl explants with collar region from in vitro seedlings of Coffea canephora Pierre ex. Frohner cv. C × R andAgrobacterium tumefaciens-mediated transformation," Plant Cell, Tissue and Organ Culture, vol. 101, no. 3, pp. 339–347, 2010.

108. T. Leroy and M. Dufour, "Coffea spp. Genetic transformation," in Transgenic Crops of the World: Essential Protocols, I. S. Curtis, Ed., pp. 159–170, Kluwer Academic Publishers, Dordrecht, The Netherlands, 2004.

109. T. Leroy, M. Royer, M. Paillard et al., "Introduction de genes d interet agronomique dans l'espece Coffea canephora Pierre par transformation avec Agrobacterium sp," in Proceedings of the 17th International Conference on Coffee Science (ASIC '97), pp. 439–446, Nairobi, Kenya, 1997.

110. J. van Boxtel, A. Eskes, and M. Berthouly, "Glufosinate as an efficient inhibitor of callus proliferation in coffee tissue," In Vitro Cellular & Developmental Biology—Plant, vol. 33, no. 1, pp. 6–12, 1997.

111. N. P. Samson, C. Campa, M. Noirot, and A. de Kochko, "Potential use of D-xylose for coffee transformation," in Proceedings of the 20th International Conference on Coffee Science (ASIC '04), pp. 707–713, Bangalore, India, 2004.

112. M. K. Mishra, S. Devi, A. McCormac et al., "Green fluorescent protein as a visual selection marker for coffee transformation," Biologia, vol. 65, no. 4, pp. 639–646, 2010.

113. B. Perthuis, T. Leroy, M. Dufour et al., "Variability in the insecticidal protein concentration within transformed Coffea canephora observed in a field experiment," in Proceedings of the 21st International Cconference on Coffee Science (ASIC '06), pp. 1390–1393, Montpellier, France, September 2006.

114. C. Montagnon, "Genetically modified coffees- the experience of CIRAD," in ICO seminar on Genetically Modified Coffee Workshop, May 2005,http://dev.ico.org/event_pdfs/gm/presentations/Christophe%20

Montagnon.pdf.

115. I. Méndez-López, R. Basurto-Ríos, and J. E. Ibarra, "Bacillus thuringiensis serovar israelensis is highly toxic to the coffee berry borer, Hypothenemus hampei Ferr. (Coleoptera: Scolytidae)," FEMS Microbiology Letters, vol. 226, no. 1, pp. 73–77, 2003.

116. M. F. Grossi de Sa, R. A. Pereira, E. V. S. A. Barros et al., "Uso de inibidores de alfa-amilases no controle da broca-do-cafe," in Anais do workshop Internacional de Manejo da Broca-do-cafe, Paraná, Brazil, 2004.

117. V. P. Campos, P. Sivapalan, and N. C. Gnanapragasam, "Nematode parasites of coffee, cocoa and tea," inPlant Parasitic Nematodes in Subtropical and Tropical Agriculture, M. Luc, R. A. Sikora, and J. Bridge, Eds., pp. 113–126, CAB International, Wallingford, UK, 1990.

118. B. Bertrand, F. Anthony, and P. Lashermes, "Breeding for resistance to Meloidogyne exigua in Coffea arabica by introgression of resistance genes of Coffea canephora," Plant Pathology, vol. 50, no. 5, pp. 637–643, 2001.

119. M. M. Chaves, J. P. Maroco, and J. S. Pereira, "Understanding plant responses to drought—from genes to the whole plant," Functional Plant Biology, vol. 30, no. 3, pp. 239–264, 2003.

120. M. M. Chaves and M. M. Oliveira, "Mechanisms underlying plant resilience to water deficits: prospects for water-saving agriculture," Journal of Experimental Botany, vol. 55, no. 407, pp. 2365–2384, 2004. ·

121. I. Coraggio and R. Tuberosa, "Molecular basis of plant adaptation to abiotic stress and approaches to enhance tolerance to hostile environment," in Hand Book of Plant Biotechnology, P. Christou and H. Klee, Eds., pp. 413–466, John Wiley & Sons, West Sussex, UK, 2004.

122. H. A. M. van der Vossen, "Agronomy I: coffee breeding practices," in Coffee: Recent Developments, R. J. Clarke and O. G. Vitzthum, Eds., pp. 184–201, Blackwell Science, London, UK, 2001.

123. M. S. Sreenivasan, "Breeding for leaf rust resistance in India," in Coffee Rust: Epidemiology, Resistance and Management, A. C. Kushalappa and A. B. Eskes, Eds., pp. 316–323, CRC Press, Boca Raton, Fla, USA, 1989.

124. B. J. Staskawicz, F. M. Ausubel, B. J. Baker, J. G. Ellis, and J. D. G. Jones, "Molecular genetics of plant disease resistance," Science, vol. 268, no. 5211, pp. 661–667, 1995.

125. B. J. Feys and J. E. Parker, "Interplay of signaling pathways in plant disease resistance," Trends in Genetics, vol. 16, no. 10, pp. 449–455,

2000.

126. F. L. W. Takken and M. H. A. J. Joosten, "Plant resistance genes: their structure, function and evolution," European Journal of Plant Pathology, vol. 106, no. 8, pp. 699–713, 2000.·

127. X. Dong, "SA, JA, ethylene, and disease resistance in plants," Current Opinion in Plant Biology, vol. 1, no. 4, pp. 316–323, 1998.

128. M. P. Does and B. J. C. Cornelissen, "Emerging strategies to control fungal diseases using transgenic plants," in in proceedings of International Crop Science Congress, V. L. Chopra, R. B. Singh, and A. Verma, Eds., pp. 233–244, Oxford and IBH, New Delhi, India, 1998.

129. D. Fernandez, P. Santos, C. Agostini et al., "Coffee (Coffea arabica L.) genes early expressed during infection by the rust fungus (Hemileia vastatrix)," Molecular Plant Pathology, vol. 5, no. 6, pp. 527–536, 2004.

130. B. de Nardi, R. Dreos, L. Del Terra et al., "Differential responses of Coffea arabica L. leaves and roots to chemically induced systemic acquired resistance," Genome, vol. 49, no. 12, pp. 1594–1605, 2006.

131. K. David, "Fun without the buzz: decaffeination process and issues," 2002,http://www.virtualcoffee.com/sept_2002/decafe.html.

132. M. N. Clifford, "Chemical and physical aspects of green coffee and coffee products," in Coffee: Botany Biochemistry and Production of Beans and Beverages, M. N. Clifford and K. C. Williamson, Eds., pp. 304–374, AVI, Westport, Conn, USA, 1985.

133. C. A. B. de Maria, L. C. Trugo, F. R. Aquino Neto, R. F. A. Moreira, and C. S. Alviano, "Composition of green coffee water-soluble fractions and identification of volatiles formed during roasting," Food Chemistry, vol. 55, no. 3, pp. 203–207, 1996.

134. M. N. Clifford, "Chlorogenic acids and other cinnamates—nature, occurrence and dietary burden,"Journal of the Science of Food and Agriculture, vol. 79, no. 3, pp. 362–372, 1999. ·

135. C. Campa, M. Noirot, M. Bourgeois et al., "Genetic mapping of a caffeoyl-coenzyme A 3-0-methyltransferase gene in coffee trees. Impact on chlorogenic acid content," Theoretical and Applied Genetics, vol. 107, no. 4, pp. 751–756, 2003.

136. K. Hahlbrock and D. Scheel, "Physiology and molecular biology of phenyl-propanoid metabolism,"Annual Review of Plant Physiology and Plant Molecular Biololgy, vol. 40, pp. 347–369, 1989.

137. V. Mahesh, J. J. Rakotomalala, L. L. Gal et al., "Isolation and genetic mapping of a Coffea canephoraphenylalanine ammonia-lyase gene

(CcPAL1) and its involvement in the accumulation of caffeoyl quinic acids," Plant Cell Reports, vol. 25, no. 9, pp. 986–992, 2006. ·

138. G. Parvatam, V. Mahesh, G. A. Ravishankar, C. Campa, and A. de Kochko, "Functional validation ofCoffea PAL genes using genetic engineering," in Proceedings of the 21st International Conference on Coffee Science (ASIC '06), pp. 702–705, Montpellier, France, September 2006.

139. L. F. Protasio Pereira, R. M. Galvão, A. K. Kobayashi, S. M. B. Cação, and L. G. Esteves Vieira, "Ethylene production and ACC oxidase gene expression during fruit ripening of Coffea arabica L," Brazilian Journal of Plant Physiology, vol. 17, no. 3, pp. 283–289, 2005.

140. K. R. Neupane, S. Moisyadi, and J. Stiles, "Cloning and characterization of fruit expressed ACC synthase and ACC oxidase from coffee," in In Proceedings of the 18th International Conference on Coffee Science (ASIC '99), pp. 322–326, Helsinki, Finland, 1999.

141. A. F. Ribas, L. F. P. Pereira, and L. G. E. Vieira, "Genetic transformation of coffee," Brazilian Journal of Plant Physiology, vol. 18, no. 1, pp. 83–94, 2006.

Chapter 5

RECENT ADVANCES IN UTILIZING TRANSCRIPTION FACTORS TO IMPROVE PLANT ABIOTIC STRESS TOLERANCE BY TRANSGENIC TECHNOLOGY

Hongyan Wang[1], Honglei Wang[1], Hongbo Shao[2,3] and Xiaoli Tang[3]

[1]Institute of Technology, Yantai Academy of China Agriculture University, Yantai, China

[2]Jiangsu Key Laboratory for Bioresources of Saline Soils, Provincial Key Laboratory of Agrobiology, Institute of Biotechnology, Jiangsu Academy of Agricultural Sciences, Nanjing, China

[3]Key Laboratory of Coastal Biology and Bioresources Utilization, Yantai Institute of Coastal Zone Research, Chinese Academy of Sciences, Yantai, China

Agricultural production and quality are adversely affected by various abiotic stresses worldwide and this will be exacerbated by the deterioration of global climate. To feed a growing world population, it is very urgent to breed stress-tolerant crops with higher yields and improved qualities against multiple environmental stresses. Since conventional breeding approaches had marginal success due to the complexity of stress tolerance traits, the transgenic approach is now being popularly used to breed stress-tolerant crops. So identifying and characterizing the critical genes involved in plant stress responses is an essential prerequisite for engineering stress-tolerant crops. Far beyond the manipulation of single functional gene, engineering certain regulatory genes has emerged as an effective strategy now for controlling the expression of many stress-responsive genes. Transcription factors (TFs) are good candidates for genetic engineering to breed stress-tolerant crop because of their role as master regulators of many stress-responsive genes. Many TFs belonging to families AP2/EREBP, MYB, WRKY, NAC, bZIP have been found to be involved in various abiotic stresses and some TF genes have also been engineered to improve stress tolerance in model and crop plants. In this review, we take five large families of TFs as examples and review the recent progress of TFs involved in plant abiotic stress responses and their potential utilization to improve multiple stress tolerance of crops in the field conditions.

INTRODUCTION

Agricultural production and quality are adversely affected by a broad range of abiotic stresses including drought, salinity, heat, and cold. Especially when these stresses occur in combination, it can have devastating effects on plant growth and productivity. It is estimated that more than 50% of worldwide yield loss for major crop are caused by abiotic stresses (Shao et al., 2009; Ahuja et al., 2010; Lobell et al., 2011). According to the current climate prediction models, the deterioration of global climate will inevitably cause an increased frequency of drought, heat wave, and salinization (Easterling et al., 2000; Ipcc, 2007, 2008). This means that agricultural productivity will face a greater challenge in fighting against environmental stresses. Meanwhile, the growing world population will reach close to ten billion by the year 2050 and then almost two times of current agricultural productivity is needed to feed the large population (Bengtsson et al., 2006; United Nations, 2015). Moreover, such a tremendous increase of crop productivity must be achieved with no increase in arable land and in the face of multiple environmental stresses. Where is the way to solve this problem? Many scholars and experts worldwide have reached a consensus that breeding stress-tolerant crops with higher yields and improved qualities against multiple environmental stresses is an effective strategy, as well as one of the greatest challenges faced by modern agriculture (Takeda and Matsuoka, 2008; Newton et al., 2011; Liu J.-H. et al., 2014). In the past few decades, a great deal of efforts has been devoted to breeding crop cultivars with various stress-tolerant traits. Two main approaches have been employed to this process. One is traditional breeding methods such as wide-cross hybridization and mutation breeding, which often brings about unpredictable results. Another is modern transgenic technology by introducing novel exogenous genes or altering the expression levels of endogenous genes to improve stress tolerance. Since conventional breeding approaches have marginal success due to the complexity of stress tolerance traits, the transgenic approach is now being popularly used to develop transgenic crops tolerant to abiotic stresses (Yamaguchi and Blumwald, 2005). Therefore, deciphering the molecular mechanisms by which plants perceive and transduce stress signals to cellular machinery to initiate adaptive responses is an essential prerequisite for identification of the key genes and pathways to engineer stress-tolerant crop plants (Ray et al., 2009; Heidarvand and Amiri, 2010; Sanchez et al., 2011).

Substantial progress has been made to unravel the molecular mechanisms of abiotic stress responses in plants by means of high throughput sequencing and functional genomics tools. To date, a number of critical genes involved in abiotic stress tolerance have been identified and validated, which are generally classified into two types: functional genes and regulatory genes (Shinozaki et al., 2003). The former encodes important enzymes and metabolic proteins (functional proteins), such as detoxification enzyme, water channel, ion transporter, heat shock protein (HSP), and late embryogenesis abundant (LEA) protein, which directly function to protect cells from stresses. The latter encodes various regulatory proteins including TFs, protein kinases and protein phosphatases, which regulate signal transduction and gene expression in the stress responses. Although there have been numerous studies on functional genes, most of these studies pay more attention to single gene or several genes encoding enzymes and protective proteins by imposing a given stress. Due to the complexity of stress responses regulated by multi-genes, little success has been achieved by a single functional gene approach to significantly enhance plant stress tolerance (Mittler and Blumwald, 2010; Varshney et al., 2011). Given the complexity and variability of field conditions where crops are often simultaneously subjected to multiple abiotic stresses or some in combination (Ahuja et al., 2010), more and more studies have paid close attention to regulatory genes and found that some regulatory genes including TFs play essential roles in multiple abotic stress responses by regulating a large spectrum of downstream stress-responsive genes. Thus, genetically modifying the expression of certain regulatory genes can greatly influence plant stress tolerance because it mimics or enhances stress signals to regulate many downstream stress-responsive genes at a time (Century et al., 2008; Yang et al., 2011). Among the regulatory genes, stress-responsive TFs have attracted particular attention due to their important roles in plant stress responses (Chen and Zhu, 2004; Xu et al., 2008a). In this paper, we mainly review the recent progress of TFs involved in plant abiotic stress responses and their potential utilization to improve multiple stress tolerance of crops in the field conditions.

The Generic Signaling Pathway Involved in Plant Abiotic Stress Responses

As sessile organisms, plants have evolved various defense mechanisms at multiple levels to respond to unfavorable environment including diverse abiotic stresses. As stated before, it is imperative to dissect regulatory mechanisms of stress response and identify the key regulators involved in this process to breed or genetically engineer stress-tolerant plants. With the availability of plant genomes and various omics tools including genomics, transcriptomics, and proteomics tools, major progress has been made in deciphering the stress signaling pathways and relevant components involved in plant abiotic stress response, but there is still much more to be determined (Liu J.-H. et al., 2014). According to our current knowledge about stress signaling pathways, the generic signaling pathway for any given abiotic stress can be divided into the following major steps: signal perception, signal transduction, stress responsive gene expression, in turn the activation of physiological, and metabolic responses (Chaves et al., 2003; Yamaguchi-Shinozaki and Shinozaki, 2006; Pérez-Clemente et al., 2013). In this process, plant cells first perceive stress stimulus through sensors or receptors located in the cell wall or membrane. Then the captured extracellular signals are converted into intracellular ones through second messengers including calcium ions, inositol phosphate, reactive oxygen species (ROS), cyclic nucleotides (cAMP and cGMP), sugars, and nitric oxide. Subsequently, these second messengers initiate the corresponding signaling pathways to transduce the signals (Chaves et al., 2009; Bhargava and Sawant, 2013). In many signal transduction pathways, the phosphorylation, and dephosphorylation of proteins mediated by protein kinase and phosphatases, respectively, is an important and effective mechanism for signal relay (Singh et al., 2003). For example, the mitogen activated protein kinases (MAPKs) pathway and calcium-dependent protein kinases (CDPKs) pathway are known to be involved in plant abiotic stress responses (Schaller et al., 2008; Huang G.T. et al., 2012). At the end of the phosphorylation cascade, TFs are activated or suppressed by protein kinases or phosphatases, and they further bind specifically to cis-elements in the promoters of stress-responsive genes and regulate their transcription (Danquah et al., 2014). Meanwhile, TFs themselves are regulated at the transcription level by other upstream components (Hirayama and Shinozaki, 2010) and also subjected to various tiers of modifications at the post-transcription level, such as ubiquitination and sumoylation, thus forming a complex regulatory network to modulate the expression of stress responsive genes, which in turn determine the activation of physiological and metabolic responses (Dong et al., 2006;Miura et al., 2007; Mizoi et al., 2013). All the components mentioned above, from the foremost receptors to the downstream functional genes, constitute the generic pathway for plant abiotic stress signal transduction (Figure 1). As one of the most important regulators, TFs function as terminal transducers and directly regulate the expression of an array of downstream genes by interacting with the specific cis-elements in their promoter region (Yamaguchi-Shinozaki and Shinozaki, 2006). In the last few decades, considerable research has been conducted to identify and characterize various TFs involved in plant abiotic stress responses either in abscisic acid (ABA)-dependent pathway or ABA-independent pathway, such as *AP2/EREBP, MYB, WRKY, NAC, bZIP*, and so on (Vinocur and Altman, 2005; Umezawa et al., 2006; Golldack et al., 2011). Numerous efforts have been also made to improve plant stress tolerance by engineering these TF genes, and some promising results have been reported in succession (Table 1). In the following sections, we mainly summarize current information on several major TF families including their features, roles, and biotechnological uses for improving the abiotic stress tolerance in plants.

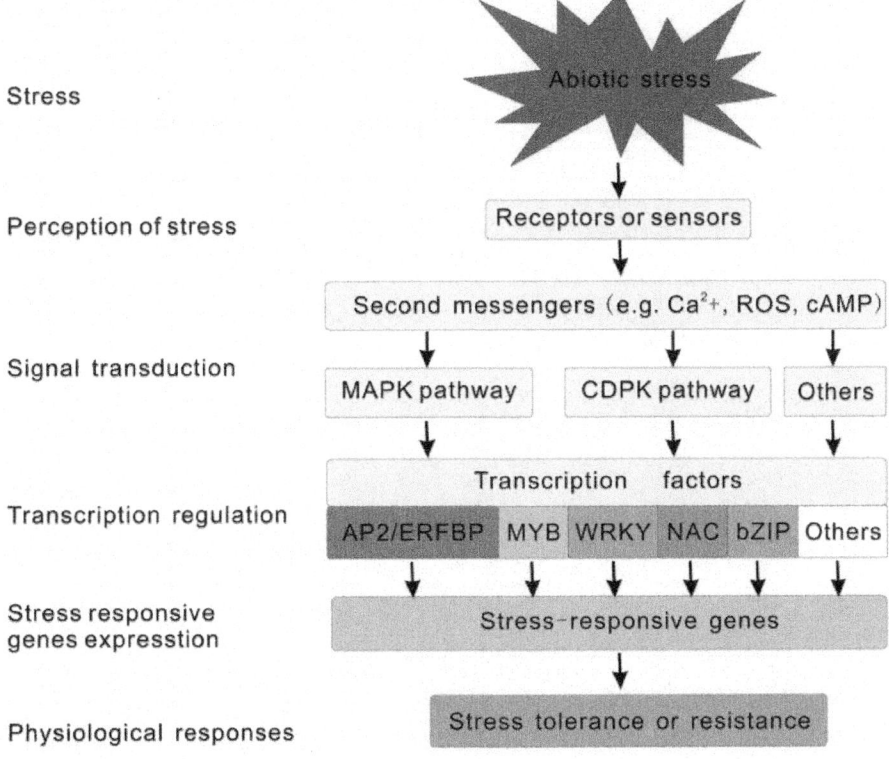

Figure 1. Generic signaling pathway involved in plant abiotic stress responses.

Table 1. Some examples of transgenic plants over-expressing transcription factor genes in recent years

Family	Gene	Donor	Acceptor	Enhanced tolerance	References
AP2/ERFBP	LcDREB3a	Leymus chinensis	Arabidopsis	Drought and salinity↑	Peng et al., 2013
	LcDREB2	Leymus chinensis	Arabidopsis	Salinity↑	Peng et al., 2013
	LcERF054	Lotus corniculatus	Arabidopsis	Salinity↑	Sun et al., 2014
	VrDREB2A	Vigna radiata	Arabidopsis	Drought and salinity↑	Chen et al., 2015
	GmERF3	Glycine max	Tobacco	Drought and salinity↑	Zhang et al., 2009
	GmERF7	Glycine max	Tobacco	Salinity↑	Zhai et al., 2013
	SsDREB	Suaeda salsa	Tobacco	Drought and salinity↑	Zhang X. et al., 2015
	JERF3	Solanum lycopersicum	Rice	Drought↑	Zhang et al., 2010
	OsDREB2A	Oryza sativa	Rice	Drought and salinity↑	Mallikarjuna et al., 2011
	OsERF4a	Oryza sativa	Rice	Drought↑	Joo et al., 2013
	AtDREB1A	Arabidopsis	Rice	Drought↑	Ravikumar et al., 2014
	TaERF3	Triticum aestivum	Wheat	Drought and salinity↑	Rong et al., 2014
	TaPIE1	Triticum aestivum	Wheat	Cold↑	Zhu et al., 2014
	EaDREB2	Erianthus arundinaceus	Sugarcane	Drought and salinity↑	Augustine et al., 2015
	StDREB1	Solanum tuberosum	Potato	Salinity↑	Bouaziz et al., 2013
MYB	AtMYB15	Arabidopsis	Arabidopsis	Drought and salinity↑	Ding et al., 2009
	LcMYB1	Leymus chinensis	Arabidopsis	Salinity↑	Cheng et al., 2013b
	GmMYBJ1	Glycine max	Arabidopsis	Drought and cold↑	Su et al., 2014
	TaMYB3R1	Triticum aestivum	Arabidopsis	Drought and salinity↑	Cai et al., 2015
	TaPIMP1	Triticum aestivum	Tobacco	Drought and salinity↑	Liu et al., 2011b
	SbMYB2	Scutellaria baicalensis	Tobacco	NaCl, mannitol, and ABA stresses↑	Qi et al., 2015
	SbMYB7	Scutellaria baicalensis	Tobacco	NaCl, mannitol, and ABA stresses↑	Qi et al., 2015
	LeAN2	Lycopersicum esculentum	Tobacco	Chilling and oxidative stresses↑	Meng et al., 2014
	LeAN2	Lycopersicum esculentum	Tomato	Heat↑	Meng et al., 2015
	AtMYB44	Arabidopsis	Soybean	Drought and salinity↑	Seo et al., 2012
	OsMYB2	Oryza sativa	Rice	Drought, cold, and salinity↑	Yang et al., 2012
	OsMYB91	Oryza sativa	Rice	Salinity↑	Zhu et al., 2015
	MdSIMYB1	Malus × domestica	Apple	Drought, cold, and salinity↑	Wang et al., 2014
WRKY	AtWRKY28	Arabidopsis	Arabidopsis	Salinity↑	Babitha et al., 2013
	OsWRKY45	Oryza sativa	Arabidopsis	Drought and salinity↑	Qiu and Yu, 2009
	TaWRKY79	Triticum aestivum	Arabidopsis	Drought↑	Qin et al., 2013
	VvWRKY11	Vitis vinifera	Arabidopsis	Drought↑	Liu et al., 2011a
	ZmWRKY33	Zea may	Arabidopsis	Salinity↑	Li et al., 2013
	GhWRKY34	Gossypium hirsutum	Arabidopsis	Salinity↑	Zhou et al., 2015
	GsWRKY20	Glycine soja	Arabidopsis	Drought↑	Luo et al., 2013
	TaWRKY79	Triticum aestivum	Arabidopsis	Salinity and ionic stress↑	Qin et al., 2013
	TaWRKY93	Triticum aestivum	Arabidopsis	Salinity, drought, and low temperature↑	Qin et al., 2015
	TaWRKY10	Triticum aestivum	Tobacco	Drought and salinity↑	Wang et al., 2013
	GhWRKY39	Gossypium hirsutum	Tobacco	Salinity↑	Shi et al., 2014
	BdWRKY36	Brachypodium distachyon	Tobacco	Drought↑	Sun et al., 2015
	ZmWRKY58	Zea may	Rice	Drought and salinity↑	Cai et al., 2014
	MtWRKY76	Medicago truncatula	Medicago truncatula	Drought and salinity↑	Liu et al., 2016
NAC	ANAC019	Arabidopsis	Arabidopsis	Cold↑	Jensen et al., 2010
	ONAC063	Oryza sativa	Arabidopsis	Salinity and osmotic tolerance↑	Yokotani et al., 2009
	GmNAC20	Glycine max	Arabidopsis	Salinity and freezing tolerance↑	Hao et al., 2011
	ZmSNAC1	Zea may	Arabidopsis	Cold, salinity, and drought↑	Lu et al., 2012
	BnNAC5	Brassica napus	Arabidopsis	Salinity↑	Zhong et al., 2012

Family	Gene	Donor	Acceptor	Enhanced tolerance	References
	TaNAC67	Triticum aestivum	Arabidopsis	Cold, salinity, and drought↑	Mao et al., 2014
	TaNAC29	Triticum aestivum	Arabidopsis	Drought and salinity↑	Huang et al., 2015
	MLNAC5	Miscanthus lutarioriparius	Arabidopsis	Drought and cold↑	Yang et al., 2015
	TaNAC2a	Triticum aestivum	Tobacco	Drought	Tang et al., 2012
	AhNAC3	Arachis hypogaea	Tobacco	Drought↑	Liu et al., 2013
	SNAC1	Oryza sativa	Wheat	Drought and salinity↑	Saad et al., 2013
	OsNAP	Oryza sativa	Rice	Cold, salinity, and drought↑	Chen et al., 2014
bZIP	ABP9	Zea may	Arabidopsis	Drought, salinity, and cold↑	Zhang et al., 2011
	GmbZIP1	Glycine max	Arabidopsis	Drought, salinity, and cold↑	Gao et al., 2011
	ZmbZIP72	Zea may	Arabidopsis	Drought and salinity↑	Ying et al., 2012
	TabZIP60	Triticum aestivum	Arabidopsis	Drought, salt, and freezing tolerance↑	Zhang L. et al., 2015
	PtrABF	Poncirus trifoliata	Tobacco	Drought↑	Huang et al., 2010
	GmbZIP1	Glycine max	Tobacco	Drought, salinity, and cold↑	Gao et al., 2011
	LrbZIP	Nelumbo nucifera	Tobacco	Salinity↑	Cheng et al., 2013a
	OsbZIP71	Oryza sativa	Rice	Drought and salinity↑	Liu C. et al., 2014

AP2/EREBP Transcription Factors

AP2/ERFBP family includes a large group of plant-specific TFs and is characterized by the presence of the highly conserved *AP2*/ethylene-responsive element-binding factor (ERF) DNA-binding domain that directly interact with GCC box and/or dehydration-responsive element (DRE)/C-repeat element (CRT) cis-acting elements at the promoter of downstream target genes (Riechmann and Meyerowitz, 1998). AP2/ERFBP TFs perform a variety of roles in plant developmental processes and stress responses, such as vegetative and reproductive development, cell proliferation, abiotic and biotic stress responses, and plant hormone responses (Nakano et al., 2006; Licausi et al., 2010; Sharoni et al., 2011). Presently, a multitude of AP2/ERFBP members have been identified in various species by means of genome-wide analysis, such as 145 in *Arabidopsis* (Riechmann and Meyerowitz, 1998), 163 in rice (Sharoni et al., 2011), 200 in poplar (Zhuang et al., 2008), 291 in *Chinese cabbage* (Song et al., 2013), 171 in *foxtail millet* (Lata et al., 2014), 116 in moso bamboo (Wu et al., 2015). Based on the number and similarity of *AP2/ERF* domains, these *AP2/EREBP* TFs are grouped into four major subfamilies: *AP2* (Apetala 2), *RAV* (related to *ABI3/VP1*), *DREB* (dehydration-responsive element-binding protein), and *ERF* (Sakuma et al., 2002; Sharoni et al., 2011). Among these, both *ERF* and *DREB* subfamilies have been extensively studied due to their involvement in plant responses to biotic and abiotic stresses.

The DREB subfamily can regulate the expression of multiple dehydration/cold-regulated (*RD/COR*) genes by interacting with *DRE/CRT* cis-elements (A/GCCGAC) located in the promoters of *RD/COR* genes that are responsive to water deficit and low-temperature, such as *COR15A, RD29A/COR78*, and *COR6.6* (Stockinger et al., 1997; Liu et al., 1998; Lucas et al., 2011). A lot of DREB-type TFs have been tested in many plants including *Arabidopsis*, wheat, tomato, soybean, rice, maize, and barley (Agarwal et al., 2006; Lata and

Prasad, 2011;Mizoi et al., 2012). According to the variation in some conserved motifs and biological functions in divergent species, *DREB* TFs are further classified into two major subgroups: *DREB1*/C-repeat-binding factor (*DREB1/ CBF*) and *DREB2*, and each of them is involved in separate signal transduction pathway under abiotic stresses (Dubouzet et al., 2003). Commonly, *DREB1/ CBF* genes are involved in low temperature stress responses in *Arabidopsis* and rice, while *DREB2* genes respond to dehydration, high salinity and heat shock (Liu et al., 1998; Sakuma et al., 2002; Lucas et al., 2011). For example, three major *DREB1/CBF* members in *Arabidopsis, DREB1A/CBF3, DREB1B/ CBF1* and *DREB1C/CBF2* are rapidly induced in response to cold stress (Stockinger et al., 1997; Gilmour et al., 1998; Liu et al., 1998; Shinwari et al., 1998). Over-expressing any one of these three *DREB1s/CBFs* displayed significantly improved tolerance to freezing, drought, and high salinity in transgenic *Arabidopsis* (Gilmour et al., 1998; Jaglo-Ottosen et al., 1998; Liu et al., 1998). Further, over-expressing *Arabidopsis DREB1/CBF* genes improved freezing tolerance in oilseed rape (Jaglo et al., 2001) and chilling tolerance in tomato, tobacco and rice (Tsai-Hung et al., 2002; Kasuga et al., 2004; Ito et al., 2006). Some *DREB1/CBF* homologous genes have also been isolated from many other plant species including tomato, oilseed rape, wheat, rye, rice, and maize, and some of them have been used to produce transgenic plants with improved tolerance to abiotic stress (Jaglo et al., 2001; Dubouzet et al., 2003; Qin et al., 2004). In contrast, *DREB2* genes have been studied in a limited number of plant species, but the existing studies have shown that*DREB2* genes are also involved in abiotic stress responses in plants. In *Arabidopsis, DREB2A* and *DREB2B* are major *DREB2s* induced by dehydration, high salinity, and heat, while *DREB2C* is induced by heat later than them (Liu et al., 1998; Nakashima et al., 2000; Sakuma et al., 2006b; Lim et al., 2007). Over-expression of the constitutively active form of *AtDREB2A* from *Arabidopsis* improved the tolerance to drought and osmotic stress in transgenic *Arabidopsis* plants (Sakuma et al., 2006a; Xu et al., 2008b). Over-expression of*ZmDREB2A* from maize enhanced drought tolerance in transgenic *Arabidopsis* plants (Qin et al., 2007). The transgenic plants harboring *GmDREB2* from soybean also showed enhanced tolerance to drought and high salinity without any growth retardation (Chen et al., 2007), as was observed in transgenic rice by over-expressing *OsDREB2A* under control of stress-inducible *RD29A* promoter (Mallikarjuna et al., 2011).

The ERF subfamily is the largest group of the AP2/EREBP TF family (Dietz et al., 2010) and functions in plant stress tolerance by regulating the stress-responsive genes through interacting with the cis-element GCC boxes with core sequence of AGCCGCC (Ohme-Takagi and Shinshi, 1995; Hao et al., 1998). An array of ERF genes are induced by various abiotic stresses, such as drought, high salinity, osmotic stress, and cold (Xu et al., 2008a). Over-expression of these ERF genes resulted in improvement of abiotic stress tolerance in transgenic plants, as summarized in Table 1. It is worth mentioning that some ERFs function in both biotic and abiotic stress tolerance, and this is partly due to their involvement in various hormonal signaling pathways including ethylene, JA, or SA (Liang et al., 2008). For example, over-expressing *TaPIE1* in wheat significantly enhanced resistance to both pathogen and freezing stress (Zhu et al., 2014). Over-expressing *GmERF3* in tobacco not only enhanced resistance against infection by pathogen and tobacco mosaic virus (TMV) but also improved tolerance to high salinity and dehydration (Zhang et al., 2009). So far, functions of a limited number of *ERFs* have been well characterized, but most of *ERF* family members have yet to be identified.

MYB Transcription Factors

The MYB TFs are widely distributed in plants and form a large family characterized by a highly conserved MYB domain for DNA-binding, which contains from 1 to 4 imperfect repeats (*MYB* repeat) at the N-terminus. In contrast, the activation domain is located at the C-terminus and varies significantly among MYBs, leading to versatile regulatory roles of *MYB* family. According to the number of MYB domain repeats, the MYB TFs are divided into four groups: 1R-MYB (MYB-related type), R2R3-MYB, R1R2R3-MYB, and 4R-MYB, containing one, two, three, and four *MYB* repeats, respectively. Among them, the R2R3-MYBs are more prevalent in plants (Dubos et al., 2010; Ambawat et al., 2013; Li et al., 2015). So far, large numbers of MYB members have been identified in different plant species, such as 198 in *Arabidopsis* (Yanhui et al., 2006), 183 in rice (Yanhui et al., 2006), 229 in apple (Cao et al., 2013), 177 in sweet orange (Hou et al., 2014), 209 in *foxtail millet* (Muthamilarasan et al., 2014).

Numerous MYB TFs have been found to function in many significant physiological and biochemical processes including cell development and cell cycle, primary and secondary metabolism, hormone synthesis and signal transduction, as well as in plant responses to various biotic and abiotic stresses (Allan et al., 2008; Dubos et al., 2010; Ambawat et al., 2013). Recently, some abiotic stress-responsive *MYB* TFs in *Arabidopsis* and other plants have been well summarized by Li (Li et al., 2015). For example, *AtMYB15* improved freezing tolerance by regulating *CBF* expression (Agarwal et al., 2006); *AtMYB44, AtMYB60,* and *AtMYB61* improved drought tolerance by regulating stomatal movement (Cominelli et al., 2005; Liang et al., 2005; Jung et al., 2008). Especially, *AtMYB96* improved drought tolerance either by integrating ABA and auxin signals (Seo et al., 2009) or by activating cuticular wax biosynthesis (Seo et al., 2011), and also improved freezing and drought tolerance by regulating a lipid-transfer protein LTP3. This fact shows that a MYB factor can regulate diverse target genes involved in various physiological processes under abiotic stresses. In addition, *OsMYB2* from rice was induced by salt, cold, and dehydration stress. The transgenic plants with over-expressing *OsMYB2*exhibited enhanced tolerance to various stresses by the alteration of expression levels of numerous genes involving diverse functions in stress response (Yang et al., 2012). Salt and freezing tolerance in *Arabidopsis* was significantly enhanced by over-expressing either *GmMYB76* or *GmMYB177* from soybean (Liao et al., 2008).

WRKY Transcription Factors

The WRKY family is also extensively distributed in plants and contains many members. WRKY TFs are characterized by the presence of one or two highly conserved WRKY domains of about 60 amino acid residues, which contains a conserved WRKYGQK motif at the N-terminus and a C2H2 or C2HC zinc-finger motif at the C-terminus The WRKY domains can specifically bind to W-box cis-elements with a core sequence of TTGACC/T, located at the promoters of many target genes. According to the number of WRKY domains and the feature of the zinc finger motif, the WRKY TFs can be categorized into three groups. Group I members contain two WRKY domains and a C2H2 zinc-finger motif, whereas group II and III members contain one WRKY domain with a C2H2 and C2HC zinc-finger motif, respectively (Eulgem et al., 2000; Ulker and Somssich, 2004; Pandey and Somssich, 2009; Rushton et al., 2010). Since the cloning of the first cDNA encoding a WRKY protein (SPF1) from sweet potato (Ishiguro and Nakamura, 1994), an increasing number of WRKY TFs have been identified in various plants, such as 74 in *Arabidopsis* (Ulker and Somssich, 2004), 102 in rice (Wu et al., 2005), 104 in poplar (He et al., 2012), 86 in *Brachypodium distachyon* (Wen et al., 2014), 182 in soybean (Bencke-Malato et al., 2014), and 116 and 102 genes in two different species of cotton (Dou et al., 2014).

WRKY TFs have been shown to participate in various processes in plants, including plant growth, seed development, leaf senescence, and responses to biotic and abiotic stresses (Rushton et al., 2010). Accumulating evidence has demonstrated that WRKY TFs play key roles in plant responses to a variety of abiotic stresses such as drought, salt, heat, cold, and osmotic pressure, and these topics have been extensively reviewed recently (Chen et al., 2012; Rushton et al., 2012; Tripathi et al., 2014; Banerjee and Roychoudhury, 2015). Over-expression of some stress-responsive *WRKY* genes showed enhanced tolerance to abiotic stresses in transgenic plants. For example, transgenic rice plants harboring *OsWRKY11* gene showed significant heat and drought tolerance (Wu et al., 2009). Transgenic *Arabidopsis* plants over-expressing *GmWRKY21* gene exhibited improved tolerance to cold stress, while over-expressing *GmWRKY54* gene improved tolerance to drought and salt stress (Zhou et al., 2008). Transgenic *Arabidopsis* plants over-expressing *VvWRKY11*improved to tolerance mannitol-induced osmotic stress (Liu et al., 2011a). Although some WRKYs in several plants have been functionally characterized, the vast majority of WRKYs in many plants, especially in non-model plants, are far from being functionally elucidated.

NAC Transcription Factors

Like the transcription factor families mentioned above, the NAC TFs also comprise a large plant-specific superfamily present in a wide range of plant species. The typical features of a NAC TF contain a highly conserved NAC domain in the N-terminal region and a variable transcriptional regulatory region in the C-terminal region. The NAC domain is associated with DNA binding, nucleus-oriented localization, and the formation of homodimers or heterodimers with other NAC proteins, while the transcriptional regulatory functions as a transcriptional activator or repressor (Olsen et al., 2005; Puranik et al., 2012). NAC TFs can regulate the transcription of downstream target genes by interacting with *NAC* recognition sequence (NACRS) with the CACG core-DNA binding motif in the promoter of these genes. NAC TFs have been found to participate in various processes including flower development, formation of secondary walls and cell division, shoot apical meristem formation, leaf senescence, as well as biotic and abiotic stress responses (Olsen et al., 2005; Tran et al., 2010; Nakashima et al., 2012; Nuruzzaman et al., 2013; Banerjee and Roychoudhury, 2015). To date, a lot of putative NAC TFs have been identified in many sequenced species at genome-wide scale, such as 117 in *Arabidopsis* and 151 in rice (Nuruzzaman et al., 2010), 152 in soybean (Le et al., 2011), 204 in Chinese cabbage (Liu T.K. et al., 2014), 152 in maize (Shiriga et al., 2014), tomato (Su et al., 2015). Moreover, quite a lot of them have been found to be involved in abiotic stress responses. For instance, 33 *NAC* genes changed significantly in *Arabdopsis* under salt stress (Jiang and Deyholos, 2006), 40 *NAC* genes changed under drought or salt stress in rice (Fang et al., 2008), 38 *NAC* genes changed in soybean under drought stress (Le et al., 2011), 32 *NAC* genes responded to at least two kinds of treatments in *Chrysanthemum lavandulifolium*(Huang H. et al., 2012). These stress-responsive *NAC* genes showed differential expression patterns such as tissue-specific, developmental stage- or stress-specific expression, indicating their involvement in the complex signaling networks during plant stress responses.

Some of these stress-responsive *NAC* genes have been over-expressed in *Arabidopsis*, rice and other plants and displayed positive effects, summarized in Table 1.

bZIP Transcription Factors

The basic leucine zipper (bZIP) family contains a conserved bZIP domain which is composed of a highly basic region for nuclear localization and DNA binding at the N-terminus and a leucine-rich motif for dimerization at the C-terminus (Landschulz et al., 1988; Hurst, 1994). Like other TFs, the bZIP TFs not only play pivotal roles in developmental processes but also respond to various abiotic stresses such as drought, high salinity, and cold stresses (Jakoby et al., 2002). Now, many members of the bZIP TF family have been identified or predicted at genome-wide level in some species. For example, it has been reported 75 in *Arabidopsis* (Jakoby et al., 2002), 89 in rice (Nijhawan et al., 2008), 125 in maize (Wei et al., 2012), 89 in barley (Pourabed et al., 2015), 55 in grapevine (Liu J. et al., 2014), 96 in *Brachypodium distachyon* (Liu and Chu, 2015). However, only a small portion of bZIP TFs has been well studied and most studies on their involvement in stress responses have shown that bZIP TFs are induced by ABA and regulate the expression of stress-related genes in ABA-dependent manner through interaction with specific ABA-responsive cis-acting elements (ABRE) in their promoter region (Uno et al., 2000; Kim et al., 2006; Zou et al., 2008). A lot of efforts have been made to improve abiotic stress tolerance in transgenic plants by over-expressing some stress-responsive bZIP genes and some successful example have been achieved, as listed in Table 1.

CONCLUSIONS AND PERSPECTIVES

Taking five large families of TFs as examples, this review emphasizes the promising roles of TFs as tools to improve plant responses to multiple abiotic stresses. In addition to the above-mentioned several TF families, there are still other TF families such as DNA binding with one finger (Dof) TFs, basic helix-loop-helix (bHLH) TFs, homeodomain-leucine zipper (HD-Zip) TFs, heat shock TFs (HSFs), etc. How to select the key TFs in such a huge gene families and fully display its potential is still an important issue before us. Although a great deal of information about TFs has been accumulated on their involvement in response to diverse abiotic stresses and a good number of promising candidate TF genes have been validated, but there are still some problems to be solved. First, functional redundancy between different TF members may hinder the dissection of the functions of an individual member. Second, most of transgenic studies based on TFs focused on plant growth and tolerance to a given stress at a given developmental stage rather than whole stage. Moreover, the evaluation of transgenic plants was conducted in controlled laboratory or greenhouse conditions rather than field conditions. Third, the constitutive over-expression of some TF genes may improve the stress tolerance, but occasionally lead to negative effects in transgenic plants such as dwarfing, late flowering, and lower yields. Finally, the complete regulation mechanism of individual transcription factor including its upstream and downstream co-regulators, as well as their interactions remains largely unknown.

Abiotic stress response in plants is an extremely complicated process because of the huge gene families and the complex interactions between TFs and cis-elements on the promoters of target genes. Moreover, one transcription factor may regulate a vast array of target genes with the corresponding cis-elements on the promoters, whereas one gene with several types of cis-elements may be regulated by different families of TFs. Thus, the stress-responsive TFs not only function independently but also cross-talk between each other in response to various abiotic stress responses, which indicates the complexity of signaling networks involved in plant stress responses. In the future research, we should first identify multiple stress-responsive TF genes by comparing their expression patterns and the identification of commonly regulated genes which have been proposed to be required for universal stress responses or represent points of cross-talk between signaling pathways (Prasch and Sonnewald, 2015). Genetic manipulation of these identified genes should be a more powerful approach for improving plant tolerance to multiple stresses than manipulation of individual functional genes. Then, the selected TF genes should be validated not only in model plant species but also in crop plants by use of stress-inducible promoter which can minimize the negative effects caused by over-expressing some TF genes. Furthermore, the critical field trials are required to evaluate the transgenetic plants, especially focusing on their growth and tolerance in the whole life period. which is a necessary step in many strategies to develop stress-tolerant crops. Taken together, we still need to struggle for a complete understanding the precise regulatory mechanisms involved in plant abiotic stress responses, which helps to obtain the promising candidate TF genes for breeding multiple abiotic stress-tolerant crops with better yields and qualities.

AUTHOR CONTRIBUTIONS

HYW, HLW, XT wrote the paper. HS provided the paper frame and revised the final paper. All authors reviewed the final manuscript.

ACKNOWLEDGMENTS

This work has been jointly supported by the National Natural Science Foundation of China (41171216), Jiangsu Autonomous Innovation of Agricultural Science and Technology [CX(15)1005], National Basic Research Program of China (2013CB430403; 2013CB430400), Open Foundation of Jiangsu Key Laboratory for Bioresources of Saline Soils (JKLBS2014005), Jiangsu Province Science and Technology Support Plan (BE2013429), and Yantai Double-hundred Talent Plan (XY-003-02), and Shuangchuang Talent Plan of Jiangsu Province.

REFERENCES

1. Agarwal, M., Hao, Y., Kapoor, A., Dong, C. H., Fujii, H., Zheng, X., et al. (2006). A R2R3 type MYB transcription factor is involved in the cold regulation of CBF genes and in acquired freezing tolerance. *J. Biol. Chem.* 281, 37636–37645. doi: 10.1074/jbc.M605895200

2. Ahuja, I., de Vos, R. C., Bones, A. M., and Hall, R. D. (2010). Plant molecular stress responses face climate change. *Trends Plant Sci.* 15, 664–674. doi: 10.1016/j.tplants.2010.08.002

3. Allan, A. C., Hellens, R. P., and Laing, W. A. (2008). MYB transcription factors that colour our fruit. *Trends Plant Sci.* 13, 99–102. doi: 10.1016/j.tplants.2007.11.012

4. Ambawat, S., Sharma, P., Yadav, N. R., and Yadav, R. C. (2013). MYB transcription factor genes as regulators for plant responses: an overview. *Physiol. Mol. Biol. Plants* 19, 307–321. doi: 10.1007/s12298-013-0179-1

5. Augustine, S. M., Ashwin Narayan, J., Syamaladevi, D. P., Appunu, C., Chakravarthi, M., Ravichandran, V., et al. (2015). Overexpression of EaDREB2 and pyramiding of EaDREB2 with the pea DNA helicase gene (PDH45) enhance drought and salinity tolerance in sugarcane (Saccharum spp. hybrid). *Plant Cell Rep.* 34, 247–263. doi: 10.1007/s00299-014-1704-6

6. Babitha, K. C., Ramu, S. V., Pruthvi, V., Mahesh, P., Nataraja, K. N., and Udayakumar, M. (2013). Co-expression of AtbHLH17 and AtWRKY28 confers resistance to abiotic stress in *Arabidopsis*. *Transgenic Res.* 22, 327–341. doi: 10.1007/s11248-012-9645-8

7. Banerjee, A., and Roychoudhury, A. (2015). WRKY proteins: signaling and regulation of expression during abiotic stress responses. *Sci. World J.* 2015:807560. doi: 10.1155/2015/807560

8.   Bencke-Malato, M., Cabreira, C., Wiebke-Strohm, B., Bücker-Neto, L., Mancini, E., Osorio, M. B., et al. (2014). Genome-wide annotation of the soybean WRKY family and functional characterization of genes involved in response to *Phakopsora pachyrhizi* infection. *BMC Plant Biol.* 14:236. doi: 10.1186/s12870-014-0236-0

9.   Bengtsson, M., Shen, Y., and Oki, T. (2006). A SRES-based gridded global population dataset for 1990–2100. *Popul. Environ.* 28, 113–131. doi: 10.1007/s11111-007-0035-8

10. Bhargava, S., and Sawant, K. (2013). Drought stress adaptation: metabolic adjustment and regulation of gene expression. *Plant Breeding* 132, 21–32. doi: 10.1111/pbr.12004

11. Bouaziz, D., Pirrello, J., Charfeddine, M., Hammami, A., Jbir, R., Dhieb, A., et al. (2013). Overexpression of StDREB1 transcription factor increases tolerance to salt in transgenic potato plants. *Mol. Biotechnol.* 54, 803–817. doi: 10.1007/s12033-012-9628-2

12. Cai, H., Tian, S., Dong, H., and Guo, C. (2015). Pleiotropic effects of TaMYB3R1 on plant development and response to osmotic stress in transgenic *Arabidopsis. Gene* 558, 227–234. doi: 10.1016/j.gene.2014.12.066

13. Cai, R., Zhao, Y., Wang, Y., Lin, Y., Peng, X., Li, Q., et al. (2014). Overexpression of a maize WRKY58 gene enhances drought and salt tolerance in transgenic rice. *Plant Cell Tiss. Org. Cult.* 119, 565–577. doi: 10.1007/s11240-014-0556-7

14. Cao, Z. H., Zhang, S. Z., Wang, R. K., Zhang, R. F., and Hao, Y. J. (2013). Genome wide analysis of the apple MYB transcription factor family allows the identification of MdoMYB121 gene confering abiotic stress tolerance in plants. *PLoS ONE* 8:e69955. doi: 10.1371/journal.pone.0069955

15. Century, K., Reuber, T. L., and Ratcliffe, O. J. (2008). Regulating the regulators: the future prospects for transcription-factor-based agricultural biotechnology products. *Plant Physiol.* 147, 20–29. doi: 10.1104/pp.108.117887

16. Chaves, M. M., Flexas, J., and Pinheiro, C. (2009). Photosynthesis under drought and salt stress: regulation mechanisms from whole plant to cell. *Ann. Bot.* 103, 551–560. doi: 10.1093/aob/mcn125

17. Chaves, M. M., Maroco, J., and Joao, S. P. (2003). Understanding plant responses to drought - from genes to the whole plant. *Funct. Plant Biol.* 30, 239–264. doi: 10.1071/FP02076

18. Chen, H., Liu, L., Wang, L., Wang, S., and Cheng, X. (2015). VrDREB2A,

a DREB-binding transcription factor from *Vigna radiata*, increased drought and high-salt tolerance in transgenic *Arabidopsis thaliana*. *J. Plant Res*. doi: 10.1007/s10265-015-0773-0. [Epub ahead of print].

19. Chen, L., Song, Y., Li, S., Zhang, L., Zou, C., and Yu, D. (2012). The role of WRKY transcription factors in plant abiotic stresses. *Biochim. Biophys. Acta* 1819, 120–128. doi: 10.1016/j.bbagrm.2011.09.002

20. Chen, M., Wang, Q. Y., Cheng, X. G., Xu, Z. S., Li, L. C., Ye, X. G., et al. (2007). GmDREB2, a soybean DRE-binding transcription factor, conferred drought and high-salt tolerance in transgenic plants. *Biochem. Biophys. Res. Commun*. 353, 299–305. doi: 10.1016/j.bbrc.2006.12.027

21. Chen, W. J., and Zhu, T. (2004). Networks of transcription factors with roles in environmental stress response. *Trends Plant Sci*. 9, 591–596. doi: 10.1016/j.tplants.2004.10.007

22. Chen, X., Wang, Y., Lv, B., Li, J., Luo, L., Lu, S., et al. (2014). The NAC family transcription factor OsNAP confers abiotic stress response through the ABA pathway. *Plant Cell Physiol*. 55, 604–619. doi: 10.1093/pcp/pct204

23. Cheng, L., Li, S., Hussain, J., Xu, X., Yin, J., Zhang, Y., et al. (2013a). Isolation and functional characterization of a salt responsive transcriptional factor, LrbZIP from lotus root (*Nelumbo nucifera* Gaertn). *Mol. Biol. Rep*. 40, 4033–4045. doi: 10.1007/s11033-012-2481-3

24. Cheng, L., Li, X., Huang, X., Ma, T., Liang, Y., Ma, X., et al. (2013b). Overexpression of sheepgrass R1-MYB transcription factor LcMYB1 confers salt tolerance in transgenic *Arabidopsis*. *Plant Physiol. Biochem*. 70, 252–260. doi: 10.1016/j.plaphy.2013.05.025

25. Cominelli, E., Galbiati, M., Vavasseur, A., Conti, L., Sala, T., Vuylsteke, M., et al. (2005). A guard-cell-specific MYB transcription factor regulates stomatal movements and plant drought tolerance. *Curr. Biol*. 15, 1196–1200. doi: 10.1016/j.cub.2005.05.048

26. Danquah, A., de Zelicourt, A., Colcombet, J., and Hirt, H. (2014). The role of ABA and MAPK signaling pathways in plant abiotic stress responses. *Biotechnol. Adv*. 32, 40–52. doi: 10.1016/j.biotechadv.2013.09.006

27. Dietz, K. J., Vogel, M. O., and Viehhauser, A. (2010). AP2/EREBP transcription factors are part of gene regulatory networks and integrate metabolic, hormonal and environmental signals in stress acclimation and retrograde signalling. *Protoplasma* 245, 3–14. doi: 10.1007/s00709-010-0142-8

28. Ding, Z., Li, S., An, X., Liu, X., Qin, H., and Wang, D. (2009). Transgenic expression of MYB15 confers enhanced sensitivity to abscisic acid and

improved drought tolerance in *Arabidopsis thaliana. J. Genet. Genomics* 36, 17–29. doi: 10.1016/S1673-8527(09)60003-5

29. Dong, C. H., Agarwal, M., Zhang, Y., Xie, Q., and Zhu, J. K. (2006). The negative regulator of plant cold responses, HOS1, is a RING E3 ligase that mediates the ubiquitination and degradation of ICE1. *Proc. Natl. Acad. Sci. U.S.A.* 103, 8281–8286. doi: 10.1073/pnas.0602874103

30. Dou, L., Zhang, X., Pang, C., Song, M., Wei, H., Fan, S., et al. (2014). Genome-wide analysis of the WRKY gene family in cotton. *Mol. Genet. Genomics* 289, 1103–1121. doi: 10.1007/s00438-014-0872-y

31. Dubos, C., Stracke, R., Grotewold, E., Weisshaar, B., Martin, C., and Lepiniec, L. (2010). MYB transcription factors in *Arabidopsis. Trends Plant Sci.* 15, 573–581. doi: 10.1016/j.tplants.2010.06.005

32. Dubouzet, J. G., Sakuma, Y., Ito, Y., Kasuga, M., Dubouzet, E. G., Miura, S., et al. (2003). OsDREB genes in rice, *Oryza sativa* L., encode transcription activators that function in drought-, high-salt- and cold-responsive gene expression. *Plant J.* 33, 751–763. doi: 10.1046/j.1365-313X.2003.01661.x

33. Easterling, D. R., Meehl, G. A., Parmesan, C., Changnon, S. A., Karl, T. R., and Mearns, L. O. (2000). Climate extremes: observations, modeling, and impacts. *Science* 289, 2068–2074. doi: 10.1126/science.289.5487.2068

34. Eulgem, T., Rushton, P. J., Robatzek, S., and Somssich, I. E. (2000). The WRKY superfamily of plant transcription factors. *Trends Plant Sci.* 5, 199–206. doi: 10.1016/S1360-1385(00)01600-9

35. Fang, Y., You, J., Xie, K., Xie, W., and Xiong, L. (2008). Systematic sequence analysis and identification of tissue-specific or stress-responsive genes of NAC transcription factor family in rice. *Mol. Genet. Genomics* 280, 547–563. doi: 10.1007/s00438-008-0386-6

36. Gao, S. Q., Chen, M., Xu, Z. S., Zhao, C. P., Li, L., Xu, H. J., et al. (2011). The soybean GmbZIP1 transcription factor enhances multiple abiotic stress tolerances in transgenic plants. *Plant Mol. Biol.* 75, 537–553. doi: 10.1007/s11103-011-9738-4

37. Gilmour, S. J., Zarka, D. G., Stockinger, E. J., Salazar, M. P., Houghton, J. M., and Thomashow, M. F. (1998). Low temperature regulation of the*Arabidopsis* CBF family of AP2 transcriptional activators as an early step in cold-induced COR gene expression. *Plant J.* 16, 433–442. doi: 10.1046/j.1365-313x.1998.00310.x

38. Golldack, D., Luking, I., and Yang, O. (2011). Plant tolerance to drought and salinity: stress regulating transcription factors and their functional

significance in the cellular transcriptional network. *Plant Cell Rep.* 30, 1383–1391. doi: 10.1007/s00299-011-1068-0

39. Hao, D., Ohme-Takagi, M., and Sarai, A. (1998). Unique mode of GCC box recognition by the DNA-binding domain of ethylene-responsive element-binding factor (ERF domain) in plant. *J. Biol. Chem.* 273, 26857–26861. doi: 10.1074/jbc.273.41.26857

40. Hao, Y. J., Wei, W., Song, Q. X., Chen, H. W., Zhang, Y. Q., Wang, F., et al. (2011). Soybean NAC transcription factors promote abiotic stress tolerance and lateral root formation in transgenic plants. *Plant J.* 68, 302–313. doi: 10.1111/j.1365-313X.2011.04687.x

41. He, H., Dong, Q., Shao, Y., Jiang, H., Zhu, S., Cheng, B., et al. (2012). Genome-wide survey and characterization of the WRKY gene family in*Populus trichocarpa. Plant Cell Rep.* 31, 1199–1217. doi: 10.1007/s00299-012-1241-0

42. Heidarvand, L., and Amiri, R. M. (2010). What happens in plant molecular responses to cold stress? *Acta Physiol. Plant.* 32, 419–431. doi: 10.1007/s11738-009-0451-8

43. Hirayama, T., and Shinozaki, K. (2010). Research on plant abiotic stress responses in the post-genome era: past, present and future. *Plant J.* 61, 1041–1052. doi: 10.1111/j.1365-313X.2010.04124.x

44. Hou, X. J., Li, S. B., Liu, S. R., Hu, C. G., and Zhang, J. Z. (2014). Genome-wide classification and evolutionary and expression analyses of citrus MYB transcription factor families in sweet orange. *PLoS ONE* 9:e112375. doi: 10.1371/journal.pone.0112375

45. Huang, G. T., Ma, S. L., Bai, L. P., Zhang, L., Ma, H., Jia, P., et al. (2012). Signal transduction during cold, salt, and drought stresses in plants. *Mol. Biol. Rep.* 39, 969–987. doi: 10.1007/s11033-011-0823-1

46. Huang, H., Wang, Y., Wang, S., Wu, X., Yang, K., Niu, Y., et al. (2012). Transcriptome-wide survey and expression analysis of stress-responsive NAC genes in *Chrysanthemum lavandulifolium. Plant Sci.* 193–194, 18–27. doi: 10.1016/j.plantsci.2012.05.004

47. Huang, Q., Wang, Y., Li, B., Chang, J., Chen, M., Li, K., et al. (2015). TaNAC29, a NAC transcription factor from wheat, enhances salt and drought tolerance in transgenic *Arabidopsis. BMC Plant Biol.* 15:268. doi: 10.1186/s12870-015-0644-9

48. Huang, X. S., Liu, J. H., and Chen, X. J. (2010). Overexpression of PtrABF gene, a bZIP transcription factor isolated from *Poncirus trifoliata*, enhances dehydration and drought tolerance in tobacco via scavenging ROS and modulating expression of stress-responsive genes. *BMC Plant Biol.* 10:230. doi: 10.1186/1471-2229-10-230

49. Hurst, H. C. (1994). Transcription factors. 1: bZIP proteins. *Protein Profile* 1, 123–168.

50. Ipcc (2007). "Climate change 2007: the physical science basis," *in Contribution of Working Group I to the Fourth Assessment Report of the Intergovernmental Panel on Climate Change*, eds S. Solomon, D. Qin, M. Manning, Z. Chen, M. Marquis, K. B. Averyt, M. Tignor, and H. L. Miller (Geneva: Ipcc Secretariat).

51. Ipcc (2008). "Climate change and water," *in Technical Paper of the Intergovernmental Panel on Climate Change*, eds B. C. Bates, Z. W. Kundzewicz, J. Palutikof, and S. Wu (Geneva: Ipcc Secretariat).

52. Ishiguro, S., and Nakamura, K. (1994). Characterization of a cDNA encoding a novel DNA-binding protein, SPF1, that recognizes SP8 sequences in the 5' upstream regions of genes coding for sporamin and beta-amylase from sweet potato. *Mol. Gen. Genet.* 244, 563–571. doi: 10.1007/BF00282746

53. Ito, Y., Katsura, K., Maruyama, K., Taji, T., Kobayashi, M., Seki, M., et al. (2006). Functional analysis of rice DREB1/CBF-type transcription factors involved in cold-responsive gene expression in transgenic rice. *Plant Cell Physiol.* 47, 141–153. doi: 10.1093/pcp/pci230

54. Jaglo, K. R., Kleff, S., Amundsen, K. L., Zhang, X., Haake, V., Zhang, J. Z., et al. (2001). Components of the *Arabidopsis* C-repeat/dehydration-responsive element binding factor cold-response pathway are conserved in *Brassica napus* and other plant species. *Plant Physiol.* 127, 910–917. doi: 10.1104/pp.010548

55. Jaglo-Ottosen, K. R., Gilmour, S. J., Zarka, D. G., Schabenberger, O., and Thomashow, M. F. (1998). *Arabidopsis* CBF1 overexpression induces COR genes and enhances freezing tolerance. *Science* 280, 104–106. doi: 10.1126/science.280.5360.104

56. Jakoby, M., Weisshaar, B., Dröge-Laser, W., Vicente-Carbajosa, J., Tiedemann, J., Kroj, T., et al. (2002). bZIP transcription factors in*Arabidopsis*. *Trends Plant Sci.* 7, 106–111. doi: 10.1016/S1360-

1385(01)02223-3

57. Jensen, M. K., Kjaersgaard, T., Nielsen, M. M., Galberg, P., Petersen, K., O›shea, C., et al. (2010). The *Arabidopsis thaliana* NAC transcription factor family: structure-function relationships and determinants of ANAC019 stress signalling. *Biochem. J.* 426, 183–196. doi: 10.1042/BJ20091234

58. Jiang, Y., and Deyholos, M. K. (2006). Comprehensive transcriptional profiling of NaCl-stressed *Arabidopsis* roots reveals novel classes of responsive genes. *BMC Plant Biol.* 6:25. doi: 10.1186/1471-2229-6-25

59. Joo, J., Choi, H. J., Lee, Y. H., Kim, Y. K., and Song, S. I. (2013). A transcriptional repressor of the ERF family confers drought tolerance to rice and regulates genes preferentially located on chromosome 11. *Planta* 238, 155–170. doi: 10.1007/s00425-013-1880-6

60. Jung, C., Seo, J. S., Han, S. W., Koo, Y. J., Kim, C. H., Song, S. I., et al. (2008). Overexpression of AtMYB44 enhances stomatal closure to confer abiotic stress tolerance in transgenic *Arabidopsis*. *Plant Physiol.* 146, 623–635. doi: 10.1104/pp.107.110981

61. Kasuga, M., Miura, S., Shinozaki, K., and Yamaguchi-Shinozaki, K. (2004). A combination of the *Arabidopsis* DREB1A gene and stress-inducible rd29A promoter improved drought- and low-temperature stress tolerance in tobacco by gene transfer. *Plant Cell Physiol.* 45, 346–350. doi: 10.1093/pcp/pch037

62. Kim, Y. S., Kim, S. G., Park, J. E., Park, H. Y., Lim, M. H., Chua, N. H., et al. (2006). A membrane-bound NAC transcription factor regulates cell division in *Arabidopsis*. *Plant Cell* 18, 3132–3144. doi: 10.1105/tpc.106.043018

63. Landschulz, W. H., Johnson, P. F., and McKnight, S. L. (1988). The leucine zipper: a hypothetical structure common to a new class of DNA binding proteins. *Science* 240, 1759–1764. doi: 10.1126/science.3289117

64. Lata, C., Mishra, A. K., Muthamilarasan, M., Bonthala, V. S., Khan, Y., and Prasad, M. (2014). Genome-wide investigation and expression profiling of AP2/ERF transcription factor superfamily in foxtail millet (*Setaria italica* L.). *PLoS ONE* 9:e113092. doi: 10.1371/journal.pone.0113092

65. Lata, C., and Prasad, M. (2011). Role of DREBs in regulation of abiotic stress responses in plants. *J. Exp. Bot.* 62, 4731–4748. doi: 10.1093/jxb/err210

66. Le, D. T., Nishiyama, R., Watanabe, Y., Mochida, K., Yamaguchi-Shinozaki, K., Shinozaki, K., et al. (2011). Genome-wide survey and expression analysis of the plant-specific NAC transcription factor family

in soybean during development and dehydration stress. *DNA Res.* 18, 263–276. doi: 10.1093/dnares/dsr015

67. Li, C., Ng, C. K. Y., and Fan, L.-M. (2015). MYB transcription factors, active players in abiotic stress signaling. *Environ. Exp. Bot.* 114, 80–91. doi: 10.1016/j.envexpbot.2014.06.014

68. Li, H., Gao, Y., Xu, H., Dai, Y., Deng, D., and Chen, J. (2013). ZmWRKY33, a WRKY maize transcription factor conferring enhanced salt stress tolerances in *Arabidopsis*. *Plant Growth Regul.* 70, 207–216. doi: 10.1007/s10725-013-9792-9

69. Liang, H., Lu, Y., Liu, H., Wang, F., Xin, Z., and Zhang, Z. (2008). A novel activator-type ERF of *Thinopyrum intermedium*, TiERF1, positively regulates defence responses. *J. Exp. Bot.* 59, 3111–3120. doi: 10.1093/jxb/ern165

70. Liang, Y. K., Dubos, C., Dodd, I. C., Holroyd, G. H., Hetherington, A. M., and Campbell, M. M. (2005). AtMYB61, an R2R3-MYB transcription factor controlling stomatal aperture in *Arabidopsis thaliana*. *Curr. Biol.* 15, 1201–1206. doi: 10.1016/j.cub.2005.06.041

71. Liao, Y., Zou, H. F., Wang, H. W., Zhang, W. K., Ma, B., Zhang, J. S., et al. (2008). Soybean GmMYB76, GmMYB92, and GmMYB177 genes confer stress tolerance in transgenic *Arabidopsis* plants. *Cell Res.* 18, 1047–1060. doi: 10.1038/cr.2008.280

72. Licausi, F., Giorgi, F. M., Zenoni, S., Osti, F., Pezzotti, M., and Perata, P. (2010). Genomic and transcriptomic analysis of the AP2/ERF superfamily in *Vitis vinifera*. *BMC Genomics* 11:719. doi: 10.1186/1471-2164-11-719

73. Lim, C. J., Hwang, J. E., Chen, H., Hong, J. K., Yang, K. A., Choi, M. S., et al. (2007). Over-expression of the *Arabidopsis* DRE/CRT-binding transcription factor DREB2C enhances thermotolerance. *Biochem. Biophys. Res. Commun.* 362, 431–436. doi: 10.1016/j.bbrc.2007.08.007

74. Liu, C., Mao, B., Ou, S., Wang, W., Liu, L., Wu, Y., et al. (2014). OsbZIP71, a bZIP transcription factor, confers salinity and drought tolerance in rice. *Plant Mol. Biol.* 84, 19–36. doi: 10.1007/s11103-013-0115-3

75. Liu, H., Yang, W., Liu, D., Han, Y., Zhang, A., and Li, S. (2011a). Ectopic expression of a grapevine transcription factor VvWRKY11 contributes to osmotic stress tolerance in *Arabidopsis*. *Mol. Biol. Rep.* 38, 417–427. doi: 10.1007/s11033-010-0124-0

76. Liu, H., Zhou, X., Dong, N., Liu, X., Zhang, H., and Zhang, Z. (2011b). Expression of a wheat MYB gene in transgenic tobacco enhances resistance to *Ralstonia solanacearum*, and to drought and salt stresses. *Funct. Integr. Genomics* 11, 431–443. doi: 10.1007/s10142-011-0228-1

77. Liu, J., Chen, N., Chen, F., Cai, B., Dal Santo, S., Tornielli, G. B., et al. (2014). Genome-wide analysis and expression profile of the bZIP transcription factor gene family in grapevine (*Vitis vinifera*). *BMC Genomics* 15:281. doi: 10.1186/1471-2164-15-281

78. Liu, J.-H., Peng, T., and Dai, W. (2014). Critical cis-acting elements and interacting transcription factors: key players associated with abiotic stress responses in plants. *Plant Mol. Biol. Report.* 32, 303–317. doi: 10.1007/s11105-013-0667-z

79. Liu, L., Zhang, Z., Dong, J., and Wang, T. (2016). Overexpression of MtWRKY76 increases both salt and drought tolerance in *Medicago truncatula*. *Environ. Exp. Bot.* 123, 50–58. doi: 10.1016/j. envexpbot.2015.10.007

80. Liu, T. K., Song, X. M., Duan, W. K., Huang, Z. N., Liu, G. F., Li, Y., et al. (2014). Genome-wide analysis and expression patterns of NAC transcription factor family under different developmental stages and abiotic stresses in Chinese Cabbage. *Plant Mol. Biol. Report.* 32, 1041–1056. doi: 10.1007/s11105-014-0712-6

81. Liu, Q., Kasuga, M., Sakuma, Y., Abe, H., Miura, S., Yamaguchi-Shinozaki, K., et al. (1998). Two transcription factors, DREB1 and DREB2, with an EREBP/AP2 DNA binding domain separate two cellular signal transduction pathways in drought- and low-temperature-responsive gene expression, respectively, in *Arabidopsis. Plant Cell* 10, 1391–1406. doi: 10.1105/tpc.10.8.1391

82. Liu, X., and Chu, Z. (2015). Genome-wide evolutionary characterization and analysis of bZIP transcription factors and their expression profiles in response to multiple abiotic stresses in *Brachypodium distachyon. BMC Genomics* 16:227. doi: 10.1186/s12864-015-1457-9

83. Liu, X., Liu, S., Wu, J., Zhang, B., Li, X., Yan, Y., et al. (2013). Overexpression of *Arachis hypogaea* NAC3 in tobacco enhances dehydration and drought tolerance by increasing superoxide scavenging. *Plant Physiol. Biochem.* 70, 354–359. doi: 10.1016/j.plaphy.2013.05.018

84. Lobell, D. B., Schlenker, W., and Costa-Roberts, J. (2011). Climate trends and global crop production since 1980. *Science* 333, 616–620. doi: 10.1126/science.1204531

85. Lu, M., Ying, S., Zhang, D. F., Shi, Y. S., Song, Y. C., Wang, T. Y., et al. (2012). A maize stress-responsive NAC transcription factor, ZmSNAC1, confers enhanced tolerance to dehydration in transgenic *Arabidopsis. Plant Cell Rep.* 31, 1701–1711. doi: 10.1007/s00299-012-1284-2

86. Lucas, S., Durmaz, E., Akpinar, B. A., and Budak, H. (2011). The drought

response displayed by a DRE-binding protein from *Triticum dicoccoides*. *Plant Physiol. Biochem.* 49, 346–351. doi: 10.1016/j.plaphy.2011.01.016

87. Luo, X., Bai, X., Sun, X., Zhu, D., Liu, B., Ji, W., et al. (2013). Expression of wild soybean WRKY20 in *Arabidopsis* enhances drought tolerance and regulates ABA signalling. *J. Exp. Bot.* 64, 2155–2169. doi: 10.1093/jxb/ert073

88. Mallikarjuna, G., Mallikarjuna, K., Reddy, M. K., and Kaul, T. (2011). Expression of OsDREB2A transcription factor confers enhanced dehydration and salt stress tolerance in rice (*Oryza sativa* L.). *Biotechnol. Lett.* 33, 1689–1697. doi: 10.1007/s10529-011-0620-x

89. Mao, X., Chen, S., Li, A., Zhai, C., and Jing, R. (2014). Novel NAC transcription factor TaNAC67 confers enhanced multi-abiotic stress tolerances in *Arabidopsis*. *PLoS ONE* 9:e84359. doi: 10.1371/journal.pone.0084359

90. Meng, X., Wang, J. R., Wang, G. D., Liang, X. Q., Li, X. D., and Meng, Q. W. (2015). An R2R3-MYB gene, LeAN2, positively regulated the thermo-tolerance in transgenic tomato. *J. Plant Physiol.* 175, 1–8. doi: 10.1016/j.jplph.2014.09.018

91. Meng, X., Yin, B., Feng, H. L., Zhang, S., Liang, X. Q., and Meng, Q. W. (2014). Overexpression of R2R3-MYB gene leads to accumulation of anthocyanin and enhanced resistance to chilling and oxidative stress. *Biol. Plant.* 58, 121–130. doi: 10.1007/s10535-013-0376-3

92. Mittler, R., and Blumwald, E. (2010). Genetic engineering for modern agriculture: challenges and perspectives. *Annu. Rev. Plant Biol.* 61, 443–462. doi: 10.1146/annurev-arplant-042809-112116

93. Miura, K., Jin, J. B., Lee, J., Yoo, C. Y., Stirm, V., Miura, T., et al. (2007). SIZ1-mediated sumoylation of ICE1 controls CBF3/DREB1A expression and freezing tolerance in *Arabidopsis*. *Plant Cell* 19, 1403–1414. doi: 10.1105/tpc.106.048397

94. Mizoi, J., Ohori, T., Moriwaki, T., Kidokoro, S., Todaka, D., Maruyama, K., et al. (2013). GmDREB2A;2, a canonical DEHYDRATION-RESPONSIVE ELEMENT-BINDING PROTEIN2-type transcription factor in soybean, is posttranslationally regulated and mediates dehydration-responsive element-dependent gene expression. *Plant Physiol.* 161, 346–361. doi: 10.1104/pp.112.204875

95. Mizoi, J., Shinozaki, K., and Yamaguchi-Shinozaki, K. (2012). AP2/ERF family transcription factors in plant abiotic stress responses. *Biochim. Biophys. Acta* 1819, 86–96. doi: 10.1016/j.bbagrm.2011.08.004

96. Muthamilarasan, M., Khandelwal, R., Yadav, C. B., Bonthala, V. S., Khan,

Y., and Prasad, M. (2014). Identification and molecular characterization of MYB Transcription Factor Superfamily in C4 model plant foxtail millet (*Setaria italica* L.). *PLoS ONE* 9:e109920. doi: 10.1371/journal. pone.0109920

97. Nakano, T., Suzuki, K., Fujimura, T., and Shinshi, H. (2006). Genome-wide analysis of the ERF gene family in *Arabidopsis* and rice. *Plant Physiol.* 140, 411–432. doi: 10.1104/pp.105.073783

98. Nakashima, K., Shinwari, Z. K., Sakuma, Y., Seki, M., Miura, S., Shinozaki, K., et al. (2000). Organization and expression of two *Arabidopsis*DREB2 genes encoding DRE-binding proteins involved in dehydration- and high-salinity-responsive gene expression. *Plant Mol. Biol.* 42, 657–665. doi: 10.1023/A:1006321900483

99. Nakashima, K., Takasaki, H., Mizoi, J., Shinozaki, K., and Yamaguchi-Shinozaki, K. (2012). NAC transcription factors in plant abiotic stress responses. *Biochim. Biophys. Acta* 1819, 97–103. doi: 10.1016/j. bbagrm.2011.10.005

100. Newton, A. C., Johnson, S. N., and Gregory, P. J. (2011). Implications of climate change for diseases, crop yields and food security. *Euphytica* 179, 3–18. doi: 10.1007/s10681-011-0359-4

101. Nijhawan, A., Jain, M., Tyagi, A. K., and Khurana, J. P. (2008). Genomic survey and gene expression analysis of the basic leucine zipper transcription factor family in rice. *Plant Physiol.* 146, 333–350. doi: 10.1104/pp.107.112821

102. Nuruzzaman, M., Manimekalai, R., Sharoni, A. M., Satoh, K., Kondoh, H., Ooka, H., et al. (2010). Genome-wide analysis of NAC transcription factor family in rice. *Gene* 465, 30–44. doi: 10.1016/j.gene.2010.06.008

103. Nuruzzaman, M., Sharoni, A. M., and Kikuchi, S. (2013). Roles of NAC transcription factors in the regulation of biotic and abiotic stress responses in plants. *Front. Microbiol.* 4:248. doi: 10.3389/fmicb.2013.00248

104. Ohme-Takagi, M., and Shinshi, H. (1995). Ethylene-inducible DNA binding proteins that interact with an ethylene-responsive element. *Plant Cell*7, 173–182. doi: 10.1105/tpc.7.2.173

105. Olsen, A. N., Ernst, H. A., Leggio, L. L., and Skriver, K. (2005). NAC transcription factors: structurally distinct, functionally diverse. *Trends Plant Sci.* 10, 79–87. doi: 10.1016/j.tplants.2004.12.010

106. Pandey, S. P., and Somssich, I. E. (2009). The role of WRKY transcription factors in plant immunity. *Plant Physiol.* 150, 1648–1655. doi: 10.1104/ pp.109.138990

107. Peng, X., Zhang, L., Zhang, L., Liu, Z., Cheng, L., Yang, Y., et al. (2013). The transcriptional factor LcDREB2 cooperates with LcSAMDC2 to contribute to salt tolerance in *Leymus chinensis*. *Plant Cell Tissue Organ Cult*. 113, 245–256. doi: 10.1007/s11240-012-0264-0

108. Pérez-Clemente, R. M., Vives, V., Zandalinas, S. I., López-Climent, M. F., Munoz, V., and Gomez-Cadenas, A. (2013). Biotechnological approaches to study plant responses to stress. *Biomed Res. Int*. 2013:654120. doi: 10.1155/2013/654120

109. Pourabed, E., Ghane Golmohamadi, F., Soleymani Monfared, P., Razavi, S. M., and Shobbar, Z. S. (2015). Basic leucine zipper family in barley: genome-wide characterization of members and expression analysis. *Mol. Biotechnol*. 57, 12–26. doi: 10.1007/s12033-014-9797-2

110. Prasch, C. M., and Sonnewald, U. (2015). Signaling events in plants: stress factors in combination change the picture. *Environ. Exp. Bot*. 114, 4–14. doi: 10.1016/j.envexpbot.2014.06.020

111. Puranik, S., Sahu, P. P., Srivastava, P. S., and Prasad, M. (2012). NAC proteins: regulation and role in stress tolerance. *Trends Plant Sci*. 17, 369–381. doi: 10.1016/j.tplants.2012.02.004

112. Qi, L., Yang, J., Yuan, Y., Huang, L., and Chen, P. (2015). Overexpression of two R2R3-MYB genes from *Scutellaria baicalensis* induces phenylpropanoid accumulation and enhances oxidative stress resistance in transgenic tobacco. *Plant Physiol. Biochem*. 94, 235–243. doi: 10.1016/j.plaphy.2015.06.007

113. Qin, F., Kakimoto, M., Sakuma, Y., Maruyama, K., Osakabe, Y., Tran, L. S., et al. (2007). Regulation and functional analysis of ZmDREB2A in response to drought and heat stresses in *Zea mays* L. *Plant J*. 50, 54–69. doi: 10.1111/j.1365-313X.2007.03034.x

114. Qin, F., Sakuma, Y., Li, J., Liu, Q., Li, Y. Q., Shinozaki, K., et al. (2004). Cloning and functional analysis of a novel DREB1/CBF transcription factor involved in cold-responsive gene expression in *Zea mays* L. *Plant Cell Physiol*. 45, 1042–1052. doi: 10.1093/pcp/pch118

115. Qin, Y., Tian, Y., Han, L., and Yang, X. (2013). Constitutive expression of a salinity-induced wheat WRKY transcription factor enhances salinity and ionic stress tolerance in transgenic *Arabidopsis thaliana*. *Biochem. Biophys. Res. Commun*. 441, 476–481. doi: 10.1016/j.bbrc.2013.10.088

116. Qin, Y., Tian, Y., and Liu, X. (2015). A wheat salinity-induced WRKY transcription factor TaWRKY93 confers multiple abiotic stress tolerance in*Arabidopsis thaliana*. *Biochem. Biophys. Res. Commun*. 464, 428–433. doi: 10.1016/j.bbrc.2015.06.128

117. Qiu, Y., and Yu, D. (2009). Over-expression of the stress-induced OsWRKY45 enhances disease resistance and drought tolerance in *Arabidopsis*. Environmental and experimental botany 65. *Environ. Exp. Bot.* 65, 35–47. doi: 10.1016/j.envexpbot.2008.07.002

118. Ravikumar, G., Manimaran, P., Voleti, S. R., Subrahmanyam, D., Sundaram, R. M., Bansal, K. C., et al. (2014). Stress-inducible expression of AtDREB1A transcription factor greatly improves drought stress tolerance in transgenic indica rice. *Transgenic Res.* 23, 421–439. doi: 10.1007/s11248-013-9776-6

119. Ray, S., Dansana, P. K., Bhaskar, A., Giri, J., Kapoor, S., Khurana, J. P., et al. (2009). "Emerging trends in functional genomics for stress tolerance in crop plants," in *Plant Stress Biology*, ed H. Hirt (Weinheim: Wiley-VCH Verlag GmbH & Co. KGaA), 37–63.

120. Riechmann, J. L., and Meyerowitz, E. M. (1998). The AP2/EREBP family of plant transcription factors. *Biol. Chem.* 379, 633–646.

121. Rong, W., Qi, L., Wang, A., Ye, X., Du, L., Liang, H., et al. (2014). The ERF transcription factor TaERF3 promotes tolerance to salt and drought stresses in wheat. *Plant Biotechnol. J.* 12, 468–479. doi: 10.1111/pbi.12153

122. Rushton, D. L., Tripathi, P., Rabara, R. C., Lin, J., Ringler, P., Boken, A. K., et al. (2012). WRKY transcription factors: key components in abscisic acid signalling. *Plant Biotechnol. J.* 10, 2–11. doi: 10.1111/j.1467-7652.2011.00634.x

123. Rushton, P. J., Somssich, I. E., Ringler, P., and Shen, Q. J. (2010). WRKY transcription factors. *Trends Plant Sci.* 15, 247–258. doi: 10.1016/j.tplants.2010.02.006

124. Saad, A. S., Li, X., Li, H. P., Huang, T., Gao, C. S., Guo, M. W., et al. (2013). A rice stress-responsive NAC gene enhances tolerance of transgenic wheat to drought and salt stresses. *Plant Sci.* 203–204, 33–40. doi: 10.1016/j.plantsci.2012.12.016

125. Sakuma, Y., Liu, Q., Dubouzet, J. G., Abe, H., Shinozaki, K., and Yamaguchi-Shinozaki, K. (2002). DNA-binding specificity of the ERF/AP2 domain of *Arabidopsis* DREBs, transcription factors involved in dehydration- and cold-inducible gene expression. *Biochem. Biophys. Res. Commun.* 290, 998–1009. doi: 10.1006/bbrc.2001.6299

126. Sakuma, Y., Maruyama, K., Osakabe, Y., Qin, F., Seki, M., Shinozaki, K., et al. (2006a). Functional analysis of an *Arabidopsis* transcription factor, DREB2A, involved in drought-responsive gene expression. *Plant Cell* 18, 1292–1309. doi: 10.1105/tpc.105.035881

127. Sakuma, Y., Maruyama, K., Qin, F., Osakabe, Y., Shinozaki, K., and Yamaguchi-Shinozaki, K. (2006b). Dual function of an *Arabidopsis*transcription factor DREB2A in water-stress-responsive and heat-stress-responsive gene expression. *Proc. Natl. Acad. Sci. U.S.A.* 103, 18822–18827. doi: 10.1073/pnas.0605639103

128. Sanchez, D. H., Pieckenstain, F. L., Szymanski, J., Erban, A., Bromke, M., Hannah, M. A., et al. (2011). Comparative functional genomics of salt stress in related model and cultivated plants identifies and overcomes limitations to translational genomics. *PLoS ONE* 6:e17094. doi: 10.1371/journal.pone.0017094

129. Schaller, G. E., Kieber, J. J., and Shiu, S. H. (2008). Two-component signaling elements and histidyl-aspartyl phosphorelays. *Arabidopsis Book*6:e0112. doi: 10.1199/tab.0112

130. Seo, J., Sohn, H., Noh, K., Jung, C., An, J., Donovan, C., et al. (2012). Expression of the *Arabidopsis* AtMYB44 gene confers drought/salt-stress tolerance in transgenic soybean. *Mol. Breed.* 29, 601–608. doi: 10.1007/s11032-011-9576-8

131. Seo, P. J., Lee, S. B., Suh, M. C., Park, M. J., Go, Y. S., and Park, C. M. (2011). The MYB96 transcription factor regulates cuticular wax biosynthesis under drought conditions in *Arabidopsis*. *Plant Cell* 23, 1138–1152. doi: 10.1105/tpc.111.083485

132. Seo, P. J., Xiang, F., Qiao, M., Park, J. Y., Lee, Y. N., Kim, S. G., et al. (2009). The MYB96 transcription factor mediates abscisic acid signaling during drought stress response in *Arabidopsis*. *Plant Physiol.* 151, 275–289. doi: 10.1104/pp.109.144220

133. Shao, H. B., Chu, L. Y., Jaleel, C. A., Manivannan, P., Panneerselvam, R., and Shao, M. A. (2009). Understanding water deficit stress-induced changes in the basic metabolism of higher plants - biotechnologically and sustainably improving agriculture and the ecoenvironment in arid regions of the globe. *Crit. Rev. Biotechnol.* 29, 131–151. doi: 10.1080/07388550902869792

134. Sharoni, A. M., Nuruzzaman, M., Satoh, K., Shimizu, T., Kondoh, H., Sasaya, T., et al. (2011). Gene structures, classification and expression models of the AP2/EREBP transcription factor family in rice. *Plant Cell Physiol.* 52, 344–360. doi: 10.1093/pcp/pcq196

135. Shi, W., Liu, D., Hao, L., Wu, C. A., Guo, X., and Li, H. (2014). GhWRKY39, a member of the WRKY transcription factor family in cotton, has a positive role in disease resistance and salt stress tolerance. *Plant Cell Tissue Organ Cult.* 118, 17–32. doi: 10.1007/s11240-014-

0458-8

136. Shinozaki, K., Yamaguchi-Shinozakiy, K., and Sekiz, M. (2003). Regulatory network of gene expression in the drought and cold stress responses.*Curr. Opin. Plant Biol.* 6, 410–417. doi: 10.1016/S1369-5266(03)00092-X

137. Shinwari, Z. K., Nakashima, K., Miura, S., Kasuga, M., Seki, M., Yamaguchi-Shinozaki, K., et al. (1998). An *Arabidopsis* gene family encoding DRE/CRT binding proteins involved in low-temperature-responsive gene expression. *Biochem. Biophys. Res. Commun.* 250, 161–170. doi: 10.1006/bbrc.1998.9267

138. Shiriga, K., Sharma, R., Kumar, K., Yadav, S. K., Hossain, F., and Thirunavukkarasu, N. (2014). Genome-wide identification and expression pattern of drought-responsive members of the NAC family in maize. *Meta Gene* 2, 407–417. doi: 10.1016/j.mgene.2014.05.001

139. Singh, R., Usha, Rizvi, S. M. H., and Sonia, Jaiwal, P. (2003). "Genetic engineering for enhancing abiotic stress tolerance," in *Improvement Strategies of Leguminosae Biotechnology*, eds P. Jaiwal and R. Singh (Dordrecht: Springer), 223–243.

140. Song, X., Li, Y., and Hou, X. (2013). Genome-wide analysis of the AP2/ERF transcription factor superfamily in Chinese cabbage (*Brassica rapas*sp. pekinensis). *BMC Genomics* 14:573. doi: 10.1186/1471-2164-14-573

141. Stockinger, E. J., Gilmour, S. J., and Thomashow, M. F. (1997). *Arabidopsis thaliana* CBF1 encodes an AP2 domain-containing transcriptional activator that binds to the C-repeat/DRE, a cis-acting DNA regulatory element that stimulates transcription in response to low temperature and water deficit. *Proc. Natl. Acad. Sci. U.S.A.* 94, 1035–1040. doi: 10.1073/pnas.94.3.1035

142. Su, H. Y., Zhang, S. Z., Yin, Y. L., Zhu, D. Z., and Han, L. Y. (2015). Genome-wide analysis of NAM-ATAF1, 2-CUC2 transcription factor family in*Solanum lycopersicum. J. Plant Biochem. Biotechnol.* 24, 176–183. doi: 10.1007/s13562-014-0255-9

143. Su, L. T., Li, J. W., Liu, D. Q., Zhai, Y., Zhang, H. J., Li, X. W., et al. (2014). A novel MYB transcription factor, GmMYBJ1, from soybean confers drought and cold tolerance in *Arabidopsis thaliana. Gene* 538, 46–55. doi: 10.1016/j.gene.2014.01.024

144. Sun, J., Hu, W., Zhou, R., Wang, L., Wang, X., Wang, Q., et al. (2015). The *Brachypodium distachyon* BdWRKY36 gene confers tolerance to drought stress in transgenic tobacco plants. *Plant Cell Rep.* 34, 23–35.

doi: 10.1007/s00299-014-1684-6

145. Sun, Z. M., Zhou, M. L., Xiao, X. G., Tang, Y. X., and Wu, Y. M. (2014). Genome-wide analysis of AP2/ERF family genes from *Lotus corniculatus*shows LcERF054 enhances salt tolerance. *Funct. Integr. Genomics* 14, 453–466. doi: 10.1007/s10142-014-0372-5

146. Takeda, S., and Matsuoka, M. (2008). Genetic approaches to crop improvement: responding to environmental and population changes. *Nat. Rev. Genet.* 9, 444–457. doi: 10.1038/nrg2342

147. Tang, Y., Liu, M., Gao, S., Zhang, Z., Zhao, X., Zhao, C., et al. (2012). Molecular characterization of novel TaNAC genes in wheat and overexpression of TaNAC2a confers drought tolerance in tobacco. *Physiol. Plant.* 144, 210–224. doi: 10.1111/j.1399-3054.2011.01539.x

148. Tran, L. S., Nishiyama, R., Yamaguchi-Shinozaki, K., and Shinozaki, K. (2010). Potential utilization of NAC transcription factors to enhance abiotic stress tolerance in plants by biotechnological approach. *GM Crops* 1, 32–39. doi: 10.4161/gmcr.1.1.10569

149. Tripathi, P., Rabara, R. C., and Rushton, P. J. (2014). A systems biology perspective on the role of WRKY transcription factors in drought responses in plants. *Planta* 239, 255–266. doi: 10.1007/s00425-013-1985-y

150. Tsai-Hung, H., Jent-Turn, L., Pei-Tzu, Y., Li-Hui, C., Yee-Yung, C., Yu-Chie, W., et al. (2002). Heterology expression of the *Arabidopsis*C-repeat/dehydration response element binding Factor 1 gene confers elevated tolerance to chilling and oxidative stresses in transgenic tomato. *Plant Physiol.* 129, 1086–1094. doi: 10.1104/pp.003442

151. Ulker, B., and Somssich, I. E. (2004). WRKY transcription factors: from DNA binding towards biological function. *Curr. Opin. Plant Biol.* 7, 491–498. doi: 10.1016/j.pbi.2004.07.012

152. Umezawa, T., Fujita, M., Fujita, Y., Yamaguchi-Shinozaki, K., and Shinozaki, K. (2006). Engineering drought tolerance in plants: discovering and tailoring genes to unlock the future. *Curr. Opin. Biotechnol.* 17, 113–122. doi: 10.1016/j.copbio.2006.02.002

153. United Nations (2015). *The World Population Prospects*. New York, NY: United Nations Department of Economic and Social Affairs.

154. Uno, Y., Furihata, T., Abe, H., Yoshida, R., Shinozaki, K., and Yamaguchi-Shinozaki, K. (2000). *Arabidopsis* basic leucine zipper transcription factors involved in an abscisic acid-dependent signal transduction pathway under drought and high-salinity conditions. *Proc. Natl. Acad. Sci. U.S.A.* 97, 11632–11637. doi: 10.1073/pnas.190309197

155. Varshney, R. K., Bansal, K. C., Aggarwal, P. K., Datta, S. K., and Craufurd, P. Q. (2011). Agricultural biotechnology for crop improvement in a variable climate: hope or hype? *Trends Plant Sci.* 16, 363–371. doi: 10.1016/j.tplants.2011.03.004

156. Vinocur, B., and Altman, A. (2005). Recent advances in engineering plant tolerance to abiotic stress: achievements and limitations. *Curr. Opin. Biotechnol.* 16, 123–132. doi: 10.1016/j.copbio.2005.02.001

157. Wang, C., Deng, P., Chen, L., Wang, X., Ma, H., Hu, W., et al. (2013). A wheat WRKY transcription factor TaWRKY10 confers tolerance to multiple abiotic stresses in transgenic tobacco. *PLoS ONE* 8:e65120. doi: 10.1371/journal.pone.0065120

158. Wang, R. K., Cao, Z. H., and Hao, Y. J. (2014). Overexpression of a R2R3 MYB gene MdSIMYB1 increases tolerance to multiple stresses in transgenic tobacco and apples. *Physiol. Plant.* 150, 76–87. doi: 10.1111/ppl.12069

159. Wei, K., Chen, J., Wang, Y., Chen, Y., Chen, S., Lin, Y., et al. (2012). Genome-wide analysis of bZIP-encoding genes in maize. *DNA Res.* 19, 463–476. doi: 10.1093/dnares/dss026

160. Wen, F., Zhu, H., Li, P., Jiang, M., Mao, W., Ong, C., et al. (2014). Genome-wide evolutionary characterization and expression analyses of WRKY family genes in *Brachypodium distachyon*. *DNA Res.* 21, 327–339. doi: 10.1093/dnares/dst060

161. Wu, H., Lv, H., Li, L., Liu, J., Mu, S., Li, X., et al. (2015). Genome-wide analysis of the AP2/ERF transcription factors family and the expression patterns of DREB genes in Moso bamboo (*Phyllostachys edulis*). *PLoS ONE* 10:e0126657. doi: 10.1371/journal.pone.0126657

162. Wu, K. L., Guo, Z. J., Wang, H. H., and Li, J. (2005). The WRKY family of transcription factors in rice and *Arabidopsis* and their origins. *DNA Res.* 12, 9–26. doi: 10.1093/dnares/12.1.9

163. Wu, X., Shiroto, Y., Kishitani, S., Ito, Y., and Toriyama, K. (2009). Enhanced heat and drought tolerance in transgenic rice seedlings overexpressing OsWRKY11 under the control of HSP101 promoter. *Plant Cell Rep.* 28, 21–30. doi: 10.1007/s00299-008-0614-x

164. Xu, Z.-S., Chen, M., Li, L.-C., and Ma, Y.-Z. (2008a). Functions of the ERF transcription factor family in plants. *Botany* 86, 969–977. doi: 10.1139/B08-041

165. Xu, Z.-S., Ni, Z. Y., Liu, L., Nie, L. N., Li, L. C., Chen, M., et al. (2008b).

Characterization of the TaAIDFa gene encoding a CRT/DRE-binding factor responsive to drought, high-salt, and cold stress in wheat. *Mol. Genet. Genomics* 280, 497–508. doi: 10.1007/s00438-008-0382-x

166. Yamaguchi, T., and Blumwald, E. (2005). Developing salt-tolerant crop plants: challenges and opportunities. *Trends Plant Sci.* 10, 615–620. doi: 10.1016/j.tplants.2005.10.002

167. Yamaguchi-Shinozaki, K., and Shinozaki, K. (2006). Transcriptional regulatory networks in cellular responses and tolerance to dehydration and cold stresses. *Annu. Rev. Plant Biol.* 57, 781–803. doi: 10.1146/annurev.arplant.57.032905.105444

168. Yang, A., Dai, X., and Zhang, W. H. (2012). A R2R3-type MYB gene, OsMYB2, is involved in salt, cold, and dehydration tolerance in rice. *J. Exp. Bot.* 63, 2541–2556. doi: 10.1093/jxb/err431

169. Yang, W., Liu, X. D., Chi, X. J., Wu, C. A., Li, Y. Z., Song, L. L., et al. (2011). Dwarf apple MbDREB1 enhances plant tolerance to low temperature, drought, and salt stress via both ABA-dependent and ABA-independent pathways. *Planta* 233, 219–229. doi: 10.1007/s00425-010-1279-6

170. Yang, X., Wang, X., Lu, J., Yi, Z., Fu, C., Ran, J., et al. (2015). Overexpression of a *Miscanthus lutarioriparius* NAC gene MlNAC5 confers enhanced drought and cold tolerance in *Arabidopsis*. *Plant Cell Rep.* 34, 943–958. doi: 10.1007/s00299-015-1756-2

171. Yanhui, C., Xiaoyuan, Y., Kun, H., Meihua, L., Jigang, L., Zhaofeng, G., et al. (2006). The MYB transcription factor superfamily of *Arabidopsis*: expression analysis and phylogenetic comparison with the rice MYB family. *Plant Mol. Biol.* 60, 107–124. doi: 10.1007/s11103-005-2910-y

172. Ying, S., Zhang, D. F., Fu, J., Shi, Y. S., Song, Y. C., Wang, T. Y., et al. (2012). Cloning and characterization of a maize bZIP transcription factor, ZmbZIP72, confers drought and salt tolerance in transgenic *Arabidopsis*. *Planta* 235, 253–266. doi: 10.1007/s00425-011-1496-7

173. Yokotani, N., Ichikawa, T., Kondou, Y., Matsui, M., Hirochika, H., Iwabuchi, M., et al. (2009). Tolerance to various environmental stresses conferred by the salt-responsive rice gene ONAC063 in transgenic *Arabidopsis*. *Planta* 229, 1065–1075. doi: 10.1007/s00425-009-0895-5

174. Zhai, Y., Wang, Y., Li, Y., Lei, T., Yan, F., Su, L., et al. (2013). Isolation and molecular characterization of GmERF7, a soybean ethylene-response factor that increases salt stress tolerance in tobacco. *Gene* 513, 174–183. doi: 10.1016/j.gene.2012.10.018

175. Zhang, G., Chen, M., Li, L., Xu, Z., Chen, X., Guo, J., et al. (2009).

Overexpression of the soybean GmERF3 gene, an AP2/ERF type transcription factor for increased tolerances to salt, drought, and diseases in transgenic tobacco. *J. Exp. Bot.* 60, 3781–3796. doi: 10.1093/jxb/erp214

176. Zhang, H., Liu, W., Wan, L., Li, F., Dai, L., Li, D., et al. (2010). Functional analyses of ethylene response factor JERF3 with the aim of improving tolerance to drought and osmotic stress in transgenic rice. *Transgenic Res.* 19, 809–818. doi: 10.1007/s11248-009-9357-x

177. Zhang, L., Zhang, L., Xia, C., Zhao, G., Liu, J., Jia, J., et al. (2015). A novel wheat bZIP transcription factor, TabZIP60, confers multiple abiotic stress tolerances in transgenic *Arabidopsis*. *Physiol. Plant.* 153, 538–554. doi: 10.1111/ppl.12261

178. Zhang, X., Liu, X., Wu, L., Yu, G., Wang, X., and Ma, H. (2015). The SsDREB transcription factor from the succulent halophyte suaeda salsa enhances abiotic stress tolerance in transgenic tobacco. *Int. J. Genomics* 2015, 875497. doi: 10.1155/2015/875497

179. Zhang, X., Wang, L., Meng, H., Wen, H., Fan, Y., and Zhao, J. (2011). Maize ABP9 enhances tolerance to multiple stresses in transgenic*Arabidopsis* by modulating ABA signaling and cellular levels of reactive oxygen species. *Plant Mol. Biol.* 75, 365–378. doi: 10.1007/s11103-011-9732-x

180. Zhong, H., Guo, Q. Q., Chen, L., Ren, F., Wang, Q. Q., Zheng, Y., et al. (2012). Two *Brassica napus* genes encoding NAC transcription factors are involved in response to high-salinity stress. *Plant Cell Rep.* 31, 1991–2003. doi: 10.1007/s00299-012-1311-3

181. Zhou, L., Wang, N. N., Gong, S. Y., Lu, R., Li, Y., and Li, X. B. (2015). Overexpression of a cotton (*Gossypium hirsutum*) WRKY gene, GhWRKY34, in *Arabidopsis* enhances salt-tolerance of the transgenic plants. *Plant Physiol. Biochem.* 96, 311–320. doi: 10.1016/j.plaphy.2015.08.016

182. Zhou, Q. Y., Tian, A. G., Zou, H. F., Xie, Z. M., Lei, G., Huang, J., et al. (2008). Soybean WRKY-type transcription factor genes, GmWRKY13, GmWRKY21, and GmWRKY54, confer differential tolerance to abiotic stresses in transgenic *Arabidopsis* plants. *Plant Biotechnol. J.* 6, 486–503. doi: 10.1111/j.1467-7652.2008.00336.x

183. Zhu, N., Cheng, S., Liu, X., Du, H., Dai, M., Zhou, D. X., et al. (2015). The R2R3-type MYB gene OsMYB91 has a function in coordinating plant growth and salt stress tolerance in rice. *Plant Sci.* 236, 146–156. doi: 10.1016/j.plantsci.2015.03.023

184. Zhu, X., Qi, L., Liu, X., Cai, S., Xu, H., Huang, R., et al. (2014). The wheat ethylene response factor transcription factor pathogen-induced ERF1 mediates host responses to both the necrotrophic pathogen *Rhizoctonia cerealis* and freezing stresses. *Plant Physiol.* 164, 1499–1514. doi: 10.1104/pp.113.229575

185. Zhuang, J., Cai, B., Peng, R. H., Zhu, B., Jin, X. F., Xue, Y., et al. (2008). Genome-wide analysis of the AP2/ERF gene family in *Populus trichocarpa. Biochem. Biophys. Res. Commun.* 371, 468–474. doi: 10.1016/j.bbrc.2008.04.087

186. Zou, M., Guan, Y., Ren, H., Zhang, F., and Chen, F. (2008). A bZIP transcription factor, OsABI5, is involved in rice fertility and stress tolerance.*Plant Mol. Biol.* 66, 675–683. doi: 10.1007/s11103-008-9298-4

Chapter 6

GENETICALLY MODIFIED CROPS: INSECT RESISTANCE

A. Karthikeyan[1], R. Valarmathi[2], S. Nandini[2] and M.R. Nandhakumar[3]

[1]Centre for Plant Molecular Biology, Department of Plant Molecular Biology and Biotechnology, Tamil Nadu Agricultural University, Coimbatore-641-003, India

[2]Centre for Plant Protection Studies, Department of Entomology, Tamil Nadu Agricultural University, Coimbatore-641-003, India

[3]Department of Agronomy, Directorate of Crop Management, Tamil Nadu Agricultural University, Coimbatore-641-003, India

ABSTRACT

Insect pests have become an integral part of agricultural crops worldwide. They significantly reduce yield and affect almost every aspect of the plants. For many years major challenge for scientists has been developing the resistant varieties against pests in plants. Plant breeders have also been successful during the last century in producing a few Insect-resistant cultivars/lines of some potential crops through conventional breeding, but this again has utilized modest resources. However, this approach seems now inefficient due to a number of reasons and alternatively, genetic engineering for improving crop pest and disease resistance is being actively followed these days by the plant scientists, world-over. New tools and genes have been developed for use in the genetic engineering of plants to introduce effective resistance to biotic stresses and to understand the mechanisms of resistance. Recent advances in genetic engineering, *Bacillus thuringiensis* (Bt) has resulted in successful control of many economically important pests in food crops. This approach should allow increases in both productivity and quality of plants in an environmentally friendly manner, thereby reducing the use of and reliance on chemical control of pests.

INTRODUCTION

Agricultural productivity is highly influenced by pest and diseases, known as the most harmful factor concerning the growth and productivity of crops worldwide. Conventional breeding methods are being used to develop the varieties more resistance to biotic stresses. At the same time these methods are time taking, resource consuming and germplasm dependent. Besides it requires evaluation at hot spot area. Sometimes the screening based on natural occurrence in the hot spot areas also does not give consistent results. A combination with plant breeding approaches will likely to be needed for the improvement of crops (Roy et al., 2011). On the other hand, pest management by chemicals obviously has brought about considerable protection to crop yields over the past five decades. Regrettably, extensive and very often, indiscriminate usage of chemical pesticides has resulted in environmental degradation, adverse effects on human health and other organisms, eradication of beneficial insects and development of pest-resistant insects (Wahab, 2009). At this situation tool of genetic engineering has provided humankind with unprecedented power to manipulate and develop novel crop genotypes towards a safe and sustainable agriculture in the 21st century (Bates et al., 2005). In recent times, genetic engineering has become a source of agriculture innovations, providing a new solution to the age of -old problems (Mittler and Blumwald, 2010; Ahmad et al., 2012). Plant genes are being cloned, genetic regulatory signals deciphered and genes transferred from entirely unrelated organism to confer new agriculturally useful traits on crop plants (Wani and Sanghera, 2010; Josine et al., 2011). Recent advance in genetic engineering, Bt technology has emerged as a powerful modality for battling some of the important insect pests, It is chemical free and economically viable approach for insect pest control in plants (Hilder and Boulter, 1999; Gatehouse, 2008; DeVilliers and Hoisington, 2011; Sanahuja et al., 2011). Negotiate exchange of this transgenic technology to the developing countries at easy terms and its integration with the conventional approaches for resistance breeding will ensure evergreen revolution crucial for global food security (Dhaliwal and Uchimaya, 1999). In this review we mainly discussed on role of genetic engineering in crop improvement, Bt technology and Bt crops global status, benefits and limitations.

Major Pests in Food Crops

Before examining GM strategies for developing insect pest tolerance in plants, it is useful to consider some of the characteristics of the insects causing the damage. The first point to make is that, where as some adult insects feed off plants and can damage crops, most of the problems are caused by insect larvae.

They cause serious economic losses in many major crops by reducing yield. Food crops of the world are damaged by more than 10,000 species of insects less than 10% of the total identified pest species are generally considered as major pests (Dhaliwal et al., 2007). List of important pests of major crops are given in Table 1. The major classes of insect that cause crop damage are the orders Lepidoptera (Butterflies and moths), Diptera (flies and moths), Orthoptera (grasshoppers and crickets), Homoptera (aphids) and Coleopteran (beetles) (Dhaliwal et al., 2010). The changing scenario of insect pest problems in agriculture as a consequence of genetic engineering technology has been well documented. Detailed role of genetic engineering in crop improvement is discussed below.

Genetic Engineering of Crop Plants

Genetic engineering of plants mostly involves the addition of genetic material (single or multiple genes) that is integrated into a recipient plant, leading to the modification of the plant's genome. The plants with modified genome are known as transgenic plants or Genetically Modified (GM) plants (Pandey et al., 2011). Transfer of genes between plant species have played an important role in crop development for many decades (Carriere et al., 2010). Plant improvement whether as a result of natural selection or the efforts of plant breeder, has always relied on upon evolving, evaluating and selecting the right combination of alleles. Useful traits such as resistance to insect pests have been transferred to crop varieties from non cultivated plants, Since 1970 (Dhaliwal and Uchimaya, 1999). Success in breeding for better adapted varieties to insect pests depends upon the concerted efforts by various research domains including plant and cell physiology, molecular biology, genetics and breeding (Bhatnagar-Mathur et al., 2008; Isbat et al., 2009). Advancement field of genetic engineering have provided new technologies for gene identification and gene transfer into plants has provided the opportunity for genetically engineering insect pest resistance into agriculturally desirable cultivars without altering critical quality traits (Cassells and Doyle, 2003; Christou et al., 2006; Gulzar et al., 2011; Karthikeyan et al., 2011; Tiwari and Youngman, 2011). Moreover, transgenic research has made significant progress in crop genetic improvement and offers the prospect many advantages: not just widening the potential pool useful genes but also permitting the introduction of a number of different desirable genes at a single event and reducing the time needed to introgress introduced characters into an elite genetic background, besides introduction of molecular change by genetic engineering takes less time compared to other classical genetic methods (Behrooz et al., 2008). Hence, genetic engineering for developing insect pest tolerant plants, based on the introgression of genes that are known to be involved in insect pest response and putative tolerance, might prove to be a faster track towards improving crop varieties.

Table 1: List of important insect pests in food crops

Insect pests	Scientific name	Order	Family	Crops
American bollworm	Helicoverpa armigera	Lepidoptera	Noctuidae	Cotton
Brown plant hopper	Nilaparvata lugens	Hemiptera	Delphacidae	Rice
Diamond back moth	Plutella xylostella	Lepidoptera	Plutellidae	Cauliflower and cabbage
Fruit borer	Helicoverpa armigera	Lepidoptera	Noctuidae	Tomato
Fruit fly	Bactrocera sp.	Diptera	Tenthredinidae	Fruits and vegetables
Gall midge	Orseolia oryzae	Diptera	Cecidomyiidae	Rice
Gram pod borer	Helicoverpa armigera	Lepidoptera	Noctuidae	Chickpea and pigeon pea
Green leafhopper	Nephotettix sp.	Hemiptera	Cicadellidae	Rice
Leaf miner	Aproaerema modicella	Lepidoptera	Gelechiidae	Groundnut
Mealy bug	Several species	Hemiptera	Pseudococcidae	Several field and horticultural crops
Mustard aphid	Lipaphis erysimi	Hemiptera	Aphididae	Mustard
Pink stem borer	Sesamia inferens	Lepidoptera	Noctuidae	Wheat
Pyrilla	Pyrilla perpusilla	Hemiptera	Lophophidae	Sugarcane and rice
Shoot and fruit borer	Leucinodes orbonalis	Lepidoptera	Pyralidae	Brinjal
Thrips	Several species	Thysanoptera	Thripidae	Groundnut, cotton, chilies, roses, grapes and citrus
Top borer	Scirpophaga nivella	Lepidoptera	Pyralidae	Sugarcane
Tuber moth	Phthorimaea operculella	Lepidoptera	Gelechiidae	Potato
Yellow stem borer	Scirpophaga incertulas	Lepidoptera	Noctuidae	Rice
Whitefly	Bemisia tabaci	Hemiptera	Aleyrodidae	tobacco
Wheat aphid	Macrosiphum miscanthi	Hemiptera	Aphididae	Wheat, barley, oats

Table 2: List of important ICP proteins (Slater et al., 2009)

Cry protein	Protein size (kDa)	Susceptible insect class
Cry1A(a-i)	133	Lepidoptera
Cry1B(a-g)	140	Lepidoptera
Cry1C(a, b)	133-134	Lepidoptera
Cry1D(a, b)	131-132	Lepidoptera
Cry1E(a, b)	133-134	Lepidoptera
Cry1F(a, b)	132-134	Lepidoptera
Cry1G(a-c)	132-133	Lepidoptera
Cry1H(a, b)	133	Lepidoptera
Cry1I(a-d)	81	Lepidoptera
Cry1J(a-d)	133	Lepidoptera
Cry1Ka	137	Lepidoptera
Cry1La	133	Lepidoptera
Cry2A(a-e)	71	Lepidoptera
Cry3Aa, Cry3B(a, b), Cry3Ca	73-75	Coleoptera
Cry4Aa, Cry4Ba	135,128	Diptera
Cry5A(a, b)	142-152	Nematodes
Cry5Ac	135	Hymenoptera
Cry5Ba	140	Hymenoptera
Cry6Aa, Cry6Ba	143	Nematodes
Cry7A(a, b)	129-130	Coleopteran
Cry8Aa, Cry8B(a-c), Cry8(Ca-Ha)	131	Coleopteran
Cry9Aa, Cry9B(a, b), Cry9Ca, Cry9D(a, b), Cry9E(a, d)	130	Lepidoptera
Cry10Aa	78	Diptera
Cry11Aa	72	Diptera
Cry11B(a, b)	81	Diptera

Bacillus thuringiensis

Bt toxin gene the source of the insecticidal toxins produced in commercial transgenic plants is the soil bacterium *Bacillus thuringiensis* (Bt). It was discovered by Ishiwaki in 1901 in diseased silkworms. Further research on Bt by Steinhaus (1951) lead to renewed interest in biopesticides and as a result, the more potent products such as Thuricide a and Dipela were introduced (Bravo et al., 2007; Federici et al., 2010; Sanahuja et al., 2011). It was subsequently classified and named after its isolation from the gut of diseased flour moth larvae in thurienberg, by Ernst Berliner. The ubiquitous nature of *Bacillus thuringiensis* (Bt) is now being mirrored in major crops plants that have been engineered through recombinant DNA to carry genes responsible for producing these crystal proteins and providing host plant resistance to major pests (Ranjekar et al., 2003; DeVilliers and Hoisington, 2011). *Bacillus thuringiensis* synthesizes crystalline proteins during sporulation. These crystalline proteins are highly insecticidal at very low concentrations. Moreover, Bt strains show differing specificities of insecticidal activity toward pests and constitute a large reservoir of genes encoding insecticidal proteins, which are accumulated in the crystalline inclusion bodies produced by the bacterium on sporulation (Cry proteins, Cyt proteins) or expressed during bacterial growth (Vip proteins) (Federici et al., 2010; Sanahuja et al., 2011). The bacterium produces an insecticidal crystal protein (ICP: also called Cry proteins, encoded by cry genes). Cry proteins are one of several classes of endotoxins produced by the sporulating bacteria: Hence they were originally classified as -endotoxins, to distinguish them from the other classes of and endotoxins (Ranjekar et al., 2003; Slater et al., 2009). With the advent of molecular biology and genetic engineering, it has become possible to use Bt more effectively and rationally by introducing the ICPs of Bt in crop plants. List of important ICP proteins given in Table 2.

Bt Technology

Bacillus thuringiensis is a gram-positive aerobic, sporulating bacterium, which produces proteinaceous crystalline inclusion bodies during sporulation. There are several subspecies of this bacterium, which are effective against lepidopteran, dipteran and coleopteran insects (Tabashnik et al., 2008b). The mechanism of action of the Bt ICPs has been worked out in some detail. The molecular structure of at least three different ICPs has been studied. The crystals, upon ingestion by the insect larva, are solubilized in the highly alkaline midgut into individual protoxins which vary from 133-138 kDa in molecular weight, depending upon the type of protoxin (Slater et al., 2009). The protoxins are acted upon by midgut proteases which cleave them into two halves, the N-terminal half which is usually of 65-68 kDa is the toxin protein. The toxin protein fragment can be divided into three domains (domains I, II and III) (Ranjekar et al., 2003; DeVilliers and Hoisington, 2011). The first is involved in pore formation, the second determines receptor binding and the third is involved in protection to the toxin from proteases. The toxin protein binds to specific receptors present in the midgut epithelial membranes. Upon receptor binding, the domain I insert itself into the membrane leading to the pore formation. The disturbances in osmotic equilibrium and cell lysis lead to insect paralysis and death (DeVilliers and Hoisington, 2011). The current status of Bt technology: The first generation of insect resistant crops that were commercialized expressed single Bt Cry genes, which poses a relatively high risk that insect will evolve resistance to the toxin. In the second and third generations, scientists have mitigated this risk through stacking or pyramiding different genes such as multiple but different Cry genes and Cry genes combined with other insecticidal proteins, which target different receptors in insect pests but also provide resistance to a wider range of pests (Christou et al., 2006; Gatehouse, 2008). Alternatively, synthetic variants of Cry genes has been employed as in the case of MON863 which expresses a synthetic Bt kumamotoensis Cry3Bb1 gene against corn rootworm, which is eight times more effective than the native, non-modified version (Vaughn et al., 2005). Therefore, multiple mutations/adaptations need to be made by target pests in order to develop resistance to this robust new generation of insect resistant crops.

Bt Crops

The success of the transgenic approach led to the development of Bt crops, transgenic crops are used worldwide to control major pests of cotton, corn and soybean. Cotton (Gossypium hirsutum) tolerant to lepidopteran larvae (caterpillars), maize (Zea mays) tolerant to both lepidopteran and coleopteran larvae (rootworms) and soya bean (Glycine max) both lepidopteran and coleopteran larvae have become widely used in global agriculture and have led to reductions in pesticide usage and lower production costs (Toenniessen et al., 2003; Brookes and Barfoot, 2005). The first widely planted Bt crop cultivars were corn producing Bt toxin Cry1Ab and cotton producing Bt toxin Cry1Ac (Tabashnik et al., 2009). While most target pest populations remain susceptible to Bt crops, field-evolved resistance has been documented in some populations of five lepidopteran pests: cereal stem borer, Busseola fusca, in South Africa to Bt corn producing Cry1Ab (Kruger et al., 2009), fall armyworm, Spodoptera frugiperda, in Puerto Rico to Bt corn producing Cry1F (Marvier et al., 2008), pink bollworm, Pectinophora gossypiella, in western India to Bt cotton producing Cry1Ac (Bagla, 2010), cotton bollworm, Helicoverpa zea, in the southeastern United States to Bt cotton producing Cry1Ac and Cry2Ab (Luttrell et al., 2004; Tabashnik et al., 2008a, 2009) and bollworm, Helicoverpa punctigera, in Australia to Bt cotton producing Cry1Ac and Cry2Ab (Downes et al., 2010). Field-evolved resistance was reported to be associated with increased field damage by B. fusca, S. frugiperda, P. gossypiella and H. zea (Matten et al., 2008; Kruger et al., 2009; Tabashnik et al., 2008b, 2009; Bagla, 2010).

Global Status and Benefits of Bt Crops

Genetically, engineering crop resistance to insect pests offer the potential of a user friendly environment and consumer friendly method of crop protection to meet the demands of sustainable agriculture in the 21st century. Biotech crops, including those that are Genetically Modified (GM) with *Bacillus thuringiensis* (Bt) endotoxins for insect resistance, have been cultivated commercially and adopted in steadily increasing numbers of countries over the past 15 years. Biotech crops being cultivated globally include soybean, maize, cotton, canola, squash, papaya, sugar beet and tomato. Almost all of the global biotech crop area derives from soybean, corn, cotton and canola (Brookes and Barfoot, 2010). In 2011, biotech crops soybeans accounted for the largest share (52%), followed by corn (30%), cotton (13%) and canola (5%) (Fig. 1). GM crops have been grown commercially since 1996. In 2011, 16.7 million farmers across 29 countries (ten industrialized countries and 19 developing countries planted 160 million hectares of biotech crops. 90% or 15 million were small and resource poor farmers in developing countries (James, 2011). The US had the largest share of global biotech crop plantings in 2011 (69 million ha), followed by Brazil (30.3 M ha). The other main countries planting biotech crops in 2011 were, Argentina (23.7 M ha), India (10.6 M ha) and Canada (10.4 M ha). Global area of biotech crops in 2011: by Country (Table 3). (Brookes and Barfoot, 2010) reported 725 approvals for commercial cultivation had been granted for 155 events in 24 crops and 57 countries globally have granted regulatory approvals for biotech crops for import for food and feed use and for release in to the environment since 1996 incl.

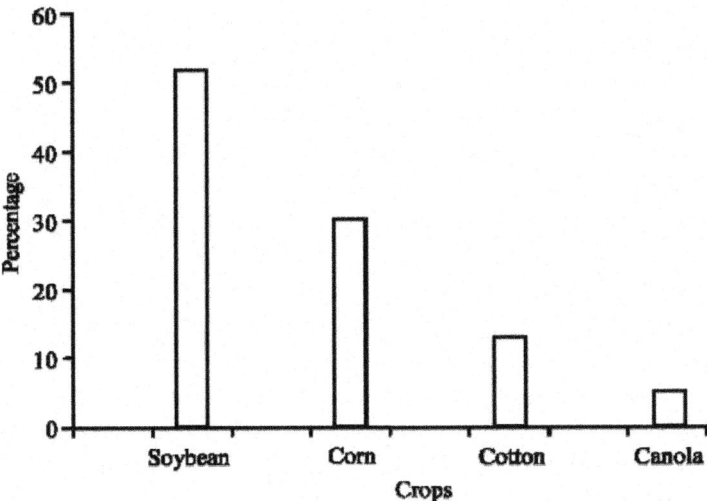

Figure 1: Global biotech crops

Table 3: Global area of biotech crops in 2011: by country (James, 2011)

Rank	Country	Total area (m ha)	Biotech crops
1	USA	69.0	Maize, soybean, cotton, canola, sugar beet, alfalfa, papaya, squash
2	Brazil	30.3	Soybean, maize, cotton
3	Argentina	23.7	Soybean, maize, cotton
4	India	10.6	Cotton
5	Canada	10.4	Canola, maize, soybean, sugar beet
6	China	3.9	Cotton, papaya, poplar, tomato, sweet pepper
7	Paraguay	2.8	Soybean
8	Pakistan	2.6	Cotton
9	South Africa	2.3	Maize, soybean, cotton
10	Uruguay	1.3	Soybean, maize
11	Bolivia	0.9	Soybean
12	Australia*	0.7	Cotton, canola
13	Philippines	0.6	Maize
14	Myanmar	0.3	Cotton
15	Burkina Faso	0.3	Cotton
16	Mexico	0.2	Cotton, soybean
17	Spain	0.1	Maize
18	Colombia	<0.1	Cotton
19	Chile	<0.1	Maize, soybean, canola
20	Honduras	<0.1	Maize
21	Portugal	<0.1	Maize
22	Czech Republic	<0.1	Maize

Japan, USA, Canada, South Korea, Mexico, Australia, Philippines, The European union, New Zealand and China. In 2011 the five lead developing countries in biotech crops are India and China in Asia, Brazil and Argentina in Latin America and South Africa on the continent of Africa, which together represent 40% of the global population, which could reach 10.1 billion by 2100. Six EU countries planted a record 114,490 hectares of biotech Bt maize, up 26% from 2010 and an additional two countries planted the biotech potato Amflora . Africa made steady progress with regulation. South Africa, Burkina Faso and Egypt, together planted a record. 2.5 million hectares; three more countries, Kenya, Nigeria and Uganda conducted field trials (James, 2011). Moreover, global scientific and regulatory authorities found biotech crops as safe as conventional crops and stated that foods from biotech crops are thoroughly evaluated through comprehensive testing for food, feed and environmental safety (James, 2009). The first generation of Bt crops has been extraordinarily successful, Bt crops offer advantages such as in-built protection against pests and other stresses, over hybrid crops and are cultivated as any other conventional crops (Brookes and Barfoot, 2010). Although, there is much debate both politically and publically concerning the environmental impact of genetically engineered crops, it is clear that Bt crops have provided immense environmental benefits. According to the recent survey of global impact of biotech crops for the period From 1996-2010, biotech crops contributed to Food Security, Sustainability and Climate Change by: increasing crop production valued at US$78.4 billion; providing a better environment, by saving 443 million kg a.i. of pesticides; a saving of 8.4 % in pesticides, Which is equivalent to a 16.1% reduction in the associated environmental impact of pesticides use on these crops, as measured by the Environmental Impact Quotient (EIQ). In 2010 alone reducing CO_2 emissions by 19 billion kg, equivalent to taking 9 million cars off the road; conserving biodiversity by saving 91 million hectares of land and helped alleviate poverty by helping 15.0 million small farmers who are some of the poorest people in the world (James, 2011). Biotech crops are essential but are not a panacea and adherence to good farming practices such as rotations and resistance management, are a must for biotech crops as they are for conventional crops, global value of biotech seed alone was valued at ~US$13 billion in 2011, with the end product of commercial grain from biotech crops valued at ~US$160 billion per year (James, 2011).

Limitations

Bt crops are not a panacea for solving all the pest problems. There are some genuine or perceived concerns. The major limitations of transgenic plants secondary pests are not controlled in the absence of sprays for the major pests, need to control the secondary pests through chemical sprays will kill the natural enemies and thus offset one of the advantages of transgenics, cost of producing and deployment of transgenics may be very high, proximity to sprayed fields will reduce the benefits of transgenics, insect migration may reduce the effectiveness of transgenics, development of resistance in insect populations may limit the usefulness of transgenics.

Perspectives

Transgenic crops play a central role in protecting the crop from its major insect pests. The production of insect tolerant plants has been major success for scientists. At the same time efficacy of transgenic crops depend on very much on whether they are viewed from the perspective of chemical pesticides or from that of no additional protective intervention. Even the best current transgenics do not perform as spectacularly as chemicals. There are many insect pests which are not susceptible to currently available range of ICP genes. Many serious pests of local, crop-specific importance have received little or no attention from this technology. There is a need to broaden the pool of genes which are available to cover these pests which are currently untreatable. Transgenic crops are used worldwide to control major pests. Development of strategies to delay the evolution of pest resistance to Bt crops requires an understanding of factors affecting responses to natural selection, which include variation in survival on Bt crops, heritability of resistance and fitness advantages associated with resistance mutations. The two main strategies adopted for delaying resistance are the refuge and pyramid strategies. Both can reduce heritability of resistance, but pyramids can also delay resistance by reducing genetic variation for resistance. One of the major challenge for scientists is accessibility of these products is relatively restricted, In some developed countries, this has been a result of vocal opposition to transgenic crops itself; but in many instances, in both developed and developing countries, it is more a case of potential economic returns not being sufficient to make the introduction of engineered crop varieties commercially viable.

CONCLUSION

Many exciting insights have emerged from recent research on plant genetic engineering. The advantages of successfully engineering plants for insect resistant response are evident. There is no doubt that the use of insect pests (Bt) resistance genes, singly and in combination, has been successful in practice, aside from social and environmental concerns. Overall, the evidence strongly suggests that in both developed and developing countries, the adoption of transgenic crops can increase the farmer's income. The increase in income to small-scale farmers in developing countries can have a direct impact on poverty alleviation and quality of life, a key component of sustainable development. In the developing world, a change in attitude by governments, non-governmental organizations and the public at large is needed for insect-resistant transgenic crops to be able to benefit all the world's population, not just a few. However, it is to be hope that encourage progress described above is maintained and developed so as to make significant contribution towards redressing the balance between world food productions and world food requirements in future.

REFERENCES

1. Ahmad, P., M. Ashraf, M. Younis, X. Hu, A. Kumar, N.A. Akram and F. Al-Qurainy, 2012. Role of transgenic plants in agriculture and biopharming. Biotechnol. Adv., 30: 524-540.

2. Bagla, P., 2010. India: Hardy cotton-munching pests are latest blow to GM crops. Science, 327: 1439-1439.

3. Bates, S.L., J.Z. Zhao, R.T. Roush and A.M. Shelton, 2005. Insect resistance management in GM crops: Past, present and future. Nat. Biotechnol., 23: 57-62.

4. Behrooz, D., S. Farajnia, M. Toorchi, S. Zakerbostanabad, S. Noeparvar and C.N. Stewart Jr., 2008. DNA-delivery methods to produce transgenic plants. Biotechnology, 7: 385-402.

5. Bhatnagar-Mathur, P., V. Vadez and K.K. Sharma, 2008. Transgenic approaches for abiotic stress tolerance in plants: Retrospect and prospects. Plant Cell Rep., 27: 411-424.

6. Bravo, A., S.S. Gill and M. Soberon, 2007. Mode of action of *Bacillus thuringiensis* Cry and Cyt toxins and their potential for insect control. Toxicon, 49: 423-435.

7. Brookes, G. and P. Barfoot, 2005. GM crops: The global economic and environmental impact-The first nine years 1996-2004. AgBioForum, 8: 187-196.

8. Carriere, Y., D.W. Crowder and B. E. Tabashnik, 2010. Evolutionary ecology of insect adaptation to Bt crops. Evol. Appl., 3: 561-573.

9. Cassells, A.C. and B.M. Doyle, 2003. Genetic engineering and mutation breeding for tolerance to abiotic and biotic stresses: Science, technology and safety. Bulg. J. Plant Physiol., 30: 52-82.

10. Christou, P., T. Capell, A. Kohli, J. Gatehouse and A. Gatehouse, 2006. Recent developments and future prospects in insect pest control in transgenic crops. Trends. Plant Sci., 11: 302-308.

11. DeVilliers, S.M. and A.D. Hoisington, 2011. The trends and future of biotechnology crops for insect pest control. Afr. J. Biotechnol., 10: 4677-4681.

12. Dhaliwal, G.S. and H. Uchimaya, 1999. Genetic engineering for disease and pest resistance in plants. Plant Biotechnol., 16: 255-261.

13. Dhaliwal, G.S., A.K. Dhawan and R. Singh, 2007. Biodiversity and ecological agriculture: Issues and perspectives. Indian J. Ecol., 34: 100-109.

14. Dhaliwal, G.S., V. Jindal and A.K. Dhawan, 2010. Insect pest problems and crop losses: Changing trends. Indian J. Ecol., 37: 1-7.

15. Downes, S., R.J. Mahon, L. Rossiter, G. Kauter, T. Leven, G. Fitt and G. Baker, 2010. Adaptive management of pest resistance by Helicoverpa species (Noctuidae) in Australia to the Cry2Ab Bt toxin in Bollgard II ® cotton. Evol. Appl., 3: 574-584.

16. Federici, A.B., H.W. Park and D.K. Bideshi, 2010. Overview of the basic biology of Bacillus thuringiensis with emphasis on genetic engineering of bacterial larvicides for mosquito control. Open Toxinol. J., 3: 83-100.

17. Gatehouse, J.A., 2008. Biotechnological prospects for engineering insect-resistant plants. Plant Physiol., 146: 881-887.

18. Gulzar, S.S., H.W. Shabir, G. Singh, L.K. Prem and N.B. Singh, 2011. Designing crop plants for biotic stresses using transgenic approach. Vegetos-Int. J. Plant Res., 24: 1-25.

19. Hilder, V.A. and D. Boulter, 1999. Genetic engineering of crop plants for insect resistance-a critical review. Crop Prot., 18: 177-191.

20. Isbat, M., N. Zeba, S.R. Kim and C.B. Hong, 2009. A BAXinhibitor-1genein Capsicumannuum is induced under various abiotic stresses and endows multi-tolerance in transgenic tobacco. J. Plant Physiol., 166: 1685-1693.

21. James, C., 2009. Global Status of Commercialized Biotech/GM Crops: 2011. Vol. 41., ISAAA Brief, Ithaca, USA..

22. James, C., 2011. Global Status of Commercialized Biotech/GM Crops: 2011. Vol. 43., ISAAA Brief, Ithaca, USA..

23. Josine, T.L., J. Ji, G. Wang and C.F. Guan, 2011. Advances in genetic engineering for plants abiotic stress control 2011. Afr. J. Biotechnol., 10: 5402-5413.

24. Direct Link |

25. Karthikeyan, A., M. Sudha, M. Pandiyan, N. Senthil, V.G. Shobana and P. Nagarajan, 2011. Screening of MYMV resistant mungbean (Vigna radiata (L.) wilczek) progenies through Agroinoculation. Int. J. Plant Pathol., 2: 115-125.

26. Kruger, M., J.B.J. van Rensburg and J. van den Berg, 2009. Perspective on the development of stem borer resistance to Bt maize and refuge compliance at the Vaalharts irrigation scheme in South Africa. Crop Prot., 28: 684-689.

27. Luttrell, R.G., I. Ali, K.C. Allen, S.Y. Young and A. Szalanski et al., 2004. Resistance to Bt in Arkansas populations of cotton bollworm. Proceedings of the 2004 Beltwide Cotton Conferences, January 5-9, 2004, National Cotton Council of America, Memphis, pp: 5-9.

28. Marvier, M., Y. Carriere, N. Ellstrand, P. Gepts and P. Kareiva et al., 2008. Harvesting data from genetically engineered crops. Science, 320: 452-453.

29. Matten, S.R., G.P. Head and H.D. Quemada, 2008. How Governmental Regulation can Help or Hinder the Integration of BT CROPS into IPM Programs. In: Integration of Insect-resistant Genetically Modified Crops within IPM Programs, Romeis, J., A.M. Shelton and G.G. Kennedy (Eds.). Springer, NewYork, pp: 27-39.

30. Mittler, R. and E. Blumwald, 2010. Genetic engineering for modern agriculture: Challenges and perspectives. Annu. Rev. Plant Biol., 61: 443-462.

31. Pandey, A., M. Kamle, L.P. Yadava, M. Muthukumar and P. Kumar et al., 2011. Genetically modified food: It Uses, future prospects and safety assessment. Biotechnology, 10: 473-487.

32. Ranjekar, P.K., A. Patankar, V. Gupta, R. Bhatnagar, J. Bentur and P.A. Kumar, 2003. Genetic engineering of crop plants for insect resistance. Curr. Sci., 84: 321-329.

33. Roy, B., S.K. Noren, A.B. Mandal and A.K. Basu, 2011. Genetic engineering for Abiotic stress tolerance in agricultural crops. Biotechnology, 10: 1-22.

34. Sanahuja, G., R. Banakar, R.M. Twyman, T. Capell and P. Christou, 2011. Bacillus thuringiensis: A century of research, development and commercial applications. Plant Biotechnol. J., 9: 283-300.

35. Slater, A., N.W. Scott and R.M. Fowler, 2009. GM starategies for insect resistance: Bacillus thuringiensis. Plant Biotechnol. Genet. Manipulation Plants, 6: 133-138.

36. Steinhaus, E.A., 1951. Possible use of Bacillus thuringiensis as an aid in the biological control of the alfalfa caterpillar. Hilgardia, 20: 359-381.

37. Tabashnik, B.E., A.J. Gassmann, D.W. Crowder and Y. Carriere, 2008. Insect resistance to Bt crops: Evidence versus theory. Nat. Biotechnol., 26: 199-202.

38. Tabashnik, B.E., A.J. Gassmann, D.W. Crowder and Y. Carriere, 2008. Field-evolved resistance to Bt toxins. Nat. Biotechnol., 26: 1074-1076.

39. Tabashnik, B.E., J.B. van Rensburg and Y. Carriere, 2009. Field-evolved insect resistance to Bt crops: Sefinition, theory and data. J. Econ. Entomol., 102: 2011-2025.

40. Tiwari, S. and R.R. Youngman, 2011. Transgenic Bt corn hybrids and pest management in the USA alternative farming systems, biotechnology, drought stress and ecological fertilisation. Sustainable Agric. Rev., 6: 15-37.

41. Toenniessen, G.H., J.C. O'Toole and J. DeVries, 2003. Advances in plant biotechnology and its adoption in developing countries. Curr. Opin. Plant Biotechnol., 6: 191-198.

42. Vaughn, T., T. Cavato, G. Brar, T. Coombe and T. DeGooyer et al., 2005. A method of controlling corn rootworm feeding using a bacillus thuringiensis protein expressed in transgenic maize. Crop Sci., 45: 931-938.

43. Wahab, S., 2009. Biotechnological approaches in the management of plant pests, diseases and weeds for Sustainable Agriculture. J. Biopesticides, 2: 115-134.

44. Wani, S.H. and G.S. Sanghera, 2010. Genetic engineering for viral disease management in plants. Not. Sci. Biol., 2: 20-28.

Chapter 7

GENETIC ENGINEERING FOR IMPROVING QUALITY AND PRODUCTIVITY OF CROPS

Asis Datta

National Institute of Plant Genome Research, New Delhi 110067, India

ABSTRACT

The importance of optimal nutrition for human health and development is well recognized. Adverse environmental conditions, such as drought, flooding, extreme heat and so on, affect crop yields more than pests and diseases. Thus, a major goal of plant scientists is to find ways to maintain high productivity under stress as well as developing crops with enhanced nutritional value. Genetically-modified (GM) crops can prove to be powerful complements to those produced by conventional methods for meeting the worldwide demand for quality foods. Crops developed by genetic engineering can not only be used to enhance yields and nutritional quality but also for increased tolerance to various biotic and abiotic stresses. Although there have been some expressions of concern about biosafety and health hazards associated with GM crops, there is no reason to hesitate in consuming genetically-engineered food crops that have been thoughtfully developed and carefully tested. Integration of modern biotechnology, with conventional agricultural practices in a sustainable manner, can fulfil the goal of attaining food security for present as well as future generations.

BACKGROUND

Food insecurity and malnutrition are currently among the most serious concerns for human health, causing the loss of countless lives in developing countries. To be healthy, our daily diet must include ample high quality foods with all of the essential nutrients, in addition to foods that provide health benefits beyond basic nutrition. Even maintaining the amount of food *per capita* what we are getting today will be a mounting job in the future because of the continuing loss of arable lands and the prevalence of unfavourable environmental conditions including drought, salinity, floods, diseases and so on. In order to ensure food security for future generations, the world must produce 50% to 100% more food than at present in spite of the predicted adverse environmental conditions [1].

During the mid-20th century's green revolution, the use of agrochemicals and high-yielding crop varieties developed through conventional plant breeding practices led to a significant boost in crop productivity in India. However, conventional plant breeding alone can no longer sustain the ever-rising global food demand. It is the time to promote sustainable agricultural practices for boosting crop productivity with the utmost conservation of all available natural resources. Agricultural biotechnology is proving to be a powerful complement to conventional methods for meeting worldwide demand for quality food. With the help of modern plant biotechnological tools, today we have access to massive gene pools that can be exploited to impart desirable traits in economically important crops. Genetically-modified (GM) crops can help us to meet the demand for high-yielding, nutritionally-balanced, biotic and abiotic stress tolerant crop varieties [2–7]. While the global area under GM crops continues to expand every year [8], concerns have been expressed regarding unintended and unpredictable pleiotropic effects of these crops on human health and the environment [9]. However, novel foods developed either by conventional or genetic engineering approaches are no different in terms of possible unintended harmful effects on human health and the environment [10]. In fact, the extent of alteration in genomes, from breeding is much more than that for GM crops.

MAIN TEXT

GM Crops *versus* Classically-Bred Crops

Classically-bred and GM crops are the outcomes of genetic modifications created through different means of gene transfer technology. Both conventional breeding and GM technology may involve changes in the genetic makeup of an organism with respect to DNA sequences and the order of genes. However, the amount of genetic changes brought about by the GM technology is small and well defined as compared to classical breeding where thousands of uncharacterized genes of an organism may be involved. Furthermore, GM crops are the outcome of very specific and targeted modification in the genome where the end products such as proteins, metabolites or the phenotype are well characterized. In traditional breeding the genomes of both the parents are mixed together and randomly re-assorted into the genome of the offspring. Thus, undesirable genes can be transferred along with the desirable genes and at the same time some genes may be lost in the offspring. To rectify these problems plant breeders carry out repeated back-crossing to the desirable parent. This is a time-consuming task and may not always be able to separate a tightly linked unsafe gene. For example, potato varieties developed using traditional breeding produce excessive amounts of naturally occurring glycoalkoloids [11]. These glycoalkoloids cause alkaloid poisoning leading to gastrointestinal, circulatory, neurological and dermatological problems. Hybrids of S. tuberosum and S. brevidens produce a toxin demissidine, which is not produced in either parent [12]. Another instance was the conventionally-bred insect-resistant high psoralens variety of celery which was found to produce skin rashes in farm workers who were involved in harvesting this crop [13]. Thus, classical (non-GM) breeding methods can have unintended effects and generate potentially hazardous new products. On the other hand, GM technology employs a precise control on the timing and location of gene products resulting in tissue/organ/development/stress-specific expression - an outcome not easy to accomplish with classical breeding. Moreover, GM techniques allow introduction of new traits at one time without involving extensive cross-breeding as in the case of classical breeding. From the scientific point of view, foods developed either by conventional breeding or by GM technology can impart the same effects on human health and the environment.

GM Crops and Food Safety

GM crops produced by introducing genes for improved agronomic performance and/or enhanced nutrition are under commercial cultivation in many countries [8]. The rigour of the food safety consideration is greatly influenced by the source of the DNA used to develop the GM crop. If the DNA is from an edible plant it will make the regulatory process before commercialization easier and it will also improve consumer acceptance; as, for example, in our laboratory where the *Ama1* gene was isolated from the edible crop Amaranthus and used to develop protein-rich GM potato. It was found to be non-allergenic and safe for consumption using the mouse model [6]. Similarly, the gene *OXDC* (Oxalate decarboxylase) isolated from the edible fungi *Collybia velutipes* was found to be non-toxic and non-allergenic [14]. When we introduced a single gene encoding C-5 sterol desaturase (FvC5SD) from *Collybia velutipes* to the tomato, we obtained a crop with multiple beneficial traits, including improved drought tolerance and fungal resistance [7, 15]. Other strategies include silencing of the host genes instead of addition of a new gene to enhance shelf life of fruits and vegetables [3]. The genes derived from plant viruses can also be considered as safe transgenes as these viruses are not known to be human pathogens. Several virus-resistant transgenics harbouring either the coat protein [16] or overexpressing siRNAs [17] have been developed and released for commercial purposes. A well-known example is the GM papaya resistant to papaya ringspot virus (PRSV) [16]. Presently, about 90% of papaya cultivated in the island of Hawaii is genetically engineered with a coat protein of PRSV. Commercial cultivation of this GM papaya resulted in a considerable increase in papaya production. To date, no conventional or organic method is available to control this rampant virus.

No harmful effects have been documented after several years of extensive cultivation of GM crops in diverse environments and consumption of GM foods by more than a billion humans and by a larger number of animals [10, 18]. However, it is important that the performance of a GM crop is closely scrutinized for several generations under field conditions and that it must go through rigorous bio-safety assessments on a case-by-case basis, before being released for commercial cultivation. Detailed studies should be carried out on various allergenicity and toxicity parameters on laboratory animals. Expressed proteins must be checked for the stability, digestibility, allergenicity and toxicity. Comparative nutritional profiling should be carried out in GM crops.

Use of Markers, a Biosafety Issue in GM Crops

Selectable and scorable marker genes (SMGs) are indispensible for the selection of transformation events for the generation of GM crops. Among the most highly used selectable markers are kanamycin and hygromycin resistance genes. The major biosafety concerns that are raised regarding SMGs relate to their toxicity or allergenicity and the possibility of horizontal gene transfer (HGT) to relevant organisms and pathogens. It has been suggested that transfer of these marker genes to other plants, may result in development of new unwanted weeds. Neomycin phosphotransferase II (*NptII*) which is the most commonly used selectable marker is most extensively evaluated for biosafety. The protein had been approved by the Food and Drug administration (FDA) in 1994. Studies have shown that NptII is non-toxic and it is not expected to result in increased weediness or invasiveness and it also does not affect the non-target organisms [19–21].

CONCLUSION

Plant biotechnology has the potential to address various problems in agriculture and society. GM strategies are being employed to minimize yield losses due to various stresses (biotic and abiotic) and are being used extensively for value addition in food crops by enrichment with quality proteins, vitamins, iron, zinc, carotenoids, anthocyanins and so on. Other ongoing efforts include the enhancement of shelf life of fruits and vegetables so as significantly to reduce the post-harvest losses of perishable crops. Fruit crops are also targeted for the production of edible vaccines to combat major diseases. While the global area under GM crops continues to expand every year, no harmful effects of these crops have been documented even after several years of extensive cultivation in diverse environments and widespread human consumption [10, 18]. Insect resistant Bt crops and/or herbicide tolerant GM crops which are currently under commercial cultivation have benefited farmers through better insect and weed management, higher yields and reduced chemical pesticide use [8, 10, 22, 23].

Thus, it can be concluded that sustainable integration of conventional agricultural practices with modern biotechnology can enable the achievement of food security for present and future generations. However, it is important that the performance of a GM crop is closely scrutinized for several generations under field conditions and goes through rigorous bio-safety assessments on a case-by-case basis, before being released for commercial cultivation. GM crops are going to be an essential part of our life and the enormous potential of biotechnology must be exploited to the benefit of humankind.

ABBREVIATIONS

DNA: Deoxyribonucleicacid

FDA: Food and drug administration

FvC5SD: C-5 Sterol desaturase

GM: Genetically modified

NptII: Neomycin phosphotransferase II

OXDC: Oxalate decarboxylase

SMG: Selectable maker genes.

REFERENCES

1.   Baulcombe D: Reaping benefits of crop research. Science. 2010, 327: 761-10.1126/science.1186705.

2.   Datta A: GM crops: dream to bring science to society. Agric Res.

2012, 1: 95-99. 10.1007/s40003-012-0014-x.

3. Meli VS, Ghosh S, Prabha TN, Chakraborty N, Chakraborty S, Datta A: Enhancement of fruit shelf life by suppressing N-glycan processing enzymes. Proc Natl Acad Sci U S A. 2010, 107: 2413-2418. 10.1073/pnas.0909329107.

4. Ghosh S, Meli VK, Kumar A, Thakur A, Chakraborty N, Chakraborty S, Datta A: The N-glycan processing enzymes α-mannosidase and β-D-1 N acetylhexosaminidase are involved in ripening-associated softening in the non climacteric fruits of capsicum. J Exp Bot. 2011, 62: 571-582. 10.1093/jxb/erq289.

5. Chakraborty S, Chakraborty N, Datta A: Increased nutritive value of transgenic potato by expressing a nonallergenic seed albumin gene from *Amaranthus hypochondriacus*. Proc Natl Acad Sci U S A. 2000, 97: 3724-3729. 10.1073/pnas.97.7.3724.

6. Chakraborty S, Chakraborty N, Agrawal L, Ghosh S, Narula K, Shekhar S, Prakash Naik S, Pande PC, Chakrborti SK, Datta A: Next generation protein rich potato by expressing a seed protein gene *AmA1* as a result of proteome rebalancing in transgenic tuber. Proc Natl Acad Sci U S A. 2010, 41: 17533-17538.

7. Kamthan A, Kamthan M, Azam M, Chakraborty N, Chakraborty S, Datta A: Expression of a fungal sterol desaturase improves tomato drought tolerance, pathogen resistance and nutritional quality. Sci Rep. 2012, 2: 951-

8. James C: Global Status of Commercialized Biotech/GM Crops. ISAAA Briefs No. 43. 2011, Ithaca, NY: ISAAA

9. Dona A, Arvanitoyannis IS: Health risks of genetically modified foods. Crit Rev Food Sci Nutr. 2009, 49: 164-175. 10.1080/10408390701855993.

10. Ronald P: Plant genetics, sustainable agriculture and global food security. Genetics. 2011, 188: 11-20. 10.1534/genetics.111.128553.

11. Hellenas KE, Branzell C, Johnsson H, Slanina P: High levels of glycoalkaloids in the established swedish potato variety magnum bonum. J Sci Food Agric. 1995, 68: 249-255. 10.1002/jsfa.2740680217.

12. Laurila J, Laakso I, Valkonen JPT, Hiltunen R, Pehu E: Formation of parental type and novel alkaloids in somatic hybrids betweenSolanum brevidens and S. tuberosum. Plant Sci. 1996, 118: 145-155. 10.1016/0168-9452(96)04435-4.

13. Berkley SF, Hightower AW, Beier RC, Fleming DW, Brokopp CD,

Ivie GW, Broome CV: Dermatitis in grocery workers associated with high natural concentrations of furanocoumarins in celery. Ann Intern Med. 1986, 105: 351-355. 10.7326/0003-4819-105-3-351.

14. Kesarwani M, Azam M, Natarajan K, Mehta A, Datta A: Oxalate decarboxylase from *Collybia velutipes*. Molecular cloning and its overexpression to confer resistance to fungal infection in transgenic tobacco and tomato. J Biol Chem. 2000, 275: 7230-7238. 10.1074/jbc.275.10.7230.

15. Kamthan A, Kamthan M, Chakraborty N, Chakraborty S, Datta A: A simple protocol for extraction, derivatization, and analysis of tomato leaf and fruit lipophilic metabolites using GC-MS. Protocol Exchange. 2012, doi: 10.1038 /protex. 2012.061

16. Gonsalves D: Control of papaya ringspot virus in papaya: a case study. Annu Rev Phytopathol. 1998, 36: 415-437. 10.1146/annurev. phyto.36.1.415.

17. Bonfim K, Faria JC, Noqueira EO, Mendes EA, Araquo FJ: RNAi mediated resistance to Bean golden mosaic virus in genetically engineered common bean (Phaseolus vulgaris). Mol Plant Microbe Interact. 2007, 20: 717-726. 10.1094/MPMI-20-6-0717.

18. Park J, McFarlane I, Phipps R, Ceddia G: The impact of the EU regulatory constraint of transgenic crops on farm income. N Biotechnol. 2011, 28: 396-406. 10.1016/j.nbt.2011.01.005.

19. Nap JP, Bijvoet J, Stiekema WJ: Biosafety of kanamycin-resistant transgenic plants. Transgenic Res. 1992, 1: 239-249. 10.1007/BF02525165.

20. Fuchs RL, Ream JE, Hammond BG, Naylor MW, Leimgruber RM, Berberich SA: Safety assessment of the neomycin phosphotransferase II (NPTII) protein. Nat Biotechnol. 1993, 11: 1543-1547. 10.1038/nbt1293-1543.

21. Petersen W, Umbeck P, Hokanson K, Halsey M: Biosafety considerations for selectable and scorable markers used in cassava (*Manihot esculenta* Crantz) biotechnology. Environ Biosafety Res. 2005, 4: 89-102. 10.1051/ebr:2005016.

22. Chaudhary B, Gaur K: The development and regulation of Bt-brinjal in India. ISAAA Brief No. 38. 2009, Ithaca, NY: ISAAA

23. Carpenter JE: Peer-reviewed surveys indicate positive impact of commercialized GM crops. Nat Biotechnol. 2010, 28: 319-321. 10.1038/nbt0410-319.

Chapter 8

GENOME-BASED ANALYSIS OF THE TRAN-SCRIPTOME FROM MATURE CHICKPEA ROOT NODULES

Fabian Afonso-Grunz[1,2], Carlos Molina[2,3], Klaus Hoffmeier[2], Lukas Rycak[2], Himabindu Kudapa[4], Rajeev K. Varshney[4], Jean-Jacques Drevon[5], Peter Winter[2] and Günter Kahl[1,2]

[1]Institute for Molecular BioSciences, Goethe University Frankfurt am Main, Frankfurt am Main, Germany

[2]GenXPro GmbH, Frankfurt Biotechnology Innovation Center (FIZ), Frankfurt am Main, Germany

[3]Plant Breeding Institute, Christian-Albrechts-University Kiel, Kiel, Germany

[4]International Crops Research Institute for the Semi-Arid Tropics, Hyderabad, India

[5]French National Institute for Agricultural Research (INRA), Eco&Sols, Montpellier-Cedex, France

Symbiotic nitrogen fixation (SNF) in root nodules of grain legumes such as chickpea is a highly complex process that drastically affects the gene expression patterns of both the prokaryotic as well as eukaryotic interacting cells. A successfully established symbiotic relationship requires mutual signaling mechanisms and a continuous adaptation of the metabolism of the involved cells to varying environmental conditions. Although some of these processes are well understood today many of the molecular mechanisms underlying SNF, especially in chickpea, remain unclear. Here, we reannotated our previously published transcriptome data generated by deepSuperSAGE (Serial Analysis of Gene Expression) to the recently published draft genome of chickpea to assess the root- and nodule-specific transcriptomes of the eukaryotic host cells. The identified gene expression patterns comprise up to 71 significantly differentially expressed genes and the expression of twenty of these was validated by quantitative real-time PCR with the tissues from five independent biological replicates. Many of the differentially expressed transcripts were found to encode proteins implicated in sugar metabolism, antioxidant defense as well as biotic and abiotic stress responses of the host cells, and some of them were already known to contribute to SNF in other legumes. The differentially expressed genes identified in this study represent candidates that can be used for further characterization of the complex molecular mechanisms underlying SNF in chickpea.

INTRODUCTION

Nitrogen, besides phosphorus and potassium, is one of the primary macronutrients for plants, and consequently a limiting factor for the growth of crops (Crawford, 1995). Although di-nitrogen (N_2) represents the most abundant gas in the atmosphere, only certain prokaryotes are able to reduce it to an organic form (ammonia), and can thereby make it available for the growth of higher plants. Among these prokaryotes, rhizobia, a class of nitrogen-fixing bacteria, establishes symbiotic partnerships within the roots of some higher plants, which in turn supply them with energy and protect the very sensitive N_2-fixation machinery from deleterious oxygen. Nutritionally and ecologically seen, symbiotic nitrogen fixation (SNF) identifies chickpea (and other legumes in general) as an important crop, and consequently a promising target for research (Saxena et al., 2012; Varshney et al., 2012). Grain legumes such as chickpea form specialized organs in response to rhizobia invasion, the root nodules. Nodules are highly complex structures that protect the oxygen-sensitive N_2-fixation machinery from its own byproducts: reactive oxygen and reactive nitrogen species (ROS and RNS, respectively). These are derived from the high rates of respiration, and the leak of electrons from redox proteins to O_2. As a consequence, nodules are particularly rich not only in quantity, but also diversity of antioxidant defense mechanisms (reviewed in Matamoros et al., 2003), and especially the nodule mitochondria as the principal source of ROS exhibit a comprehensive antioxidant repertoire (Iturbe-Ormaetxe et al., 2001).

Chickpea root nodules are indeterminate in structure, since they maintain a persistent meristematic tissue that produces new cells for growth and renewal of the nodule (Lee and Copeland, 1994). The symbiotic interaction between legumes and rhizobia relies on mutual signal recognition by both partners. Rhizobial signaling molecules called nodulation factors (Nods) are first perceived by cells in the host root epidermis, and induce the expression of early nodulation genes (eNods) in these cells (reviewed in Oldroyd and Downie, 2008). Bacterial invasion of the host cells can occur either through root hair curls or cracks in the epidermis (Gage, 2004). The latter facilitates bacterial invasion of cortical cells, and does not necessarily involve Nod signaling. In general, the formation of indeterminate nodules is accomplished by root hair invasion starting with the adhesion of bacteria to root hairs. Subsequently, the root hairs curl, and the bacteria invade the plant by a newly formed infection thread. At the same time, a nodule primordium is shaped by dividing cortical cells, and once the infection thread reaches the primordium, the bacteria are released into the cytoplasm of the host cells and surrounded by a plant-derived peribacteroid membrane, henceforth termed PBM (Mylona et al., 1995). PBM biogenesis and metabolism is governed by differential gene expression patterns of both the eukaryotic host legume and the prokaryotic rhizobia, which synergistically induce the synthesis of nodulins, bacteroidines, fatty acids, polysaccharides, and other components. The mature PBM provides selectivity for metabolite and ion transport, and facilitates signaling between both the prokaryotic bacteria and eukaryotic plant cells (Krylova et al., 2007). Although these processes are well understood many of the molecular mechanisms underlying SNF in chickpea root nodules remain unclear.

Presently, next-generation sequencing (NGS)-coupled, genome-wide transcriptome profiling techniques represent the principal tools to interrogate the molecular mechanisms of gene expression in organisms across all taxa and in a wide variety of contexts. Especially whole transcriptome shotgun sequencing (RNA-Seq) has recently emerged as a potent technique for transcriptome studies (Libault et al., 2010; Hayashi et al., 2012; Reid et al., 2012; Barros De Carvalho et al., 2013). However, tag-based approaches such as SuperSAGE have several advantages compared to techniques as e.g., RNA-Seq. Transcript abundances are determined more accurately because of the uniform tag length, which impedes an introduction of biases during PCR, and the formation of di-tags, which allows for discrimination of PCR-derived tags. Additionally, the fact that only one tag is generated out of each transcript enables an unequivocal quantification of reads from a given mRNA species, and naturally results in an increased coverage, which facilitates the study of comprehensive transcriptomes as e.g. in plants (Asmann et al., 2009). In line with this, SuperSAGE was applied for the first time to simultaneously assess the differentially expressed genes from the two interacting organisms in *Magnaporthe grisea* (blast)-infected rice leaves (Matsumura et al., 2003). RNA-Seq, on the other hand, provides additional qualitative information, such as isoform expression. The substantially increased sequencing coverage of tag-based approaches must therefore carefully be weighed against the gain of information using whole transcriptome shotgun sequencing.

To date, the adaption of SuperSAGE to NGS, termed deepSuperSAGE, has been used to assess a broad spectrum of transcriptomes in many species (Sharbel et al., 2010; Zawada et al., 2011; Lenz et al., 2013) including our own works on chickpea (Molina et al., 2008, 2011). However, in our previous work on drought- and salinity-stressed chickpea roots, deepSuperSAGE transcription profiling was hampered by the lack of a genomic sequence of chickpea that prevented a faithful functional annotation of many SuperTags. The newly reported drafts of the chickpea genome (Jain et al., 2013; Varshney et al., 2013) finally changed this situation and now enabled us to assign the majority of expressed SuperTags in roots and nodules to a genomic locus and thereby to a potential function in the context of nodulation. The approximately 50,000 duplicate and homopolymer-filtered SuperTags from our previous study in fact represent nearly 1800 genes of which at least 800 are commonly expressed in both tissues. Up to 682 are more abundant in nodule tissue, and 71 genes display a highly significant differential expression (p-value < 0.01). The underlying data integrity was previously confirmed via microarray hybridization of approximately 3000 UniTags with diverse regulation levels. Of these, 660 could be reliably measured via hybridization, and the comparison between both platforms resulted in a shared tendency toward up- or downregulation of these transcripts of 79% (Molina et al., 2011). The identified set of differentially expressed genes consequently reflects necessary adaptions of the host cell transcriptome with respect to SNF in chickpea nodules.

In the past, many aspects of nodule development in legumes have been thoroughly characterized (see Ferguson et al., 2010; Desbrosses and Stougaard, 2011; Hayashi et al., 2013). However, less emphasis was paid to the transcriptomes of both nodules and roots, especially in chickpea, since annotation had to be based on the genome sequences of other legumes (e.g., *Medicago truncatula*). Now that the draft genome sequence is available, we reanalyzed the transcriptomes of unstressed chickpea nodule-free roots and mature root nodules from our previous study, and confirmed the newly identified expression patterns via quantitative real-time PCR (qRT-PCR) using five individual biological replicates.

MATERIALS AND METHODS

Plant Material

Beja 1 (INRAT 93-1) is a salt-tolerant chickpea (*C. arietinum*) variety (L'Taief et al., 2007) that was released in 2003 from the National Institute for Agricultural Research of Tunisia (INRAT) and the International Center for Agricultural Research in the Dry Areas (ICARDA). Rhizobial inoculations and growth conditions for chickpea plants are described in the work of Molina et al. (2011). Briefly, surface-sterilized Beja 1 seeds were germinated on 0.9% agar for five days in a dark chamber at room temperature, and seedlings with a minimum root length of 5 cm were subsequently inoculated with *Mesorhizobium ciceri* strain UPMCa7 (Romdhane et al., 2007). The inoculated seedlings were hydroareoponically grown in a temperature-controlled glasshouse with a day/

night temperature regime of 28/20°C, respectively, and a 16 h photoperiod for 40 days. After this period, the six-week-old chickpea plants were harvested from the hydroponic cultures. Nodule tissue was carefully separated from the remaining root system, and subsequently both tissues were immediately stored in liquid nitrogen. Plant breeding was carried out in the greenhouse facilities of the "Soil and Symbiosis Research Unit" of the National Institute for Agricultural Research (INRA) in Montpellier, France.

Total RNA Isolation, Library Preparation and 454 Sequencing

Dissected mature nodules were used in their entirety for characterization of the nodule-specific transcriptome, while transcription profiling of the roots was performed with all the remaining root material available. Total RNA isolation, and subsequent construction and sequencing of SuperSAGE libraries from these tissues were performed as previously described using RNA from a pool of 15 plants (Molina et al., 2011). In brief, total RNA was isolated as described by Pawlowski et al. (1994) with a modified precipitation of the RNA in 3M LiCl at 4°C overnight. Then the polyadenylated fraction of the total RNA was purified using the Oligotex mRNA Mini Kit (QIAGEN, Hilden, Germany), and subsequently used for construction of SuperSAGE libraries as detailed by Matsumura et al. (2003) with a modified sequencing procedure. Instead of di-tag concatenation and subsequent cloning for Sanger sequencing, the PCR amplified di-tags were directly sequenced on the GS20 platform (454 Life Sciences, Branford, USA).

Bioinformatical Processing of deepSuperSAGE Sequencing Data

Sequencing data was analyzed with the GenXPro SuperSAGE data processing pipeline. First, distinct libraries were sorted out from the bulk of sequences according to their respective barcodes. Then, PCR-derived and all low-complexity reads containing 12 or more consecutive adenine bases were eliminated. These filtered tags were subsequently mapped against the recently published draft of the chickpea genome (Varshney et al., 2013) using the short read mapper Novoalign v2.07.13 (Novocraft Technologies) with default parameter settings. Tags mapping to more than one locus were excluded from further analysis. Finally, feature annotation for the mapped loci was performed with the mRNA sequences from the "Official Gene Set" (OGSv1.0; Varshney et al., 2013), and reads were counted using the Python package HTSeq v0.5.4p2 (EMBL Heidelberg,https://pypi.python.org/pypi/HTSeq). The numbers of unambiguously annotated SuperTags were normalized to 10,000 sequenced tags in total (tags per ten thousand; TPT) for each library to warrant comparability between the libraries. Statistical significance was assessed by

χ^2 tests (Man et al., 2000), and fold changes were determined by pair-wise comparison of the normalized tag numbers. TPT counts of zero were adjusted to 0.05 to allow for calculation of fold changes even if a given tag was only present in one of the libraries (see Table S1).

Functional annotation of the expressed genes was performed with the MapMan software (version 3.5.1) developed by Thimm et al. (2004). First, a reference mapping file comprising a draft metabolic network of chickpea was generated by classification of all protein-coding sequences from the draft genome into MapMan functional categories via the Mercator tool (Lohse et al., 2014). Approximately 60% of these sequences could be assigned to one of the 34 functional classes (Figure S1, Table S2). Subsequently, the generated reference file was used for mapping of the differentially expressed genes onto different metabolic pathways and to assign these genes to several large enzyme families.

Confirmation of the Genome-Based Reanalysis by Quantitative Real-Time PCR

Quantification of mRNA by real-time PCR was performed on the StepOne Real-Time PCR System (Applied Biosystems) using independent biological replicates that were bred in the same way as described above. Root and nodule tissues from 6 freshly grown plants at the age of six weeks were dissected and used for total RNA isolation with the InviTrap Spin Plant RNA Mini Kit (Stratec Biomedical) following the recommendations of the manufacturer for use of lysis solution DCT. While nodules were simply stripped off the snap-frozen material and afterwards used entirely for total RNA isolation, dissection of root tissue was performed with a sterile razor to obtain absolutely nodule-free root tissue for characterization of the root-specific transcriptomes. Remaining DNA fragments in the isolated total RNA were digested by DNase I (Baseline-ZERO, Epicentre) as recommended by the manufacturer. Subsequent to quantification of the total RNA (Qubit, Life Technologies), all isolates were quality-controlled on the Bioanalyzer 2100 (Agilent Technologies). Isolates with an RNA Integrity Number (RIN) of 7 or higher were reverse-transcribed with SuperScript III Reverse Transcriptase (Invitrogen) following the manufacturer's instructions for first-strand cDNA synthesis. Reverse-transcribed cDNA corresponding to 20 ng total RNA was then added to each amplification reaction on the StepOne Real-Time PCR System. All amplification reactions were carried out in 12 μl volume with the 5x HOT MOLPol EvaGreen qPCR Mix (ROX) from Molegene, complemented by the respective forward and reverse primers (Table 1) in a final concentration of 250 nM each. Initial denaturation was performed at 95°C for 15 min, followed by 40 cycles of 15 s at 95°C, 20 s at 65°C and 30 s at 72°C. A final elongation step at 72°C for 5 min allowed the polymerase to complete all unfinished strands. Subsequently, a melting curve analysis was performed to verify exclusive amplification of the expected products. Additionally, the threshold cycles (C_t values) of the negative (no template) controls from all employed assays were ensured to be higher than 35.

Table 1. List of targeted mRNAs along with the respective primer and probe sequences used for qRT-PCR quantification

Gene ID*	Alias*	Accession number[†]	Primer	Sequence	Amplicon size
Ca_04993[+]	Tubulin alpha-7 chain	XM_004501016.1	Forward	GTGGTGATCTTGCCAAGGTTCAG	222
			Reverse	GACTCAGCACCAACCTCTTCATAATC	
Ca_09743[+]	Heat shock protein Hsp90	XM_004491473.1	Forward	TGTTGAAGCTTGGACTGAGCATTG	111
		XM_004491474.1	Reverse	TCGACCTCTTCCATCTTGCTACC	
Ca_03068[+]	S-adenosylmethionine synthase	XM_003609813.1	Forward	CCTCACTATCGTGAAGAACAGCTTTG	166
			Reverse	CCCATTTAAGAGGCTTCACCACTTC	
Ca_06305	Monothiol glutaredoxin-S17	XM_004504992.1	Forward	TGCTCCAAGATGCGGCTTTAG	187
			Reverse	CCATAACAATATCGCAACCGCCTATC	
Ca_13049	Integrator complex subunit	XM_004512818.1	Forward	AAATTGCAGCTGATTTGGCTTCC	302
			Reverse	AAGCTTTGTAAGGGTCCTCTGTATG	
Ca_22734	Putative uncharacterized protein	XM_004512804.1	Forward	CAATGAAGCGTTCGGGTTTGTG	114
			Reverse	CTCCGACCGCCACAACATATC	
Ca_10340	Multidrug resistance protein	XM_004503208.1 AB024992.1	Forward	AGAGTCAGGGCATGACACTCATC	148
			Reverse	CCGTGCCATGGGATGCTTAG	
Ca_04229	Neutral alpha-glucosidase AB	XM_004502926.1	Forward	GGCACCTACTTCTGGTGGAAATG	169
			Reverse	AAGCTGGTGAATGTGCCCTTTG	
Ca_07680	Putative uncharacterized protein	XM_004494957.1	Forward	CAGGAAACAGCCGAAATCTAGGATG	132
			Reverse	ACAAGCTTCTGGCCAACTATTGC	
Ca_06862	Putative uncharacterized protein	XM_004508211.1	Forward	GCTGTCCATGAGAAAGGAGATGTG	142
			Reverse	AGCTGCTCTTGGAAACTGCTTTG	
Ca_11013	Putative uncharacterized protein	XM_004498494.1	Forward	ATGGTGCACATGGAGAATTCATGG	242
			Reverse	GCAACATAGGAAGCCCTGCATAG	
Ca_10312	Coronatine-insensitive 1	XM_004503173.1	Forward	AGGGTATGGTGCATCTCCATCTG	181
			Reverse	AATCTGATCTTTGGCCAGCAAGAG	
Ca_16834	HMG I/Y like protein	XM_004512669.1	Forward	AACAACACCTGCTAGTGCTCAAC	110
			Reverse	AAATGAGGCCTAAGCACTGCAAG	
Ca_05800	6-phosphogluconate dehydrogenase	XM_004503591.1	Forward	CTTGTTCAGGCTCAGAGGGATTTG	126
			Reverse	AATTAAGAGCAGCAACACCAGTACC	
Ca_22023	Monosaccharid transport protein	XM_004504903.1	Forward	CATGTTGCCTGAGACTAAGGGAATAC	133
			Reverse	TAACAGCTCCCTTGCCCATCTC	
Ca_00007	Squamous cell carcinoma antigen	XM_004485354.1 XM_004485355.1 XM_004485356.1 XM_004485357.1	Forward	TTTGGACGATGAGCACCTTGTTG	225
			Reverse	AATGCTCTCACTTCGTGGCTTTC	
Ca_05370	Prefoldin subunit	XM_004497178.1 XM_004497179.1 XM_004497180.1	Forward	CTCAACACGTTCTCGTCGATGTC	134
			Reverse	TTTGGGATGCCACCTCAACAAG	
Ca_15777	Serine/threonine protein kinase-like protein CCR4	XM_004506314.1	Forward	TGGACCCTGAATACTATAGGCTACAAC	332
			Reverse	ACGCCGCAAGAGCTGTTTC	
Ca_12354	ADP-ribosylation factor GTPase-activating protein	XM_004509671.1	Forward	TTCCATCTCCAGTGCCGATCTC	200
			Reverse	CAGAGAATTCGGTCTTGAAGATCTGTC	
Ca_15466	Nodulin 6	XM_004497937.1	Forward	ACTGATGCCTATGCATTTCCTGAAC	124
			Reverse	ACTTCCACAGCCTCCGGAAC	
Ca_12714	Putative uncharacterized protein	XM_004502350.1 XM_004502351.1	Forward	GGCCAATCCTGAGAAGAGAAATCAC	229
			Reverse	CCATGCCTCCTCCAACAAATTGTC	
Ca_13139	Putative uncharacterized protein	XM_004498271.1 XM_004498272.1	Forward	TGGCTGAACAAACTCATTTGGGAAG	219
			Reverse	CCTGCAACCTTGATATCTCCAGGAAC	
Ca_03442	Glutathione S-transferase	XM_004495920.1	Forward	GGAAGAGAATGAAGCCAAGTTGAACAC	217
			Reverse	TAGACCAAGCTGGTCTTGCAGTG	
Ca_16084	Leghemoglobin	XM_004490852.1	Forward	GAGATGCTACATTGGGTGCTGTTC	159
			Reverse	GCCAATCCATCATAGGCGAGTTC	

*Gene ID and alias according to OGSv1.0 (Varshney et al., 2013);

†NCBI reference sequences for all transcript variants targeted by the respective primer pair;

+ reference gene used for normalization.

The relative transcript abundances between root and nodule tissue from the different biological replicates were calculated according to the $\Delta\Delta C_t$ method using the geometric mean of three previously determined reference genes. All target and reference mRNAs were quantified in duplicates for all biological replicates, respectively. The arithmetic mean of each duplicate was then used to calculate the $\Delta\Delta C_t$ values between the tissues. Thirteen candidate reference genes were screened for their target stability using a pool of reverse-transcribed cDNAs from the five biological replicates that passed quality control (Figure S2A). The determined expression ratios were analyzed with geNorm (Vandesompele et al., 2002), and six of the best performing candidates (geNorm M < 0.5) were analyzed in more detail using three individual biological replicates (Figure S2B). The optimal number of target reference genes was determined to vary around two to three. To ensure optimal comparability, the 3 best-performing candidates from the individual test run (Ca_04993, Ca_09743, and Ca_03068) were subsequently used for normalization of the gene expression ratios from root and nodule tissue.

RESULTS AND DISCUSSION

Genome Based Reanalysis

With a duplicate- and homopolymer-filtered read number of 25,160 (roots) and 26,380 (nodules), both libraries contain a similar number of reads (±5%). Genomic mapping resulted in 17,909 (71.1%) and 20,508 (77.7%) unambiguously assigned tags (root and nodule, respectively). The remaining reads were either mapped to more than one locus or could not be mapped to the mRNA-encoding sequences of the chickpea genome at all (Figure 1). Only uniquely assigned reads were taken into consideration for further analysis.

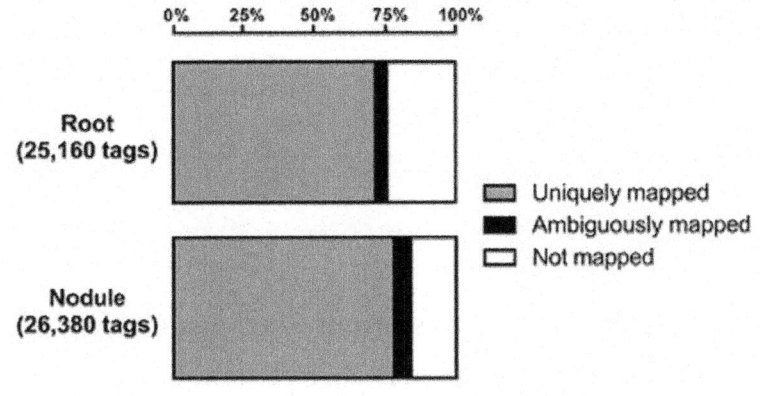

	Root		Nodule	
Uniquely mapped	17,909	71.2%	20,508	77.7%
Ambiguously mapped	1,335	5.3%	1,818	6.9%
Not mapped	5,916	23.5%	4,054	15.4%

Figure 1. Genomic mapping results of the captured transcriptomes from chickpea root and nodule tissue.

In our study in 2011, the 51,545 sequenced Beja 1 SuperTags from root and nodule tissue of unstressed chickpea plants were grouped into 11,525 UniTags (SuperTags of common origin), while the genome-based reanalysis of these SuperTags resulted in the identification of approximately 1780 expressed gene loci. These represent the transcriptionally most active genes in chickpea roots and nodules, and consequently the corresponding mRNAs are the most abundant transcripts. The 140 UniTags previously found to be more than 8-fold prevalent in nodules were reduced to 61 significantly differentially expressed loci in the genome of chickpea (Table 2). On the other hand, the number of more than 20-fold differentially expressed UniTags between both tissues increased from four transcripts to 64 that actually exhibit more than a 128-fold differential expression. Consequently, many of the slightly upregulated UniTags in nodules are expressed from the same genomic locus, and based on the genome sequence these can be combined to provide more meaningful data.

Table 2. Number of differentially expressed genes in root and nodule tissue from chickpea cultivar Beja 1

	Differentially expressed					
	>2-fold	>4-fold	>8-fold	>64-fold	>128-fold	>256-fold
Nodule upregulated	108 (953)	92 (755)	61 (692)	51 (296)	51 (51)	5 (5)
Nodule downregulated	56 (228)	42 (159)	26 (135)	13 (50)	13 (13)	3 (3)
Sum of differentially expressed genes	164 (1181)	134 (914)	87 (827)	64 (346)	64 (64)	8 (8)

Numbers on the left represent significantly differentially expressed genes (α = 0.05), while the numbers in brackets include all expressed genes regardless of any significance threshold.

Nodule-Specific Gene Expression of Chickpea Cultivar Beja 1

The identified expression profiles of Beja 1 root and nodule tissues share around 800 (~45%) expressed genes (Figure 2), while 682 (almost 40%) of the genes are heavily upregulated in nodule tissue (vs. 297 distinctly expressed genes in roots). Although the captured number of expressed genes (~1780) is relatively low compared to more recent profiling studies, the large difference in distinctly expressed genes indicates important variations in expression of the most abundant transcripts in the context of nodulation. The relatively high number of expressed genes in nodule tissue reflects an induced expression of a plethora of genes that are putatively involved in establishment and maintenance of the symbiotic relationship. A total of 71 genes were identified as highly significant (α = 0.01) differentially expressed between both tissues. The distribution of these genes across the 8 chickpea chromosomes is relatively uniform and varies around nine (±50%) differentially expressed genes per chromosome. With respect to the varying chromosome sizes, most of the differentially expressed genes are located on chromosome 8, but no particular chromosomal region was found to be significantly enriched with differentially expressed loci. Genes showing more than a 2-fold differential expression between roots and nodules are listed in Table 2. Consistent with the high number of expressed genes in nodule tissue, the number of upregulated genes in nodules is about twice as high as in roots. As expected, most genes display a relatively low differential expression, and therefore the respective significance levels vary accordingly. Sixty four significantly differentially expressed transcripts (enrichment of 128-fold or more) could be identified as highly enriched in one of the tissues. Among others, these genes include BZIP transcription factor 2 (inferred from *Phaseolus vulgaris*), nodulin 6, abscisic acid receptor PYL4, and glutathione S-transferase (inferred from *M. truncatula*) all of which are upregulated in nodules. The complete set of expressed genes is depicted in a heat map (Figure S3) that illustrates the relatively high expression ratio in nodule compared to root tissue, since numerous transcripts are found to be more abundant in nodules.

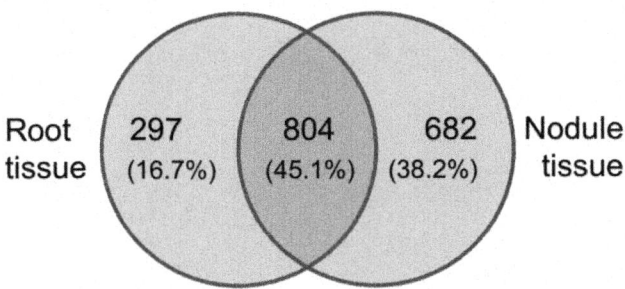

Figure 2. Tissue-specific gene expression in chickpea cultivar Beja 1. Outer numbers represent genes that are expressed in root (left) or nodule (right) tissue, while the number within the overlap represents commonly expressed genes.

Validation of the Identified Gene Expression Patterns by Quantitative Real-Time PCR

Differential expression of the ten most up and downregulated genes in chickpea nodule tissue from five biological replicates was validated via qRT-PCR and revealed a comprehensive biological variance between the different isolates (Figure 3A, Table S1). While all of the upregulated and seven out of the ten most downregulated genes in nodule tissue are found to be accordingly expressed in at least one of the replicates, some of these genes exhibit converse expression ratios in the other replicates. The expression of leghemoglobin, which is known to be expressed by legumes in response to colonization of the roots by rhizobia (Benedito et al., 2008; Libault et al., 2010), was assessed additionally to the twenty most differentially expressed genes. As expected, the mRNA encoding leghemoglobin is significantly more abundant in nodule tissue regardless of the biological replicate. The individual expression ratios, however, range from almost 25,000-fold down to 10-fold upregulated in the respective tissues. The extensive biological variance in expression of some of the twenty candidate genes becomes even more apparent after hierarchical clustering (Figure 3B). The individual replicates can be broadly classified into two groups that show similar expression ratios either for the upregulated (qPCR1 and qPCR2) or alternatively for the downregulated transcripts (qPCR3-5) compared to the pool of ten plants used for deepSuperSAGE.

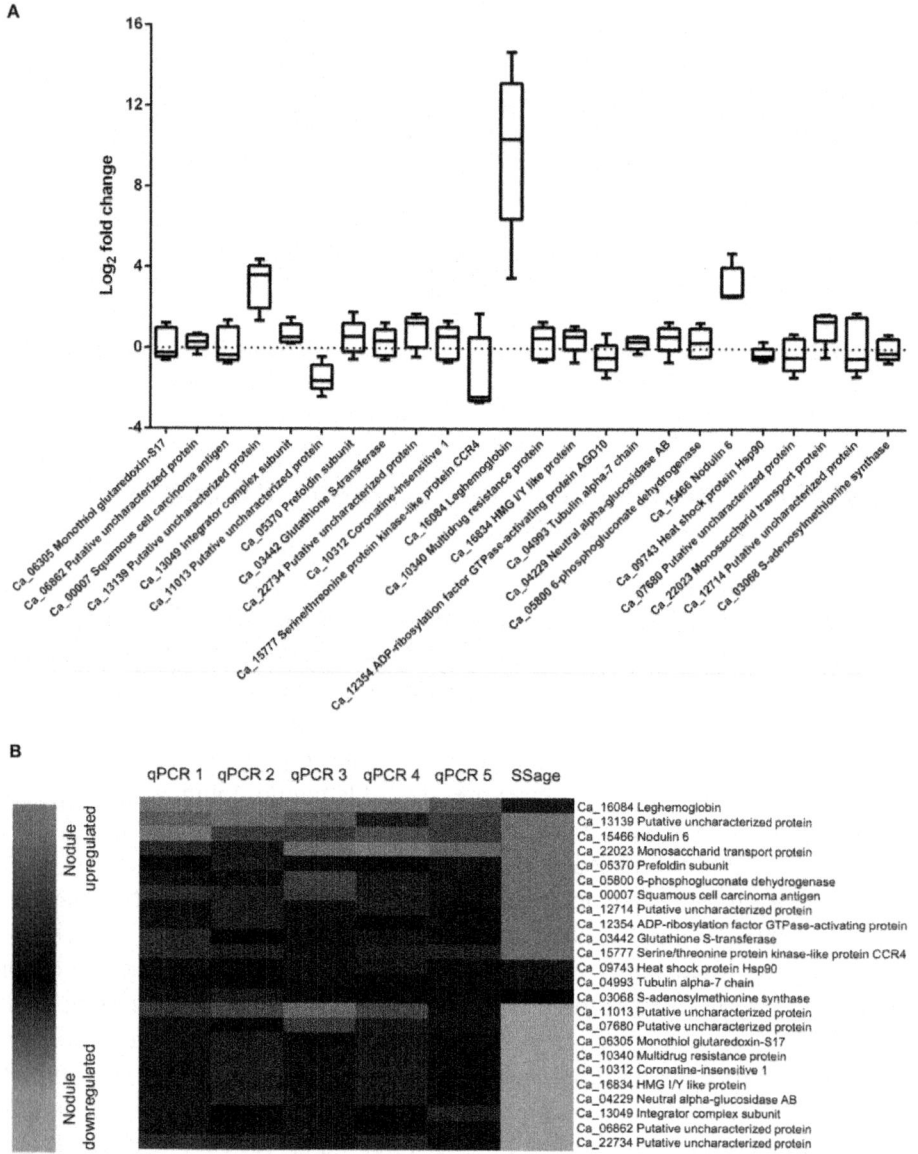

Figure 3. Biological variance in gene expression of 21 candidate and three reference genes across five biological replicates (A) and comparison of deepSuperSAGE expression patterns from ten pooled plants with the individual expression ratios determined by qRT-PCR (B). The logarithmized (base 2) expression ratios between root and nodule tissue from five biological replicates are depicted in box plots for the indicated genes. Positive values represent nodule upregulated mRNAs and negative values mRNAs that are more abundant in root tissue. The gene expression ratios of the indi-

vidual biological replicates (qPCR1-5) in comparison to the pooled (SSage) tissues are additionally shown in the heat map. Nodule upregulated transcripts are represented in red and downregulated transcripts in green. Clustering was performed with the MultiExperiment Viewer version 4.9 by hierarchical clustering of all genes and samples using Euclidean distance calculations.

Upregulated Genes in Chickpea Nodules

The 10 most upregulated genes in chickpea nodule tissue are listed in Table 3. The proteins encoded by two of the listed genes have a putative function, but the respective protein sequences could be matched with the InterPro database, which predicts a serine/threonine-protein kinase and a drug/metabolite transporter. Especially the transcript encoding the putative drug/metabolite transporter is highly upregulated in nodules (>300-fold enrichment) not only in the pooled plant tissue but also in the individual biological replicates, which suggests that the corresponding gene product is functionally important in the rhizobia-adapted metabolism of nodules. The putative drug/metabolite transporter (IPR000620) is predicted to be an integral membrane protein and may contribute to the selectivity of metabolite transport through the mature PBM.

Table 3. Proteins encoded by the 10 most upregulated genes from nodule compared to root tissue of chickpea cultivar Beja 1

ID	TrEMBL database	InterProScan	Fold change	P-value	Chr.
Ca_13139	Putative uncharacterized protein	Drug/metabolite transporter	346	0.0008	4
Ca_03442	Glutathione S-transferase	Glutathione S-transferase	346	0.0008	4
Ca_12714	Putative uncharacterized protein	Protein kinase, catalytic domain; Serine/threonine-protein kinase domain	311	0.0015	5
Ca_12354	ADP-ribosylation factor GTPase-activating protein AGD10	Arf GTPase activating protein	276	0.0028	7
Ca_15466	Nodulin 6	Amidohydrolase 2	276	0.0028	4
Ca_15777	Serine/threonine protein kinase-like protein CCR4	Protein kinase, catalytic domain; Serine/threonine-protein kinase domain	242	0.0051	6
Ca_05370	Prefoldin subunit	Prefoldin subunit	242	0.0051	4
Ca_05800	6-phosphogluconate dehydrogenase, decarboxylating	6-phosphogluconate dehydrogenase, C-terminal; NAD-binding	207	0.0095	6
Ca_22023	Monosaccharide transport protein	Sugar/inositol transporter	207	0.0095	6
Ca_00007	Squamous cell carcinoma antigen recognized by T-cells, putative	RNA recognition motif domain; RNA-processing protein, HAT helix	207	0.0095	1

Protein functions inferred from the TrEMBL and InterPro database are listed along with the respective fold changes, p-values (χ^2) and chromosomal positions. This list represents an excerpt from Table S1.

Another strongly upregulated transcript in nodule tissue encodes a glutathione S-transferase (GST) that is implicated in the antioxidant defense of legume root nodules. GSTs can directly scavenge peroxides with the help of glutathione as electron acceptor, and furthermore detoxify endogenous compounds such as peroxidized lipids via conjugation of glutathione to target substrates, which facilitates their sequestration and removal (reviewed in Becana et al., 2010). Not surprisingly, GSTs constitute a large gene family in legumes, whose N_2-fixation machinery needs protection from ROS and RNS. Nodulin 6, encoded by early nodulin gene *MtN6* in *M. truncatula*, is one of the early response genes upon rhizobia infection in legumes. The mRNAs coding for Nodulin 6 accumulate in outer cortical cells that contain pre-infection threads, and in front of growing infection threads in mature root nodules (Mathis et al., 1999). Upregulation of Nodulin 6 consequently reflects the ongoing root hair invasion in mature nodule tissue.

The transcript encoding 6-phosphogluconate dehydrogenase (6PGD) is more than 200-fold upregulated in the pooled nodule compared to root tissue. The encoded enzyme belongs to the family of oxidoreductases and is involved in the non-oxidative phase of the pentose phosphate pathway, where it catalyzes the conversion of 6-phosphogluconate to ribulose 5-phosphate. The upregulation of *6PGD* resonates with the induced expression of ribulose-phosphate 3-epimerase (~170-fold upregulated in nodules, Table S1), which in turn uses ribulose 5-phosphate as a substrate to generate the ketose sugar xylulose 5-phosphate. A primary end product of the pentose phosphate pathway is NADPH, which is needed in response to oxidative stress, besides being necessary for fatty acid synthesis. NADPH can serve as co-substrate for glutathione reductases that reduce oxidized glutathione e.g. subsequent to its oxidation via GSTs (Table 3) or other glutathione peroxidases (up to 3-fold upregulated in nodules, Table S1). Further end products of the pathway include ribose-5-phosphate, used in the synthesis of nucleic acids, and erythrose-4-phosphate, implicated in the synthesis of aromatic amino acids. Against this backdrop, the induced expression of several members of the pentose phosphate pathway is not exclusively linked to an increased anabolism in nodule tissue, but rather ensures a steady supply with reducing equivalents (in form of NADPH).

The heavily upregulated monosaccharide transport protein (MTP) represents an ortholog of the hexose transporter Mtst1 in *M. truncatula*. An alignment of the mRNA sequence of Mtst1 from *M. truncatula* with the genomic sequence of *MTP* resulted in 44% matching bases, although the *MTP* sequence still retains introns (Figure S5). Interestingly, expression of Mtst1 was associated with a successfully established symbiosis of *M. truncatula* and the vesicular-arbuscular mycorrhizal fungus *Glomus versiforme* (Harrison, 1996). *In situ* hybridization revealed high Mtst1 mRNA levels in phloem fiber cells of the vascular tissue, cells of the root tip, and in cortical cells of the mycorrhizal root, especially in highly invaded areas. The author therefore suggests that increased expression levels of Mtst1 are correlated with internal growth of the fungus and with a functioning symbiosis, because the affected cells exhibit an increased metabolism, which in turn requires an intensified energy supply. To our knowledge an ortholog of Mtst1 in chickpea has not yet been described. Analogous to the function of Mtst1 in the context of vesicular-arbuscular mycorrhizal associations, MTP potentially ensures the sugar supply for root cells directly involved in the symbiotic association of rhizobia and legumes. The strong upregulation of *MTP* in nodules might furthermore contribute to the salt-tolerant trait of chickpea cultivar Beja 1, since hexose sugars also increase the cytosolic solute concentration, and thus can contribute to maintain the osmotic pressure (Hasegawa et al., 2000). One of the major bottlenecks of SNF in plants is the sensitivity of the symbiotic interaction, which renders the nodules very susceptible to abiotic stresses. The activity of enzymes directly involved in SNF, for example, was drastically decreased in salt-stressed nodules (Cordovilla et al., 1994). The nodule-specific induction of *MTP* along with *6PGD* thus indicates that several of the upregulated genes in nodules are not only linked to an increased metabolism, but also involved in the salt-tolerant trait of Beja 1. This is in line with our previous finding that several relatively low expressed UniTags in unstressed nodules become highly abundant in salt-stressed nodule tissue (Molina et al., 2011). 6PGD was additionally identified as a stress-responsive protein in a comparative proteomic analysis of *A. thaliana* roots that had been exposed to 150 mM NaCl (Jiang et al., 2007). The differential expression of this gene with the onset of salt stress in *A. thaliana* seems to be tightly linked to the accumulation of compatible osmolytes, and the present findings emphasize the importance of this mechanism in the context of SNF in legumes.

Functional Classification of Differentially Expressed Genes in Chickpea Nodule Tissue

The identified expression patterns of chickpea root and nodule tissues display comprehensive differences. Many gene products involved in sugar metabolism, antioxidant defense as well as biotic and abiotic stress responses of the host cells show drastically altered levels after adaption to the symbiotic relationship (Figure S4). This is in line with a recent report of nodulation-relevant genes identified in soybean roots 10 days after inoculation with *Bradyrhizobium japonicum* (Barros De Carvalho et al., 2013). The upregulated genes in nodules were found to be primarily involved in host cell metabolism, cell wall modifications and the antioxidant defense system. Interestingly, glycolysis and the citric acid cycle were the most active metabolic pathways in the context of SNF. The strong induction of several members of the pentose phosphate shunt in Beja 1 root nodules, however, reflects a shifted nitrogen balance in nodule tissue that is linked to an increased anabolism. Naturally, the interacting host cells have to ensure a steady supply of metabolites for the nitrogen-fixing rhizobia on the one hand, but also provide nitrogenous compounds for the rest of the plant on the other. Accordingly, the gene encoding asparagine synthetase is heavily upregulated in nodule tissue (~70-fold, see Table S1, Figure S4B), since asparagine represents one of the primary nitrogen transport products in plants.

The comprehensive expression of genes encoding GSTs in nodule tissue is accompanied by an upregulation of several oxidoreductases involved in different metabolic pathways as well as the two isoflavone 7-O-methyltransferase isoforms 8 and 9 (Figure 4). A globally induced expression of genes encoding GSTs was previously identified in nodule tissue from *M. truncatula* by Benedito et al. (2008), and the present findings confirm the importance of ROS scavenging for SNF in chickpea.

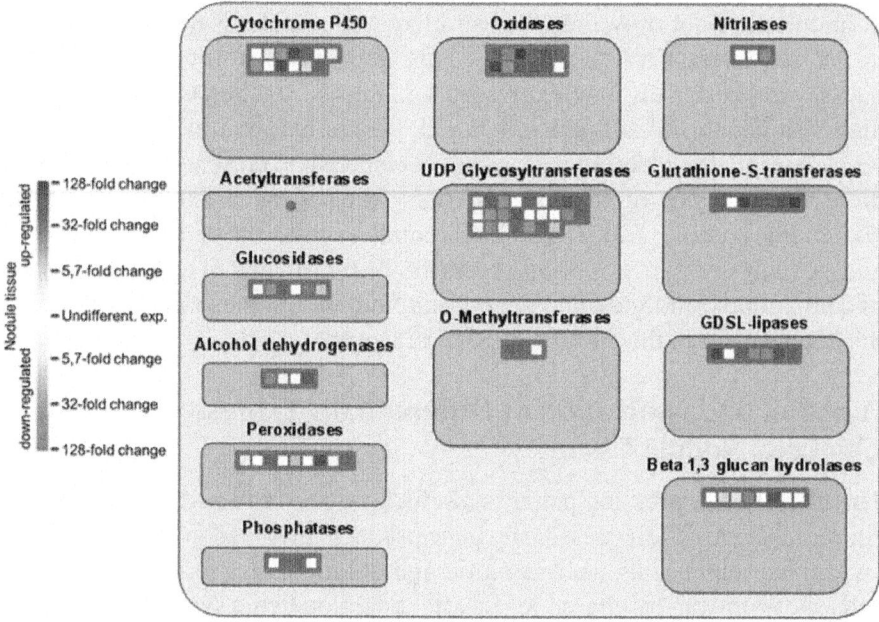

Figure 4. MapMan-based classification of the expressed genes into large enzyme families.Each field represents the expression of a particular gene. Upregulated genes in nodule tissue are shown in red, genes with reduced expression in green, and undifferentially expressed ones in white. Dark gray fields indicate that none of the expressed genes could be assigned to the respective class.

Interestingly, overexpression of isoflavone 7-O-methyltransferase 8 was shown to enhance disease resistance in *M. sativa* (He and Dixon, 2000), while the other isoform was identified as drought-responsive transcript in chickpea

(Varshney et al., 2009). Against this backdrop, upregulation of isoform 9 could well contribute to the salt-tolerant trait of Beja 1, since an *a priori* increased expression of stress-responsive genes already prepares the very sensitive interaction of eukaryotic and prokaryotic cells in root nodules for a possible onset of stress. Expression profiles of other large enzyme families such as nitrilases, glucosidases or the cytochrome P450 superfamily of monooxygenases are less consistent, but several peroxidases implicated in the response to oxidative stress as well as defense responses to fungi are heavily upregulated in nodules. These peroxidases apparently add to the comprehensive antioxidant defense repertoire that protects the oxygen-sensitive N_2-fixation machinery in chickpea nodules.

Functional annotation onto known biotic stress pathways reveals a reduced expression of disease resistance proteins, defense genes and of genes involved in mediating host cell responses to pathogens (Figure 5). This is consistent with the massive downregulation of the multidrug resistance and disease resistance response proteins in mature Beja 1 nodules, and confirms our previous observation that a sustained inhibition of at least some of the defense reactions in root nodules seems to be a prerequisite for maintenance of the symbiosis.

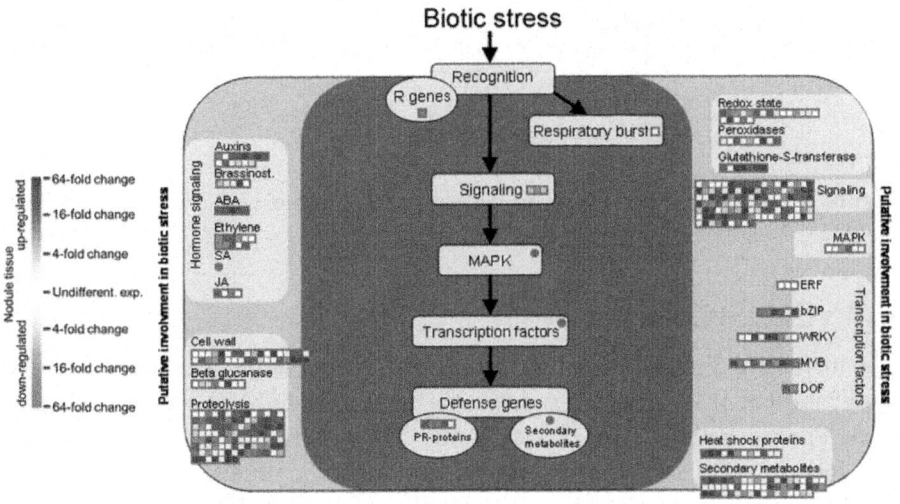

Figure 5. Gene expression relating to biotic stress pathways based on the functional annotation with MapMan. Please consult Figure 4 for further details.

The functional analysis of genes involved in generation of secondary metabolites in nodules illustrates a strong upregulation of genes implicated in phenylpropanoid and flavonoid biosynthesis although several genes encoding proteins of the shikimic acid pathway are downregulated (Figure 6).

Especially dihydroflavonols appear to be of special importance for SNF, which is likely linked to their function as antioxidants and radical scavengers (Haraguchi et al., 1996). The most upregulated gene involved in the synthesis of phenylpropanoids actually encodes a nicotinamidase that operates in the salvage pathway of NAD biosynthesis (Wang and Pichersky, 2007). The induced expression of this gene in nodule tissue underlines the importance of NAD as redox carrier in the rhizobia-adapted metabolism.

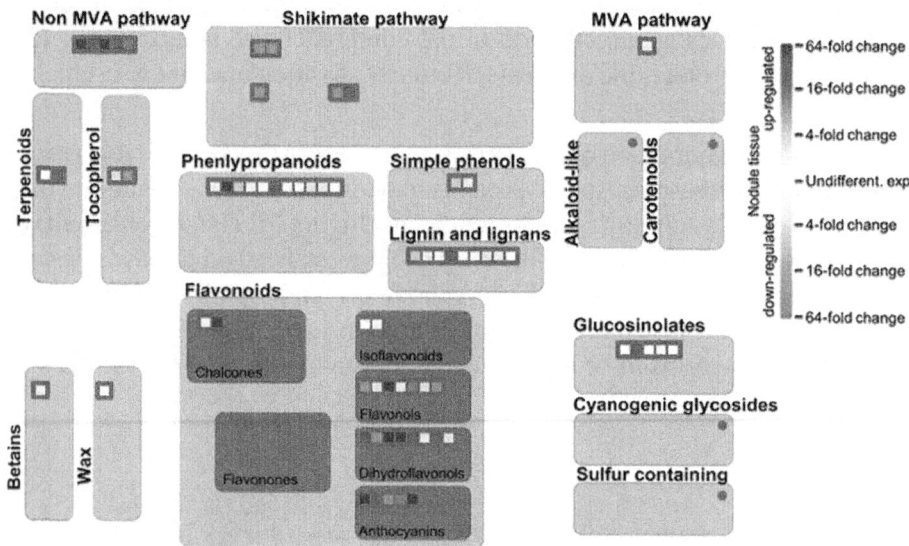

Figure 6. Differential expression of genes involved in biosynthesis of secondary metabolites. Please consult Figure 4 for further details.

CONCLUSIONS

The increasing throughput of next-generation sequencing technologies and the application of 3rd generation sequencing (such as the SMRT sequencing system from Pacific Biosciences) are currently intensifying the efforts to decipher the genome of a steadily increasing number of legumes. As in the case of chickpea, these genome sequences will allow for more detailed analyses of the differential gene expression in relation to SNF in legumes. Using the recently published chickpea genome sequence, we reanalyzed the transcriptomes of unstressed root and nodule tissue and identified a strong upregulation of genes encoding glutathione S-transferases or with implications in phenylpropanoid and flavonoid biosynthesis. The expression of twenty candidate genes from the reanalyzed dataset was validated using five individual biological replicates, revealing a comprehensive biological variance in the corresponding expression.

Additionally to the characterization of the most differentially expressed mRNAs and their potential functions in relation to SNF, we implemented a functional analysis via MapMan for the transcriptomes of both tissues. The functional classification of all protein-coding chickpea genome sequences will pave the way for future functional analyses of chickpea mRNAs, also retrospectively (as in the present study) and not only from root or nodule, but from any tissue of interest.

ACKNOWLEDGMENTS

The authors thank Catherine Pernot, INRA, for assistance in preparation of the new tissue samples as well as the anonymous reviewers for valuable advices. Research of the authors was supported by funds from Deutsche Gesellschaft für Technische Zusammenarbeit (GTZ, project 08.7860.3-001.00), BMBF IGSTC (project IND 09/515: Biotechnological approaches to improve chickpea crop productivity for farming community and industry), and Aquarhiz (project EU-FP6: Modulation of plant-bacteria interactions to enhance tolerance to water deficit for grain legumes in the Mediterranean dry lands).

REFERENCES

1. Asmann, Y. W., Klee, E. W., Thompson, E. A., Perez, E. A., Middha, S., Oberg, A. L., et al. (2009). 3' tag digital gene expression profiling of human brain and universal reference RNA using Illumina Genome Analyzer. *BMC Genomics* 10:531. doi: 10.1186/1471-2164-10-531

2. Barros De Carvalho, G. A., Batista, J. S., Marcelino-Guimaraes, F. C., Costa Do Nascimento, L., and Hungria, M. (2013). Transcriptional analysis of genes involved in nodulation in soybean roots inoculated with *Bradyrhizobium japonicum* strain CPAC 15. *BMC Genomics* 14:153. doi: 10.1186/1471-2164-14-153

3. Becana, M., Matamoros, M. A., Udvardi, M., and Dalton, D. A. (2010). Recent insights into antioxidant defenses of legume root nodules. *New Phytol.* 188, 960–976. doi: 10.1111/j.1469-8137.2010.03512.x

4. Benedito, V. A., Torres-Jerez, I., Murray, J. D., Andriankaja, A., Allen, S., Kakar, K., et al. (2008). A gene expression atlas of the model legume*Medicago truncatula*. *Plant J.* 55, 504–513. doi: 10.1111/j.1365-313X.2008.03519.x

5. Cordovilla, M. P., Ligero, F., and Lluch, C. (1994). The effect of salinity on N fixation and assimilation in *Vicia faba. J. Exp. Bot.* 45, 1483–1488.

6. Crawford, N. M. (1995). Nitrate: nutrient and signal for plant growth. *Plant Cell* 7, 859–868. doi: 10.1105/tpc.7.7.859

7. Desbrosses, G. J., and Stougaard, J. (2011). Root nodulation: a paradigm for how plant-microbe symbiosis influences host developmental pathways. *Cell Host Microbe* 10, 348–358. doi: 10.1016/j. chom.2011.09.005

8. Ferguson, B. J., Indrasumunar, A., Hayashi, S., Lin, M. H., Lin, Y. H., Reid, D. E., et al. (2010). Molecular analysis of legume nodule development and autoregulation. *J. Integr. Plant Biol.* 52, 61–76. doi: 10.1111/j.1744-7909.2010.00899.x

9. Gage, D. J. (2004). Infection and invasion of roots by symbiotic, nitrogen-fixing rhizobia during nodulation of temperate legumes. *Microbiol. Mol. Biol. Rev.* 68, 280–300. doi: 10.1128/mmbr.68.2.280-300.2004

10. Haraguchi, H., Mochida, Y., Sakai, S., Masuda, H., Tamura, Y., Mizutani, K., et al. (1996). Protection against oxidative damage by dihydroflavonols in *Engelhardtia chrysolepis*. *Biosci. Biotechnol. Biochem.* 60, 945–948. doi: 10.1271/bbb.60.945

11. Harrison, M. J. (1996). A sugar transporter from *Medicago truncatula*: altered expression pattern in roots during vesicular-arbuscular (VA) mycorrhizal associations. *Plant J.* 9, 491–503. doi: 10.1046/j.1365-313X.1996.09040491.x

12. Hasegawa, P. M., Bressan, R. A., Zhu, J. K., and Bohnert, H. J. (2000). Plant cellular and molecular responses to high salinity. *Annu. Rev. Plant Physiol. Plant Mol. Biol.* 51, 463–499. doi: 10.1146/annurev. arplant.51.1.463

13. Hayashi, S., Gresshoff, P. M., and Ferguson, B. J. (2013). Systemic signalling in legume nodulation: nodule formation and its regulation. Long-distance systemic signaling and communication in plants. *Signal. Commun. Plants* 19, 219–229. doi: 10.1007/978-3-642-36470-9_11

14. Hayashi, S., Reid, D. E., Lorenc, M. T., Stiller, J., Edwards, D., Gresshoff, P. M., et al. (2012). Transient Nod factor-dependent gene expression in the nodulation-competent zone of soybean (*Glycine max* [L.] Merr.) roots. *Plant Biotechnol. J.* 10, 995–1010. doi: 10.1111/j.1467-7652.2012.00729.x

15. He, X. Z., and Dixon, R. A. (2000). Genetic manipulation of isoflavone 7-O-methyltransferase enhances biosynthesis of 4'-O-methylated isoflavonoid phytoalexins and disease resistance in alfalfa. *Plant Cell* 12, 1689–1702. doi: 10.2307/3871183

16. Iturbe-Ormaetxe, I., Matamoros, M. A., Rubio, M. C., Dalton, D. A., and Becana, M. (2001). The antioxidants of legume nodule mitochondria.*Mol. Plant Microbe Interact.* 14, 1189–1196. doi: 10.1094/

mpmi.2001.14.10.1189

17. Jain, M., Misra, G., Patel, R. K., Priya, P., Jhanwar, S., Khan, A. W., et al. (2013). A draft genome sequence of the pulse crop chickpea (*Cicer arietinum* L.). *Plant J.* 74, 715–729. doi: 10.1111/tpj.12173

18. Jiang, Y., Yang, B., Harris, N. S., and Deyholos, M. K. (2007). Comparative proteomic analysis of NaCl stress-responsive proteins in *Arabidopsis*roots. *J. Exp. Bot.* 58, 3591–3607. doi: 10.1093/jxb/erm207

19. Krylova, V., Dubrovo, P., and Izmailov, S. (2007). Metabolite transport across the peribacteroid membrane during broad bean development. *Russ. J. Plant Physiol.* 54, 184–190. doi: 10.1134/S1021443707020045

20. Lee, H., and Copeland, L. (1994). Ultrastructure of chickpea nodules. *Protoplasma* 182, 32–38. doi: 10.1134/S1021443707020045

21. Lenz, T. L., Eizaguirre, C., Rotter, B., Kalbe, M., and Milinski, M. (2013). Exploring local immunological adaptation of two stickleback ecotypes by experimental infection and transcriptome-wide digital gene expression analysis. *Mol. Ecol.* 22, 774–786. doi: 10.1111/j.1365-294X.2012.05756.x

22. Libault, M., Farmer, A., Joshi, T., Takahashi, K., Langley, R. J., Franklin, L. D., et al. (2010). An integrated transcriptome atlas of the crop model*Glycine max*, and its use in comparative analyses in plants. *Plant J.* 63, 86–99. doi: 10.1111/j.1365-313X.2010.04222.x

23. Lohse, M., Nagel, A., Herter, T., May, P., Schroda, M., Zrenner, R., et al. (2014). Mercator: a fast and simple web server for genome scale functional annotation of plant sequence data. *Plant Cell Environ.* 37, 1250–1258. doi: 10.1111/pce.12231

24. L'Taief, B., Sifi, B., Zaman-Allah, M., Drevon, J. J., and Lachaal, M. (2007). Effect of salinity on root-nodule conductance to the oxygen diffusion in the *Cicer arietinum-Mesorhizobium ciceri* symbiosis. *J. Plant Physiol.* 164, 1028–1036. doi: 10.1016/j.jplph.2006.05.016

25. Man, M. Z., Wang, X., and Wang, Y. (2000). POWER_SAGE: comparing statistical tests for SAGE experiments. *Bioinformatics* 16, 953–959. doi: 10.1093/bioinformatics/16.11.953

26. Matamoros, M. A., Dalton, D. A., Ramos, J., Clemente, M. R., Rubio, M. C., and Becana, M. (2003). Biochemistry and molecular biology of antioxidants in the rhizobia-legume symbiosis. *Plant Physiol.* 133, 499–509. doi: 10.1104/pp.103.025619

27. Mathis, R., Grosjean, C., De Billy, F., Huguet, T., and Gamas, P. (1999).

The early nodulin gene MtN6 is a novel marker for events preceding infection of *Medicago truncatula* roots by *Sinorhizobium meliloti*. *Mol. Plant Microbe Interact.* 12, 544–555. doi: 10.1094/mpmi.1999.12.6.544

28. Matsumura, H., Reich, S., Ito, A., Saitoh, H., Kamoun, S., Winter, P., et al. (2003). Gene expression analysis of plant host-pathogen interactions by SuperSAGE. *Proc. Natl. Acad. Sci. U.S.A.* 100, 15718–15723. doi: 10.1073/pnas.2536670100

29. Molina, C., Rotter, B., Horres, R., Udupa, S. M., Besser, B., Bellarmino, L., et al. (2008). SuperSAGE: the drought stress-responsive transcriptome of chickpea roots. *BMC Genomics* 9:553. doi: 10.1186/1471-2164-9-553

30. Molina, C., Zaman-Allah, M., Khan, F., Fatnassi, N., Horres, R., Rotter, B., et al. (2011). The salt-responsive transcriptome of chickpea roots and nodules via deepSuperSAGE. *BMC Plant Biol.* 11:31. doi: 10.1186/1471-2229-11-31

31. Mylona, P., Pawlowski, K., and Bisseling, T. (1995). Symbiotic nitrogen fixation. *Plant Cell* 7, 869–885. doi: 10.1105/tpc.7.7.869

32. Oldroyd, G. E., and Downie, J. A. (2008). Coordinating nodule morphogenesis with rhizobial infection in legumes. *Annu. Rev. Plant Biol.* 59, 519–546. doi: 10.1146/annurev.arplant.59.032607.092839

33. Pawlowski, K., Kunze, R., Vries, S. D., and Bisseling, T. (1994). Isolation of total, poly(A) and polysomal RNA from plant tissues. *Plant Mol. Biol. Man.* D5, 1–13.

34. Reid, D. E., Hayashi, S., Lorenc, M., Stiller, J., Edwards, D., Gresshoff, P. M., et al. (2012). Identification of systemic responses in soybean nodulation by xylem sap feeding and complete transcriptome sequencing reveal a novel component of the autoregulation pathway. *Plant Biotechnol. J.* 10, 680–689. doi: 10.1111/j.1467-7652.2012.00706.x

35. Romdhane, S. B., Tajini, F., Trabelsi, M., Aouani, M. E., and Mhamdi, R. (2007). Competition for nodule formation between introduced strains of *Mesorhizobium ciceri* and the native populations of rhizobia nodulating chickpea *Cicer arietinum* in Tunisia. *World J. Microbiol. Biotechnol.* 23, 1195–1201. doi: 10.1007/s11274-006-9325-z

36. Saxena, R. K., Penmetsa, R. V., Upadhyaya, H. D., Kumar, A., Carrasquilla-Garcia, N., Schlueter, J. A., et al. (2012). Large-scale development of cost-effective single-nucleotide polymorphism marker assays for genetic mapping in pigeonpea and comparative mapping in legumes. *DNA Res.* 19, 449–461. doi: 10.1093/dnares/dss025

37. Sharbel, T. F., Voigt, M. L., Corral, J. M., Galla, G., Kumlehn, J., Klukas, C., et al. (2010). Apomictic and sexual ovules of *Boechera* display

heterochronic global gene expression patterns. *Plant Cell* 22, 655–671. doi: 10.1105/tpc.109.072223

38. Thimm, O., Blasing, O., Gibon, Y., Nagel, A., Meyer, S., Kruger, P., et al. (2004). MAPMAN: a user-driven tool to display genomics data sets onto diagrams of metabolic pathways and other biological processes. *Plant J.* 37, 914–939. doi: 10.1111/j.1365-313X.2004.02016.x

39. Vandesompele, J., De Preter, K., Pattyn, F., Poppe, B., Van Roy, N., De Paepe, A., et al. (2002). Accurate normalization of real-time quantitative RT-PCR data by geometric averaging of multiple internal control genes. *Genome Biol.* 3: RESEARCH0034. doi: 10.1186/gb-2002-3-7-research0034

40. Varshney, R. K., Hiremath, P. J., Lekha, P., Kashiwagi, J., Balaji, J., Deokar, A. A., et al. (2009). A comprehensive resource of drought- and salinity-responsive ESTs for gene discovery and marker development in chickpea (*Cicer arietinum* L.). *BMC Genomics* 10:523. doi: 10.1186/1471-2164-10-523

41. Varshney, R. K., Kudapa, H., Roorkiwal, M., Thudi, M., Pandey, M. K., Saxena, R. K., et al. (2012). Advances in genetics and molecular breeding of three legume crops of semi-arid tropics using next-generation sequencing and high-throughput genotyping technologies. *J. Biosci.* 37, 811–820. doi: 10.1007/s12038-012-9228-0

42. Varshney, R. K., Song, C., Saxena, R. K., Azam, S., Yu, S., Sharpe, A. G., et al. (2013). Draft genome sequence of chickpea (*Cicer arietinum*) provides a resource for trait improvement. *Nat. Biotechnol.* 31, 240–246. doi: 10.1038/nbt.2491

43. Wang, G., and Pichersky, E. (2007). Nicotinamidase participates in the salvage pathway of NAD biosynthesis in *Arabidopsis*. *Plant J.* 49, 1020–1029. doi: 10.1111/j.1365-313X.2006.03013.x

44. Zawada, A. M., Rogacev, K. S., Rotter, B., Winter, P., Marell, R. R., Fliser, D., et al. (2011). SuperSAGE evidence for CD14++CD16+ monocytes as a third monocyte subset. *Blood* 118, e50–e61. doi: 10.1182/blood-2011-01-326827

Chapter 9

GENOME-WIDE ANALYSIS AND EXPRESSION PROFILING SUGGEST DIVERSE ROLES OF GH3 GENES DURING DEVELOPMENT AND ABIOTIC STRESS RESPONSES IN LEGUMES

Vikash K. Singh, Mukesh Jain and Rohini Garg

Functional and Applied Genomics Laboratory, National Institute of Plant Genome Research, New Delhi, India

Growth hormone auxin regulates various cellular processes by altering the expression of diverse genes in plants. Among various auxin-responsive genes, *GH3* genes maintain endogenous auxin homeostasis by conjugating excess of auxin with amino acids. *GH3* genes have been characterized in many plant species, but not in legumes. In the present work, we identified members of GH3 gene family and analyzed their chromosomal distribution, gene structure, gene duplication and phylogenetic analysis in different legumes, including chickpea, soybean, *Medicago*, and *Lotus*. A comprehensive expression analysis in different vegetative and reproductive tissues/stages revealed that many of *GH3* genes were expressed in a tissue-specific manner. Notably, chickpea *CaGH3-3*, soybean *GmGH3-8* and *-25*, and *Lotus LjGH3-4*, *-5*, *-9* and *-18* genes were up-regulated in root, indicating their putative role in root development. In addition, chickpea *CaGH3-1* and *-7*, and *Medicago MtGH3-7*, *-8*, and *-9* were found to be highly induced under drought and/or salt stresses, suggesting their role in abiotic stress responses. We also observed the examples of differential expression pattern of duplicated *GH3* genes in soybean, indicating their functional diversification. Furthermore, analyses of three-dimensional structures, active site residues and ligand preferences provided molecular insights into function of *GH3* genes in legumes. The analysis presented here would help in investigation of precise function of *GH3* genes in legumes during development and stress conditions.

INTRODUCTION

Auxin is an important phytohormone which regulates various aspects of plant growth and development. Most of these processes are regulated by auxin-responsive genes, namely auxin/indole-3-acetic acid (Aux/IAA), auxin-response factor (ARF), small auxin-up RNAs (SAUR) and Gretchen Hagen3 (GH3; Hagen and Guilfoyle, 2002). Auxin-responsiveness to these genes is conferred by auxin-responsive elements (AuxREs, TGTCTC) present in their promoters (Hagen et al., 1991; Li et al., 1994; Ulmasov et al., 1995; Hagen and Guilfoyle, 2002). To understand molecular mechanism of auxin action, several auxin-responsive genes have been isolated and characterized from many plant species, such as pea, soybean, tobacco, and cucumber (Hagen and Guilfoyle, 2002).

Gretchen Hagen3 family of proteins maintain auxin level by catalyzing conjugation of amino acids with indole-3-acetic acid, salicylic acid (SA), and jasmonic acid (JA; Staswick et al., 2002, 2005). The first *GH3* gene was identified by Hagen et al. (1984) as an early auxin-responsive gene in soybean. Since then, a large number of *GH3* homologs have been identified in numerous plant species ranging from mosses to angiosperms (Jain et al., 2006; Terol et al., 2006; Ludwig-Müller et al., 2008; Kumar et al., 2012; Yuan et al., 2013). The studies on GH3 proteins have revealed their regulatory function in plant growth, organ development, light signaling, abiotic stress tolerance, and plant defense responses (Woodward and Bartel, 2005; Park et al., 2007; Jain and Khurana, 2009; Ludwig-Muller, 2011; Du et al., 2012; Kumar et al., 2012; Yuan et al., 2013). In *Arabidopsis*, GH3 gene family has been classified into three groups (I–III) based on sequence similarity and substrate specificities (Staswick et al., 2002). Group I GH3 proteins of *Arabidopsis* are JA-amido synthetases (Staswick et al., 2002, 2005). AtGH3-11, a group I GH3 protein, was characterized based on analysis of *jar1* mutant, which was insensitive to JA and was required for the formation of bioactive jasmonate JA-isoleucine (Staswick et al., 2002). A different allele of this gene (*FIN219*) was identified as a phytochrome A signaling component, having crucial role in photomorphogenesis (Hsieh et al., 2000). Group II GH3 proteins of *Arabidopsis* are involved in conjugation of IAA to various amino acids (Staswick et al., 2002, 2005). *AtGH3-2* gain-of-function mutant, *Ydk1-D*, was shown to be responsible for short primary root, reduced lateral root number, and apical dominance (Takase et al., 2004). In another report *AtGH3-6* mutant, *dfl1*, was shown to regulate shoot elongation and lateral root formation negatively, but positively regulate the light responses to hypocotyl length (Nakazawa et al., 2001). Some Group II

GH3 proteins of rice (TLD1/OsGH3-13, OsGH3-2, and OsGH3-8) have also been characterized that conjugate IAA with aspartate or alanine (Chen et al., 2009, 2010; Zhang et al., 2009). A gain-of-function mutant of rice *OsGH3-13* gene, *tld1-D*, resulted in increased tillers, enlarged leaf angles, dwarfism and improved drought tolerance (Zhang et al., 2009). AtGH3-12/PBS3 is the only characterized member of group III, which catalyzes the conjugation of glutamic acid (Glu) to 4-aminobenzoate and 4-hydroxybenzoate and is involved in SA signaling (Jagadeeswaran et al., 2007; Nobuta et al., 2007; Okrent et al., 2009). Recently, the crystal structure and mechanism of catalytic action of AtGH3-12 and JAR1/AtGH3-11 (Westfall et al., 2012) in *Arabidopsis*, and VvGH3-1 in grapevine (Peat et al., 2012) have also been reported.

Legumes are nutritionally important crop plants, which serve as a rich source of proteins and fibers. Although the first auxin-responsive gene was identified from soybean (Hagen et al., 1984), genome-wide analysis of *GH3*genes in legumes is still lacking. This may be attributed to scarcity of genomic resources for legumes until recently. However, in recent years several genomic resources have been generated for various legumes. The genome and transcriptome sequences of desi and kabuli chickpea (*Cicer arietinum*), soybean (*Glycine max*),*Medicago* (*Medicago truncatula*), and *Lotus* (*Lotus japonicus*) have been published (Sato et al., 2008;Schmutz et al., 2010; Garg et al., 2011; Young et al., 2011; Jain et al., 2013; Varshney et al., 2013). The availability of genome annotation provides an opportunity for characterization of GH3 gene family in legumes, which can help in better understanding of their function in various cellular processes. The availability of crystal structures of GH3 proteins (Peat et al., 2012; Westfall et al., 2012) provides a resource to identify substrate specificity determining motifs/residues in the GH3 proteins, which can help in understanding auxin-mediated regulation of cellular processes in legumes.

Here, we performed genome-wide identification and analysis of GH3 gene family in four legume species, including chickpea, soybean, *Medicago,* and *Lotus*. We report their genomic organization, chromosomal distribution, sequence homology, and phylogenetic relationship in/among different legumes. Comprehensive gene expression analyses in various tissues/stages and abiotic stress conditions have also been performed to gain insight into their putative function. Putative promoter sequences of the *GH3* genes were also analyzed for identification of *cis*-regulatory elements, which may be involved in various development processes and stress responses. In addition, their ligand preferences were predicted based on the protein structure and sequence analysis. These data provide a framework for further in-depth functional analyses of *GH3* genes in legumes.

MATERIAL AND METHODS

Identification of *GH3* Genes

Chickpea genome annotation was downloaded from Chickpea Genome Analysis Project (CGAP v1.0; Jain et al., 2013), soybean and *Medicago* genome annotations were downloaded from Phytozome (v9.0[1]), and *Lotus* genome annotation was taken from miyakazusa.jp database (v2.5[2]). A total of 19 protein sequences of GH3 family members of *Arabidopsis* and 13 protein sequences of rice GH3 family members were downloaded from TAIR[3] and RGAP database[4], respectively. The rice and *Arabidopsis* GH3 proteins were searched in chickpea, soybean, *Medicago* and *Lotus* proteomes individually, using BLASTP with an *e*-value cutoff of 1e-05. Further, the HMM profile of GH3 domain was downloaded from pfam database[5] and HMMER was used to search proteomes of chickpea, soybean, *Medicago,* and *Lotus* for GH3 domain. All the tentative gene lists obtained from these two searches were combined to make a non-redundant gene list for each legume, and their protein sequences were searched in pfam database to confirm the presence of conserved GH3 domain. Using the similar strategies, we investigated the chickpea transcriptome sequence (Garg et al., 2010) as well for identification of any additional GH3 gene family member that may not be represented in chickpea genome annotation.

Sequence Analysis and Phylogenetic Tree Construction

Multiple sequence alignment of all the GH3 protein sequences of chickpea, soybean, *Medicago* and *Lotus* with *Arabidopsis* GH3 protein sequences was carried out using MAFFT and phylogenetic tree was constructed by UPGMA method using CLC Genomics Workbench (v4.7.2). Bootstrap analysis was performed by taking 1,000 replicates and the generated tree was viewed using FigTree (v1.3.1).

Gene Duplication Analysis

Synteny analysis was performed using Plant Genome Duplication Database[6]. Syntenic blocks were evaluated using Circos tool. Information about the chromosome locations was obtained from Phytozome database. Genes were regarded as segmentally duplicated if they found to be coparalogs located on duplicated blocks, as proposed by Wei et al. (2007). Tandem duplication was characterized as multiple genes of one family located within the same or neighboring intergenic region (Du et al., 2013a).

Promoter Sequence Analysis

Genomic co-ordinates of coding sequences were determined using GFF files obtained from chickpea and soybean genome annotation. The regions of 2,000 bp upstream from start codon were extracted from genomic DNA sequences. *Cis*-regulatory elements on both strands of promoter sequences were scanned using PLACE web server[7].

Homology Modeling

The 3-D protein structures of AtGH3-11 (Protein Data Bank code 4EPL; Westfall et al., 2012) and Vv-GH3-1 (Protein Data Bank code 4B2G; Peat et al., 2012) were downloaded from Protein Data Bank[8]. Phyre2 (Protein Homology/ AnalogY Recognition Engine[9]) was used for predicting the protein structure by homology modeling under 'intensive' mode (Kelley and Sternberg, 2009). The protein structures modeled with >90% confidence were selected. The core of predicted protein structure or allowed area in the plot showing the preferred region for psi/phi angles pair for residues was determined through Ramachandran plot using RAMPAGE server[10] and models were viewed using Chimera (V1.9). Only those structures representing >95% of residues in favored region were considered for further analysis. For substrate binding site prediction, templates and model were superimposed using MatchMaker of Chimera (V1.9) and ligands were transferred on model from templates.

Plant Material and Stress Treatments

Chickpea (*C. arietinum* L. genotype ICC4958) seeds were grown in culture room and field for collection of various tissue samples. Mature leaf, young leaf, young pod, flower buds (FB), flower bud opened (FBO), unopened flowers (UOF), and mature flower (MF) were harvested from field grown plants. Root and shoot tissues were collected from 15-day-old chickpea seedlings grown in autoclaved mixture (1:1) of agropeat and vermiculite in plastic pots in the culture room maintained at 22 ± 1°C with a photoperiod of 14 h, as described (Garg et al., 2010). Germinating seedlings (GS) were collected after 5 days of seed germination on wet Whatman paper sheet in Petri dishes as described (Singh et al., 2013). Two stages of flower bud development (FB 4 mm and FBO 8–10 mm) were collected on the basis of size and morphological differences (Singh et al., 2013). Two stages of flower development, including young flower with closed petals (UOF) and MF with opened petals were also collected. For stress treatments, 10-day-old chickpea seedlings were kept in water for control, 150 mM solution of NaCl for salt stress, at 4°C for cold stress and between folds of tissue paper for desiccation stress. Root and shoot tissues were harvested separately after 5 h of treatments as described (Garg et al., 2010). All samples were quickly frozen into liquid nitrogen after harvesting and stored at -80°C till RNA isolation.

RNA Isolation and Quantitative RT-PCR Analysis

Total RNA was extracted using TRI reagent (Sigma Life Science, St. Louis, MO, USA) following the manufacturer's instructions. RNA quality and quantity was determined using Nanodrop 1000 spectrophotometer (Thermo Fisher Scientific, Wilmington, DE, USA). RNA samples with 260/280 ratio between 1.8 and 2.1 and 260/230 ratio between 2.0 and 2.5 were used for cDNA synthesis. Primers were designed for all genes using Primer Express (v3.0) software (Applied Biosystems, Foster City, CA, USA). Specificity of each pair of primers was determined via BLAST search. All the primer sequences used have been listed in Supplemental Table S1. For each tissue, at least two independent biological replicates and three technical replicates of each biological replicate were taken for the analysis. Real time PCR analysis was performed using the 7500 Detection System (Applied Biosystems) as described (Garg et al., 2010). The expression of *elongation factor-1 alpha* gene was used as internal control for normalization of sample input variance (Garg et al., 2010).

RNA-seq and Microarray Data Analysis

The expression patterns of chickpea and soybean *GH3* genes were analyzed using RNA-seq data from various tissue/stages of development. For chickpea, we mapped our RNA-seq data (Singh et al., 2013) on the genome using TopHat (v2.0.6), assembled with Cufflinks (v2.1.1), and merged with Cuffmerge to estimate read count in FPKM. For soybean, normalized gene expression data (RPKM) was downloaded from SoySeq[11]. *Medicago* and *Lotus GH3* gene expression data were downloaded from MtGEA[12] and LjGEA[13], respectively. Probsets corresponding to *MtGH3* and *LjGH3* genes were identified using BLASTN search with best hits.

RESULTS AND DISCUSSION

GH3 Gene Family in Legumes

The availability of genome sequences provides an opportunity to identify and analyze GH3 gene family in legumes. We investigated members of GH3 gene family in four legumes, including chickpea, soybean, *Medicago,* and *Lotus*, using two strategies, BLASTP and HMM profile search. For chickpea, we selected genome sequence of desi genotype (ICC4958), because of the availability of comprehensive expression (RNA-seq) data from various tissues/ developmental stages (Jain et al., 2013) and abiotic stress conditions (Garg et al., 2014), which can provide better insights into the functions of *GH3* genes (as described in latter sections). The GH3 gene family members identified via these two searches were combined and a unique gene list was obtained for each legume species. In total, 11, 28, 10, and 18 *GH3* gene members were identified in chickpea, soybean, *Medicago* and *Lotus*, respectively, after analyzing their protein sequences in pfam database for the presence of conserved GH3 domain. To identify additional members of GH3 gene family in chickpea, which may not be represented in the genome annotation, the published chickpea transcriptome (Garg et al., 2011) was also analyzed using similar strategies. This resulted in the identification of one additional GH3 gene family member for a total of 12 in chickpea. A list of *GH3* genes and their identifiers in different legumes along with their genomic co-ordinates is given in Supplemental Table S2.

The number of GH3 proteins identified in chickpea, *Medicago* and *Lotus* were comparable to *Arabidopsis* (10; excluding group III members, which are exclusively present in *Arabidopsis*), rice (13), tomato (15), and sorghum (16; Jain et al., 2006; Wang et al., 2010; Kumar et al., 2012). Whereas, the number of GmGH3 proteins are found to be approximately double as compared to other legume plants. The soybean genome has undergone two rounds of whole genome duplication, including an ancient duplication prior to the divergence of papilionoids (58–60 Mya) and a soybean-specific duplication that is estimated to have occurred ~13 Mya (Schmutz et al., 2010), which might have resulted into duplication of members of this gene family.

Genomic Organization and Chromosomal Distribution

All GH3 proteins identified in legumes showed the presence of characteristic GH3 domain, and sequence conservation in the core region in multiple sequence alignment of proteins. The gene structure (exon–intron organization) analysis of *CaGH3* and *GmGH3* genes revealed that number of introns varied from one to four except for *CaGH3-11* and *GmGH3-24*, which do not have any intron (Figure 1). Most of the *CaGH3* and*GmGH3* had similar intron-phasing distribution (Figure 1) and followed the pattern reported earlier for rice*GH3* genes (Jain et al., 2006). Next, we analyzed the distribution of *GH3* genes on the chromosomes in different legumes. Only two *CaGH3* genes could be located on the linkage groups, whereas others were located on scaffolds (Supplemental Table S2). This may be due to availability of incomplete draft genome sequence and unanchored scaffold as of now (Jain et al., 2013). For soybean, all the 28 *GmGH3* genes were distributed on 14 of 20 chromosomes, with six *GH3* genes located on chromosome 12, four and three being present on chromosome 6 and 13, respectively, two each on chromosome 2, 3, 15, and 17, and one each on chromosome 1, 5, 7, 10, 11, 16, and 19 (Supplemental Table S2). In soybean, many *GH3* genes were clustered, such as adjacent genes on chromosome 6 (*GmGH3-7, -8, -9,* and *-10*), chromosome 12 (*GmGH3-14* and *-15,* and *GmGH3-17, -18* and *-19*), chromosome 13 (*GmGH3-20, -21,* and *-22*), and chromosome 17 (*GmGH3-26* and *-27*).

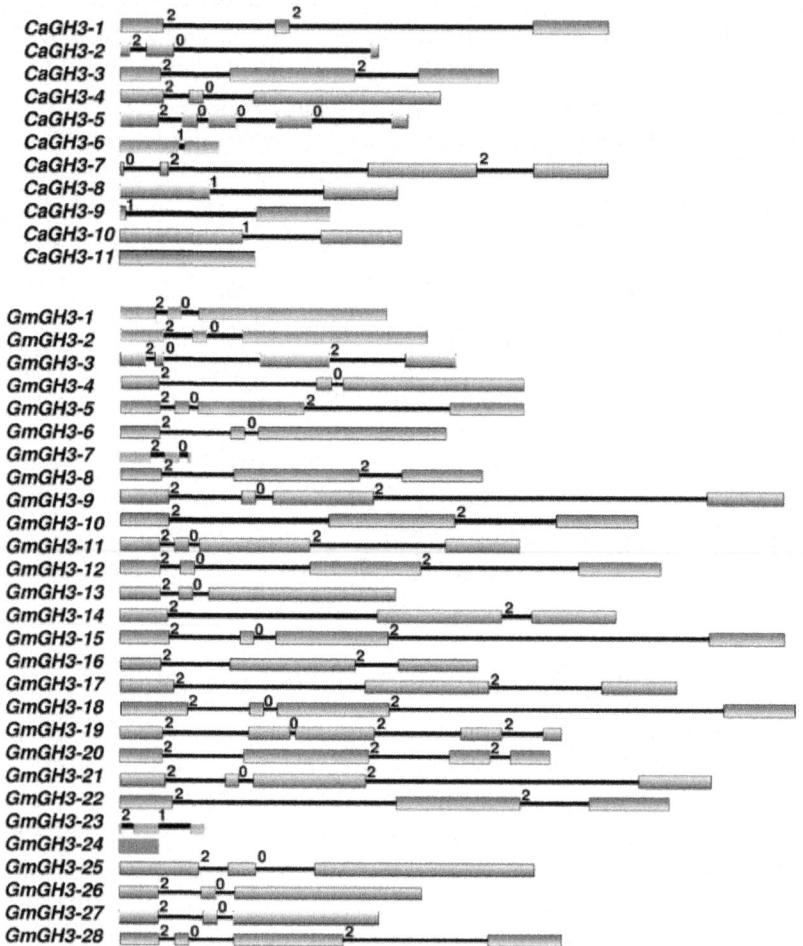

Figure 1. Exon-intron organization of chickpea and soybean *GH3* **genes.** Boxes and lines represent exons and introns, respectively. The numbers 0, 1, and 2 represent phase 0, 1, and 2 introns, respectively.

The amino acid sequences of these genes [GmGH3-7 to -10 (7–8% identity), GmGH3-14, and -15 (32% identity), GmGH3-17, -18, -19 (28–57% identity), GmGH3-20 to -22 (30–68%), and GmGH3-26 and -27 (70% identity)] showed very low (7%) to high (70%) similarity (Supplemental Table S3B), indicating that these *GmGH3* genes probably resulted from tandem duplication and some of them diverged during course of evolution. In *Medicago,* 7 of 10*GH3* genes were distributed on 5 of 8 chromosomes and three *MtGH3* genes were located on scaffolds (Supplemental Table S2). Chromosome 5 and 8 of *Medicago* harbored two *MtGH3* genes each and one each resided on chromosome 2, 3, and 7. In *Lotus*, out of 18 *LjGH3* genes, only eight were located on 4 of 6 chromosomes; three located on chromosome 3, two each on chromosomes 2 and 4, and one on chromosome 1 (Supplemental Table S2). Altogether, it appears that tandem gene duplication resulted in the amplification of GH3 gene family members in legumes and low homology between them suggested their divergence during course of evolution.

Sequence Analysis and Phylogenetic Relationship

Pairwise analysis of the full-length protein sequences of chickpea and soybean GH3 proteins showed very high homology, 76.9% between paralogous pair, CaGH3-7 and CaGH3-8 (Figure 2; Supplemental Table S3A), and 97.4% between paralogous pair, GmGH3-8 and GmGH3-16 (Figure 2; Supplemental Table S3B). Such high homologies suggest that they may perform similar functions (Jain et al., 2006).

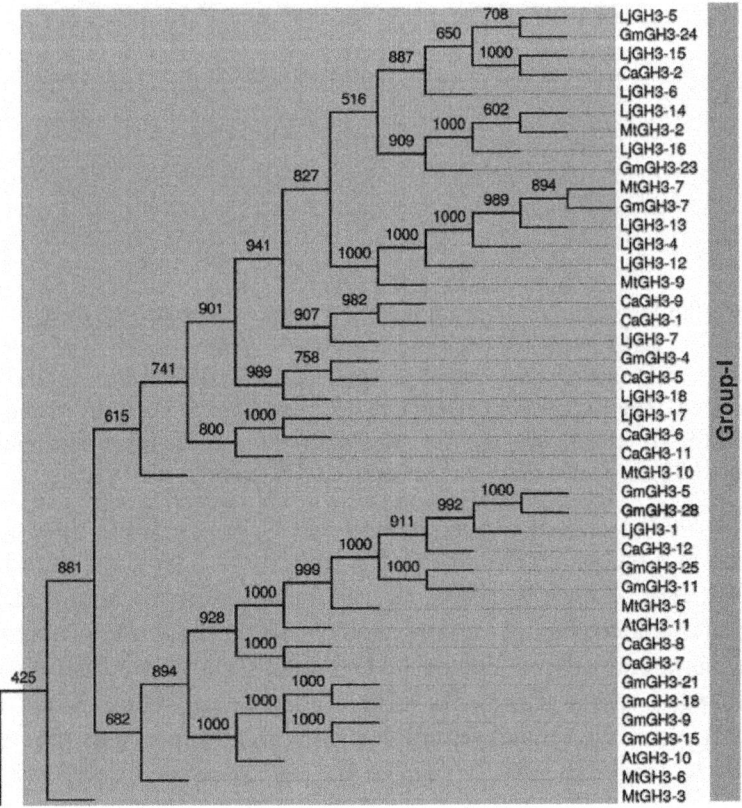

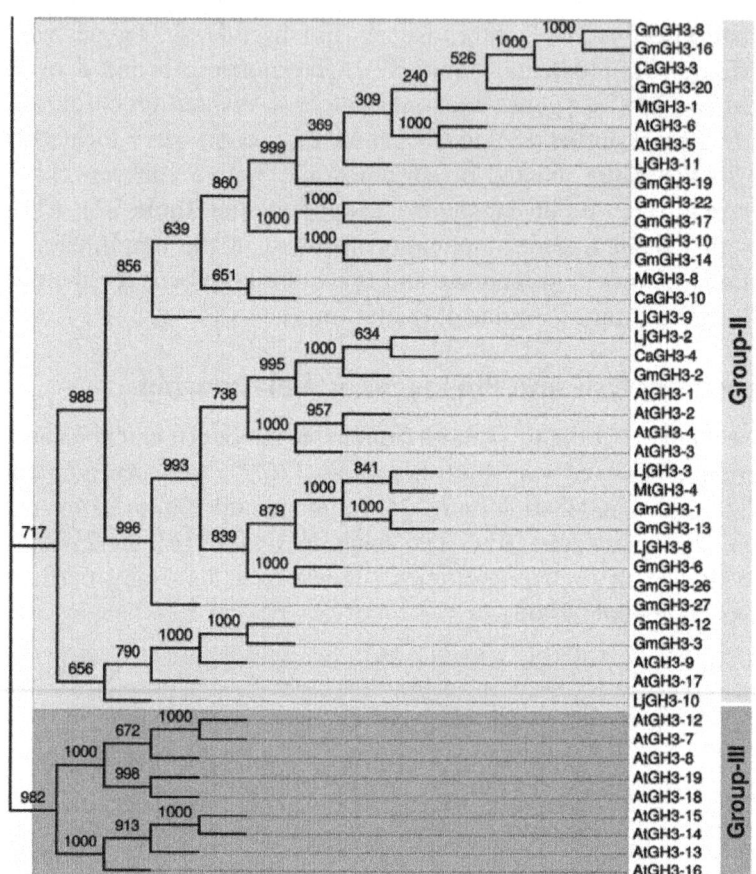

Figure 2. Phylogenetic relationship among chickpea, soybean, *Medicago, Lotus,* and*Arabidopsis* **GH3 proteins.** Multiple sequence alignment of all GH3 proteins from chickpea (CaGH3), soybean (GmGH3), *Medicago* (MtGH3), *Lotus* (LjGH3) and *Arabidopsis* (AtGH3) was performed and tree was generated by UPGMA method. FigTree was used for visualization of the tree. The value at the nodes represents bootstrap values from 1000 replicates. Different groups of GH3 proteins are labeled.

In *Arabidopsis*, GH3 proteins have been classified into three groups on the basis of sequence similarity and specificity to adenylate plant hormones (Staswick et al., 2002). We also analyzed the phylogenetic relationship among GH3 proteins identified in legumes and classified them into different groups. Phylogenetic analysis of legume GH3 proteins showed clustering into only two groups, I and II. Group III GH3 proteins were found absent in all the legumes (Figure 2). This observation is consistent to previous reports (Jain et al., 2006;Kumar et al., 2012; Yuan et al., 2013) and suggested that group III

GH3 proteins might have been lost in legumes during the course of evolution. The group I consisted of nine members of CaGH3 proteins, 12 GmGH3 proteins, seven MtGH3 proteins, and 12 LjGH3 proteins (Figure 2). Group II included three CaGH3 proteins, 16 GmGH3 proteins, three MtGH3 proteins, and six LjGH3 proteins (Figure 2).

Phylogenetic tree comprising CaGH3, GmGH3, MtGH3, LjGH3, and AtGH3 proteins showed a total of 26 sister pairs. Group I comprised of 12 sister pairs, four of GmGH3-GmGH3 proteins, two each of CaGH3-CaGH3 and CaGH3-LjGH3 proteins, and one each of CaGH3-GmGH3, GmGH3-MtGH3, GmGH3-LjGH3, MtGH3-LjGH3 proteins. Group II consisted of eleven sister pairs, including six GmGH3-GmGH3 proteins, two each of AtGH3-AtGH3 proteins, and one each of CaGH3-MtGH3, CaGH3-LjGH3, and MtGH3-LjGH3 proteins. In Group III, three sister pairs of only *Arabidopsis* GH3 proteins were present. To gain further insight into structural diversity of *GH3* genes, we compared exon/intron organization of individual *GH3* gene in chickpea and soybean. Most of the sister pairs shared similar exon/intron structures, intron numbers and intron phasing (Figure 2; Supplemental Tables S3A,B). However, all closely located *GmGH3* genes, such as *GmGH3-7, -8, -9, -10* on chromosome 6, *GmGH3-17, -18*, and *-19* on chromosome 12, *GmGH3-20, -21*, and *-22* on chromosome 13, and *GmGH3-26* and *-27* on chromosome 17, were not paired together (Figure 2). This suggested that these genes might have diverged substantially during evolution. Most of the AtGH3/GmGH3 proteins showed 1:4 orthologous relationship, such as AtGH3-10/GmGH3-9, -15, -21, and -18 (Figure 2). Presence of such orthologous relationship between AtGH3/GmGH3 pairs is also in agreement with the fact that soybean whole genome duplication happened twice in the past (Schmutz et al., 2010). Some *Arabidopsis* and legume GH3 protein pairs (AtGH3-5 and -6/CaGH3-3 and -10, AtGH3-9, and -17/GmGH3-3, and -12, AtGH3-11/MtGH3-5, LjGH3-1) exhibited n:n orthologous relationship, which suggest that members of this family have diversified both in *Arabidopsis* and legumes independently (Wang et al., 2007; Wu et al., 2011; Liu and Hu, 2013; Yuan et al., 2013).

Differential Expression Patterns of *GH3* Genes during Development

Phytohormone auxin is required for plant morphogenesis, including tropistic growth, root patterning, vascular tissue differentiation, axillary bud formation, and floral organ development (Zhao, 2010). Expression analysis of *GH3* genes in various tissue-types during different developmental stages in different plant species have suggested their diverse roles in plants (Gee et al., 1991; Nakazawa et al., 2001; Takase et al., 2004; Khan and Stone, 2007; Jain and Khurana, 2009; Zhang et al., 2009; Böttcher et al., 2010; Kuang et al., 2011; Kumar et al., 2012). Therefore, we performed expression analysis of *GH3* genes in various tissue/stages of development in legumes to know their putative functions. Availability of gene expression atlas covering various tissues/organs and stages of development (Benedito et al., 2008; Libault et al., 2010; Severin et al., 2010;Singh et al., 2013; Verdier et al., 2013), serves as resource to profile expression of candidate genes in legumes. We analyzed the expression of chickpea *GH3* genes using our RNA-seq data (Singh et al., 2013) and validated the results via qRT-PCR analysis (Figure 3).

This analysis revealed that *CaGH3* genes were differentially expressed in various tissues/stages of development. *CaGH3-3* and *CaGH3-5* genes exhibited higher expression in root, which was also confirmed via qRT-PCR, suggesting their role in chickpea root development (Figure 3).*CaGH3-3* orthologs in *Arabidopsis*, *AtGH3-2,* and *AtGH3-6,* were found to have role in root development (Nakazawa et al., 2001; Takase et al., 2004). In addition, *CaGH3-1* and *CaGH3-11* exhibited preferential expression in unopened flower, indicating that these genes might be involved in auxin homeostasis during a specific developmental stage of flower (Figure 3). In rice, *OsGH3-1, -4, -5, -8,* and *-11* genes displayed highest expression level in flower (Jain et al., 2006) and *OsGH3-8* has been reported as the downstream target of rice MADS-box transcription factor (OsMADS1), which is involved in patterning of inner whorl floral organ (Prasad et al., 2005).

Expression of *CaGH3-10* was also distinctly higher in unopened flower, suggesting its role in flower development. *CaGH3-10* was found to be in same phylogenetic clade with *AtGH3-5* and *-6,* whose orthologs in rice *OsGH3-1* and *-4* have higher expression in flower (Jain et al., 2006; Jain and Khurana, 2009), validating our observation. Paralogous gene pair, *CaGH3-7* and *-8* exhibited significantly higher expression in open flower bud, indicating their possible role in auxin homeostasis in early stages of flower development and support the notion that paralogs might have similar expression patterns and function. Transcript level of*CaGH3-2* could not be detected via qRT-PCR, suggesting it might be expressed in a specific tissue/stages of development. These findings highlight the role of *CaGH3* genes in overall plant development including various stages of reproductive development.

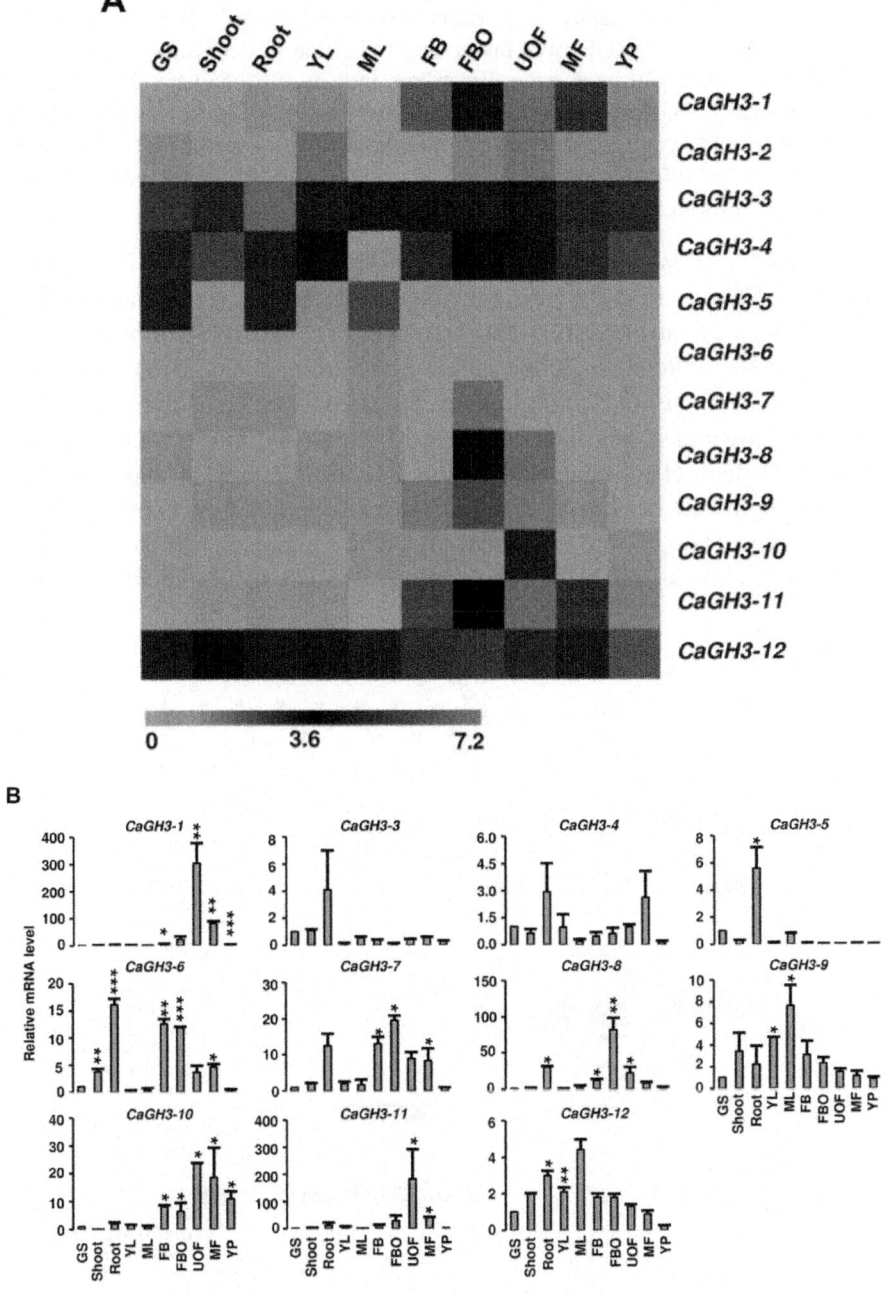

Figure 3. Expression profiles of *CaGH3* **genes during development.** (A) Heatmap showing expression profiles of *CaGH3* genes based on RNA-seq data in various tis-

sues/development stages. Heatmap was generated based on log$_2$ FPKM. (B) Real-time PCR analysis of *CaGH3* genes in various tissue/stages of development. Expression of germinating seedling (GS) was taken as a reference to determine relative mRNA level in other tissues for each gene. Error bars indicate SE of mean. *YL,* young leaf; *ML,* mature leaf; *FB,* flower bud; *UOF,* unopened flower; *FBO,* flower bud open; *MF,* mature flower; *YP,* young pod. Data points marked with asterisk (*$P \leq 0.05$, **$P \leq 0.01$, and ***$P \leq 0.001$) indicate statistically significant difference between control (GS) and other tissues.

Furthermore, we analyzed the expression profiles of *GmGH3, MtGH3,* and *LjGH3* genes in different vegetative and reproductive tissues, utilizing expression data from published RNA-seq atlas of soybean (Severin et al., 2010), *Medicago* (Benedito et al., 2008), and *Lotus* (Verdier et al., 2013), respectively. Expression analysis of*GmGH3* genes revealed their dynamic regulation in various tissues and stages of development (Figure 4A).*GmGH3-8* and *GmGH3-25* showed distinctly higher expression in root, *GmGH3-4* and *GmGH3-13* were up-regulated in nodule, *GmGH3-14* and *GmGH3-18* exhibited flower-specific expression, *GmGH3-9* showed specific expression in young leaf and *GmGH3-20* expression was higher in stages of seed development (Figure4A).

A

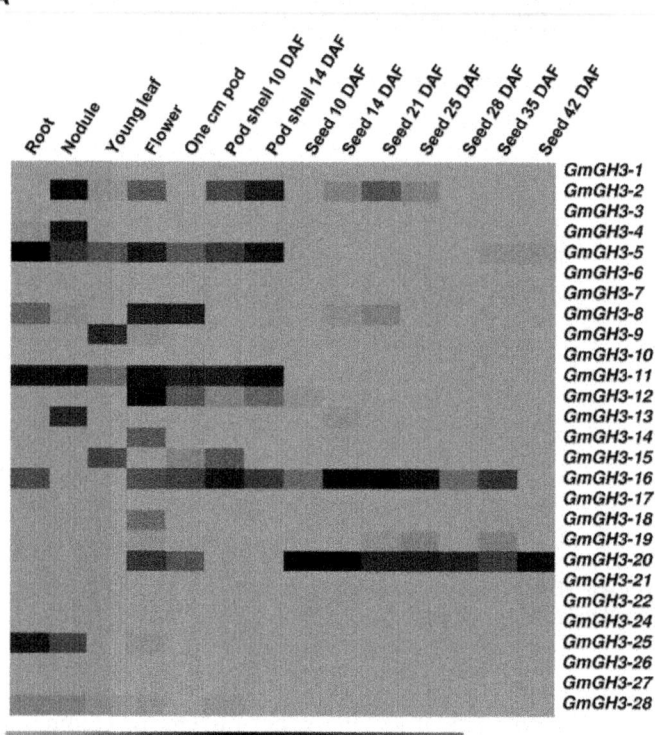

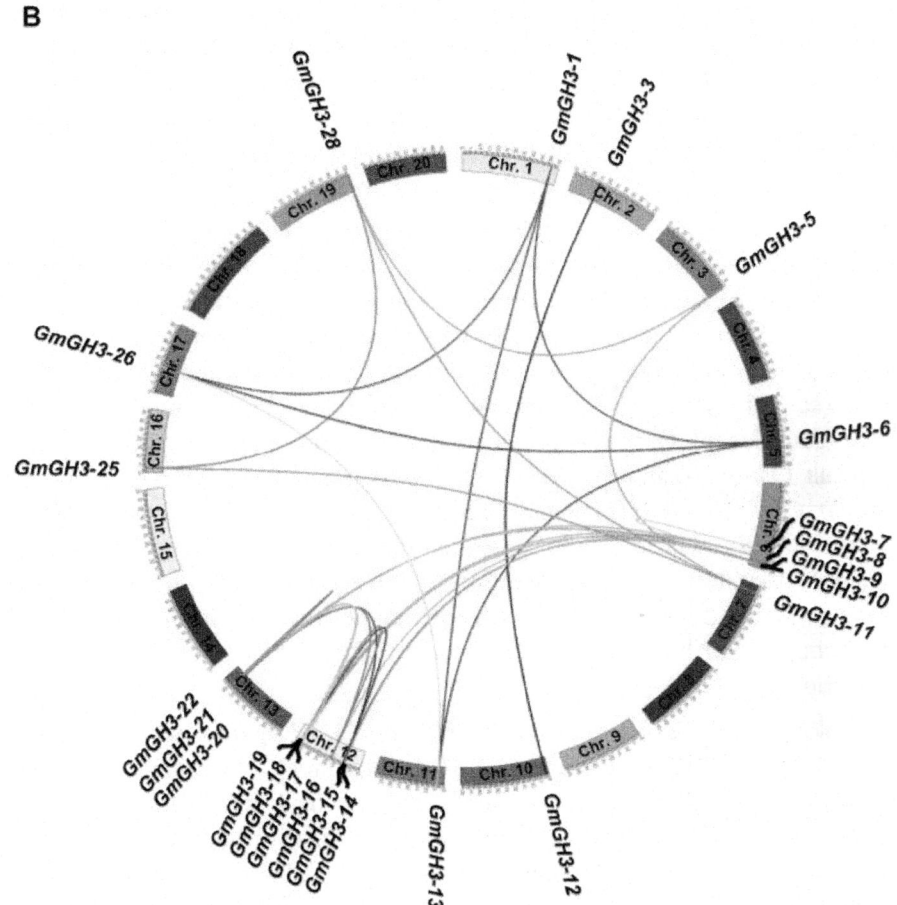

Figure 4. Expression profiles and gene duplication of *GmGH3* **genes.** (A) Heatmap showing expression profiles of soybean *GH3* genes at various stages of development. Heatmap was generated based on log$_2$ RPKM. (B) Mapping of *GmGH3* genes and duplication between them are shown on the soybean chromosomes. Duplication was determined using Plant Genome Duplication Database. Genes and their duplications were mapped on chromosomes using Circos tool. Soybean chromosomes have been arranged in circle and duplications are represented by lines.

Previously, it has been reported that *GH3* genes in soybean exhibit transient expression during floral development and higher expression in ovule and ovary at later stages of floral development (Gee et al., 1991). Reports also suggested role of *GH3* genes during seed development, for example, *GH3* gene (*YDK1*) was found to be specifically up-regulated at heart stage during embryogenesis of *Solanum chacoense* (Tebbji et al., 2010). In rice, involvement of *GH3* genes in seed development has also been reported. For instance, *OsGH3-13*overexpressing rice exhibited smaller seeds (Zhang et al., 2009) and *OsGH3-4* have higher expression during various stages of seed development (Jain and Khurana, 2009). These findings indicated that *GmGH3* genes could play an important role in seed development.

The paralogous *GmGH3* genes, *GmGH3-6* and *GmGH3-26, GmGH3-8* and *GmGH3-16, GmGH3-11* and*GmGH3-25, GmGH3-17* and *GmGH3-22,* and *GmGH3-23* and *GmGH3-24* localized on duplicated chromosomal segments, exhibited similar expression patterns in various tissues/stages of development (Figures 4A,B), suggesting their similar function. However, duplicated genes are also known to have a great degree of expression and functional divergence due to selection pressure and need for diversification (Prince and Pickett, 2002). Many duplicated *GmGH3* genes exhibited expression divergence as well, such as *GmGH3-1*and *GmGH3-13, GmGH3-3* and *GmGH3-12, GmGH3-5* and *GmGH3-28, GmGH3-9* and *GmGH3-15,* and*GmGH3-11* and *GmGH3-25* (Figures 4A,B). These results suggested that chromosomal duplication events not only facilitated the amplification of the *GmGH3* gene family members, but also resulted into expression divergence between duplicated genes, which might have contributed in the establishment of gene functional diversity during evolution.

Likewise, the expression of *MtGH3* and *LtGH3* genes was also found to be variable in various tissue/stages of development. For instance, *MtGH3-4* exhibited significantly higher expression in seed at 36 day after pollination (DAP), *MtGH3-8* showed greater expression in root and various stages of seed development (Supplemental Figure S2). *LjGH3-2* was found to be up-regulated in root, whereas *LjGH3-1, -6,* and *-12* showed distinctly higher expression in leaf, and *LjGH3-3, -4, -5,* and *-18* were seen to be up-regulated in root and nodule (Supplemental Figure S4). Expression of other *GH3* genes of legumes was also found to be variable in various tissue/stages of development elucidating their involvement in various growth and development processes (Wang et al., 2010; Kuang et al., 2011).

Furthermore, we analyzed expression patterns of paralogous/orthologous *GH3* genes to investigate their functional conservation across legumes. Although the available expression data represented diverse tissues/developmental stages in different legumes, we made an effort to define correlation in expression profiles of *GH3* genes in different legumes. Some of paralogous/orthologous *GH3* genes exhibited similar expression patterns in different legumes, such as *CaGH3-3, GmGH3-8, -16, -20, MtGH3-1* and *LjGH3-11;CaGH3-12, GmGH3-5, -11, -25, -28* and *LjGH3-1; CaGH3-5* and *GmGH3-4; CaGH3-4* and *LjGH3-2; GmGH3-13* and *LjGH3-2; MtGH3-2* and *LjGH3-14;* and *MtGH3-3* and *-6,* suggesting their conserved function across legumes (Figures 3–5; Supplemental Figures S2 and S3). Some of these paralogous/orthologous genes harbor similar cis-regulatory elements in their promoter regions (Supplemental Table S4). For instance, *CaGH3-3,GmGH3-8, -16,* and *-20* contain cis-regulatory elements, S000037, S000270, S000273, S000390, S000414, S000453, and S000461, conserved in their promoter sequences (Supplemental Table S4). An earlier study revealed similarity of gene expression profiles in various organs for a significant number of paralogous/orthologous gene pairs in *Medicago* and *Arabidopsis* (Benedito et al., 2008). Moreover, comparison of soybean transcriptome with *Medicago* and *Lotus* demonstrated similar tissue-specificity for 45% of the genes analyzed (Libault et al., 2010). Overall, these findings provide insights into the putative roles of *GH3* genes in legumes in various aspects of plant growth and development.

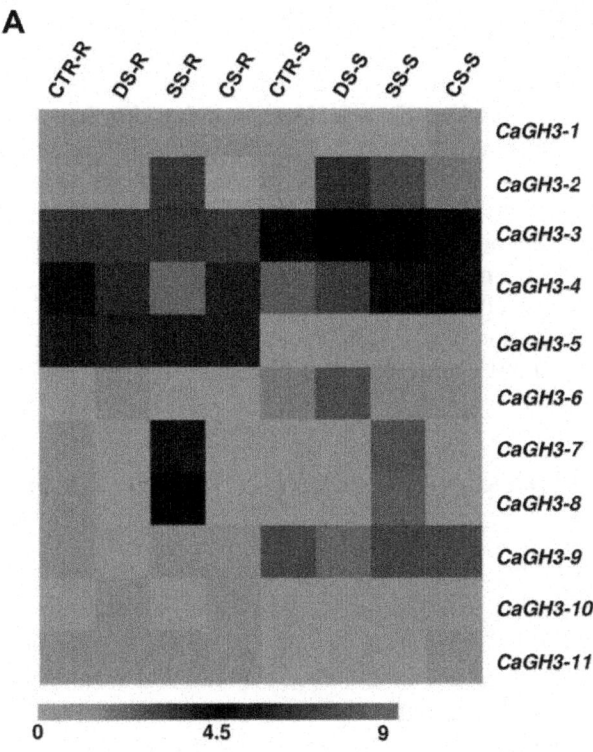

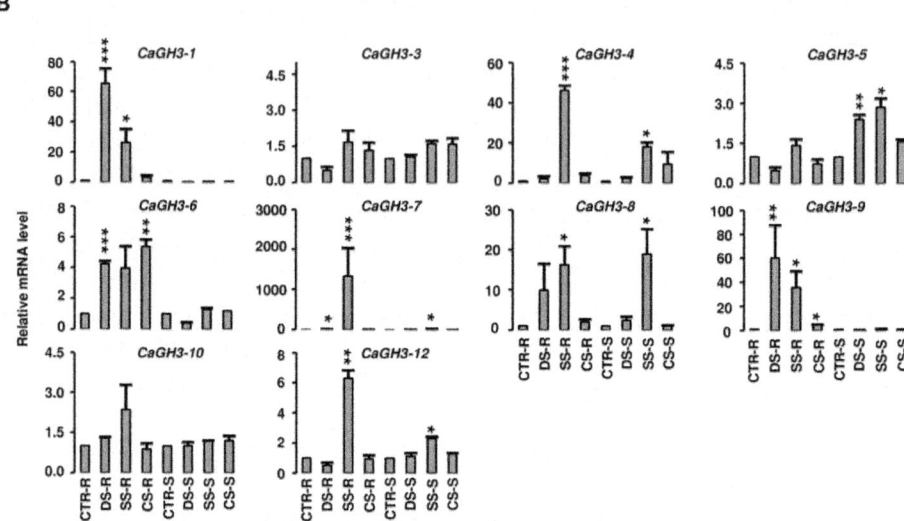

Figure 5. Expression profiles of *CaGH3* **genes under abiotic stress conditions.** (A) Heatmap showing expression of *CaGH3* genes based on RNA-seq data. Heatmap was generated based on log₂ FPKM.(B) Real-time PCR analysis of *CaGH3* genes under various stress treatments. Root control (CTR-R) and shoot control (CTR-S) was taken as a reference to determine relative mRNA level under stress conditions. Error bars indicate standard error of mean. *DS-R*: desiccation stressed root, *SS-R,* salt stressed root; *CS-R,*cold stressed root; *DS-S,* desiccation stressed shoot; *SS-S,* salt stressed shoot; *CS-S,* cold stressed shoot. Data points marked with asterisk (*$P \leq 0.05$, **$P \leq 0.01$, and ***$P \leq 0.001$) indicate statistically significant difference between control and stress treatments.

Differential Expression Patterns of *GH3* Genes under Abiotic Stresses

Plants are constantly exposed to various abiotic stresses in their life cycle. Several recent studies have implicated auxin in abiotic stress responses (Jain and Khurana, 2009; Wang et al., 2010; Du et al., 2012;Kumar et al., 2012; Yuan et al., 2013). Some studies have revealed that *GH3* genes are regulated by abiotic stresses, like drought, salt, and cold stresses (Park et al., 2007; Jain and Khurana, 2009). The transcript level of *AtGH3-5* (*WES1*) has been shown to be induced by various abiotic stress conditions, like drought, high salt, and cold (Park et al., 2007). In rice, the transcript abundance of *OsGH3-1, OsGH3-2, OsGH3-8,* and *OsGH3-13*were markedly higher in seedlings subjected to salt, drought and cold stresses (Jain and Khurana, 2009; Zhang et al., 2009; Du et al., 2012). In *Sorghum*, at least six *GH3* genes were found to be induced upon salt and drought treatments in leaf (Wang et al., 2010).

To investigate the role of legume *GH3* genes in abiotic stress responses, we performed scanning of *cis*-acting regulatory DNA elements within promoter regions (2 kb upstream from the start codon) using PLACE database. This analysis predicted several elements responsive to auxin (IAA), abscisic acid (ABA), SA, JA, drought, salinity, and disease (Supplemental Table S4), suggesting that the function of these genes may be associated with various phytohormone signals and/or environmental stresses. Considering regulatory role of *cis*-elements, we analyzed expression of *GH3* genes under abiotic stress conditions to know their function during abiotic stresses. For chickpea, we analyzed RNA-seq data from root and shoot tissues subjected to desiccation, salinity and cold conditions (Garg et al., 2014), and performed real-time PCR analysis for validation. In our analysis, paralogous gene pair, *CaGH3-1* and *-9,* showed induction under both desiccation and salinity stresses in root (Figures 5A,B), and also their promoter sequences harbor desiccation (S000414) and salinity (S000453) responsive *cis*-regulatory elements (Supplemental Table S4), indicating their role in desiccation and salinity stress. Recently, rice group-I gene, *OsGH3-12*, has also been found to be markedly induced by drought stress (Du et al., 2013b). Similarly, promoter of *CaGH3-4* harbor salinity responsive *cis*-element (S000453) and showed higher expression in root under salt stress (Figures 5A,B). Its ortholog, *AtGH3-1*, has also been found to be up-regulated under salt stress (Sani et al., 2013), corroborating our result. Group-I paralogous genes, *CaGH3-7* and *-8*, were found to be induced in root under salinity stress (Figures5A, B), implying their involvement in homeostasis of auxin under salinity stress in root. *CaGH3-5* and *-6*showed enhanced expression under desiccation, salt and cold stresses in shoot and root (Figures 5A, B), respectively, suggesting their role during multiple abiotic stress responses.

In *Medicago*, *MtGH3-8* and *-9* genes were induced under salt stress in root, and *MtGH3-7* was induced under drought stress in root (Supplemental Figure S2). Previous reports suggest that IAA, SA, JA, ethylene, and ABA regulate the protective responses of plants against both biotic and abiotic stress responses via signaling crosstalk (Bostock, 2005; Lorenzo and Solano, 2005; Mauch-Mani and Mauch, 2005; Ding et al., 2008;Domingo et al., 2009; Fu et al., 2011). In addition, orthologous genes, *CaGH3-10* and *MtGH3-8*, showed induced expression under salt stress in root (Figure 5; Supplemental Figure S2); suggesting their conserved function in both legumes. Taken together, these findings indicated that members of GH3 gene family might be involved in stress adaptation in legumes.

Homology Modeling and Substrate Preferences

The availability of crystal structures of two *Arabidopsis* GH3 proteins: AtGH3-12, which conjugate benzoate substrate and JA-specific AtGH3-11/JAR1 (Westfall et al., 2012); and grapevine IAA-amido synthetase GH3-1 (VvGH3-1) gave us an exciting opportunity to determine three-dimensional structure of GH3 members in legumes by homology modeling.

Group-I protein, CaGH3-3 and GmGH3-8 of chickpea and soybean, respectively, were modeled using structure of AtGH3-11 (Protein Data Bank code 4EPL; Westfall et al., 2012) and Group-II proteins, CaGH3-12 and GmGH3-25 were modeled using grapevine, Vv-GH3-1 (Protein Data Bank code 4B2G; Peat et al., 2012). The homology modeling revealed high degree of conservation in the protein structure of these proteins. To predict active sites, we transferred ligands from template to model by superimposing structures. Ligands for group-I proteins are JA-Ile and AMP (amino acid mono phosphate), and group-II proteins are adenosine-5′-[2-(1H-indole-3-yl)ethyl] phosphate (AIEP), which mimics the adenylated intermediate of the IAA conjugation reaction (Figures 6 and 7; Böttcher et al., 2012; Westfall et al., 2013). By comparing sequences of model and template, we also identified the residues forming acyl acid/hormone-binding site and nucleotide binding site (Figures 7and 8). Most of these residues were found to be conserved between the model and template. For example, hormone-binding residues of CaGH3-12 and GmGH3-25 with AtGH3-11 (JA-conjugating), Ca-Leu137, Gm-Leu115 to At-Leu117; Ca-Thr141, Gm-Thr119 to At-Thr121; Ca-Thr185, Gm-Thr163 to At-Thr166; Ca-Val188, Gm-Val168 to At-Val169; Ca-Ile323, Gm-Ile301 to At-Ile304; and Ca-Trp355, Gm-Trp333 to At-Trp336 are conserved (Figures 7 and 8). Similarly, conservation was found between hormone-binding residues of CaGH3-3 and GmGH3-8 with VvGH3-1(IAA-conjugating), Ca-Val167, Gm-Val167 to Vv-Val172; Ca-Leu168, Gm-Leu168 to Vv-Leu173; Ca-Ala332, Gm-Ala332 to Vv-Ala337; and Ca-Tyr337, Gm-Tyr337 to Vv-Tyr342 (Figures7 and 8). In addition, we also found nucleotide-binding residues, Ser, Thr, Phe, and Tyr conserved in all the structures (Figures 7 and 8), which is in agreement with earlier studies that nucleotide binding residues conserved in not only GH3 proteins but also the ANL superfamily (Gulick, 2009; Peat et al., 2012; Westfall et al., 2012, 2013).

CaGH3-3

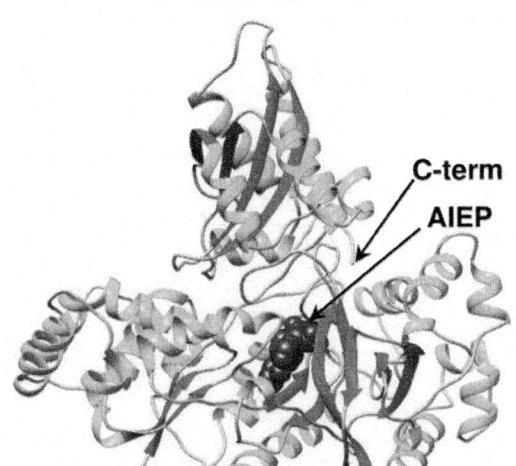

GmGH3-8

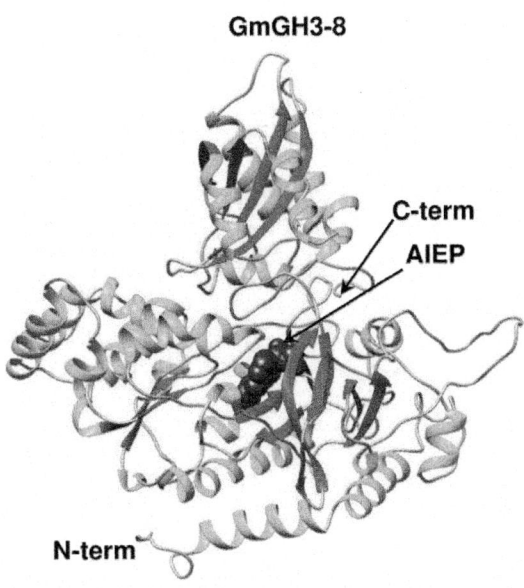

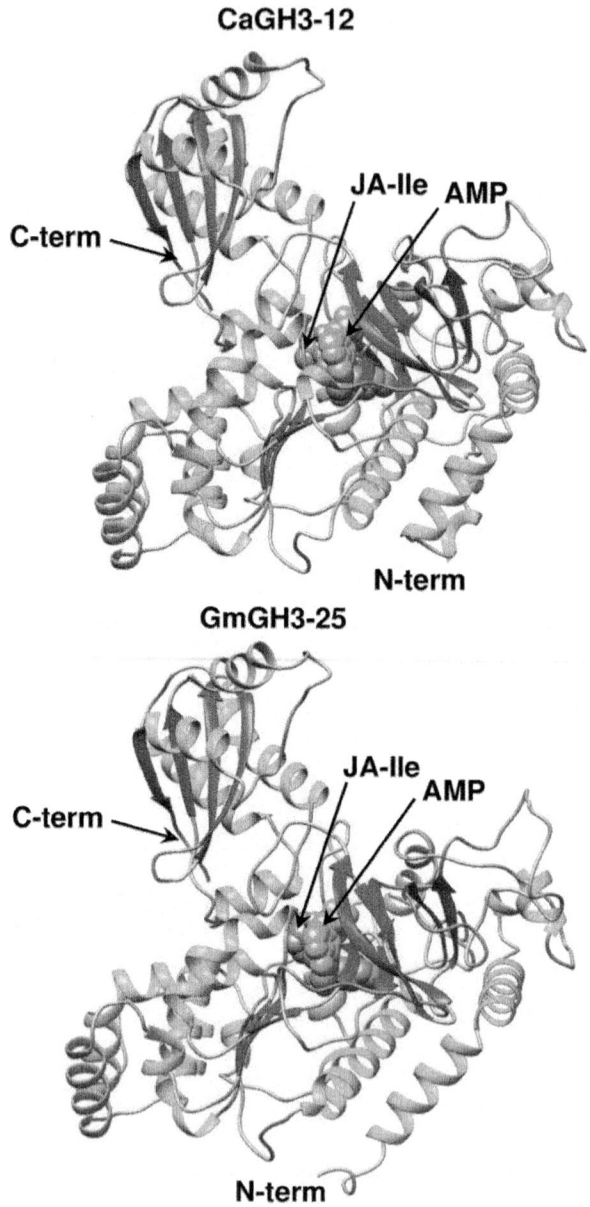

Figure 6. Predicted structures of GH3 proteins. Ribbon diagram showing the *N*- and *C*-terminal domains of chickpea (CaGH3-3 and CaGH3-8) and soybean (GmGH3-8 and GmGH3-25) GH3 protein with α-helices, β-strands and loops colored cyan, magenta, and gold, respectively. Ligands AIEP, JA-Ile, AMP are shown as space-filling model in blue, coral, and green colors, respectively.

CaGH3-3

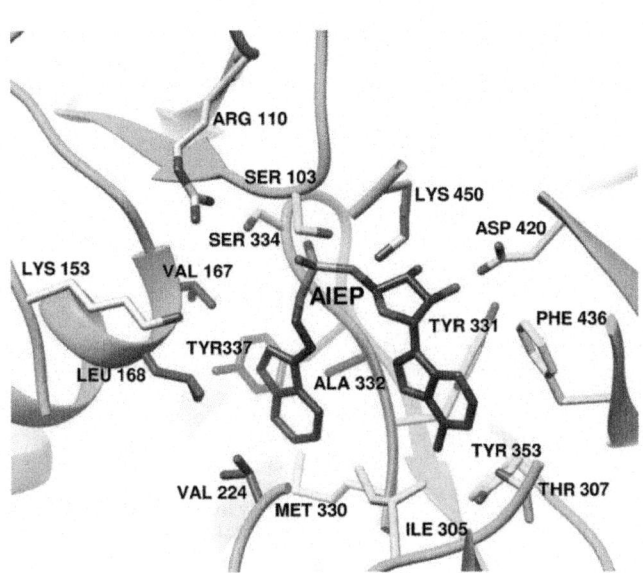

GmGH3-8

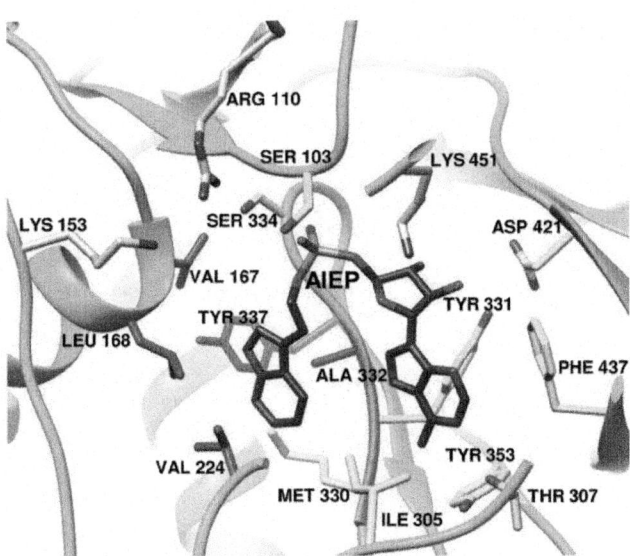

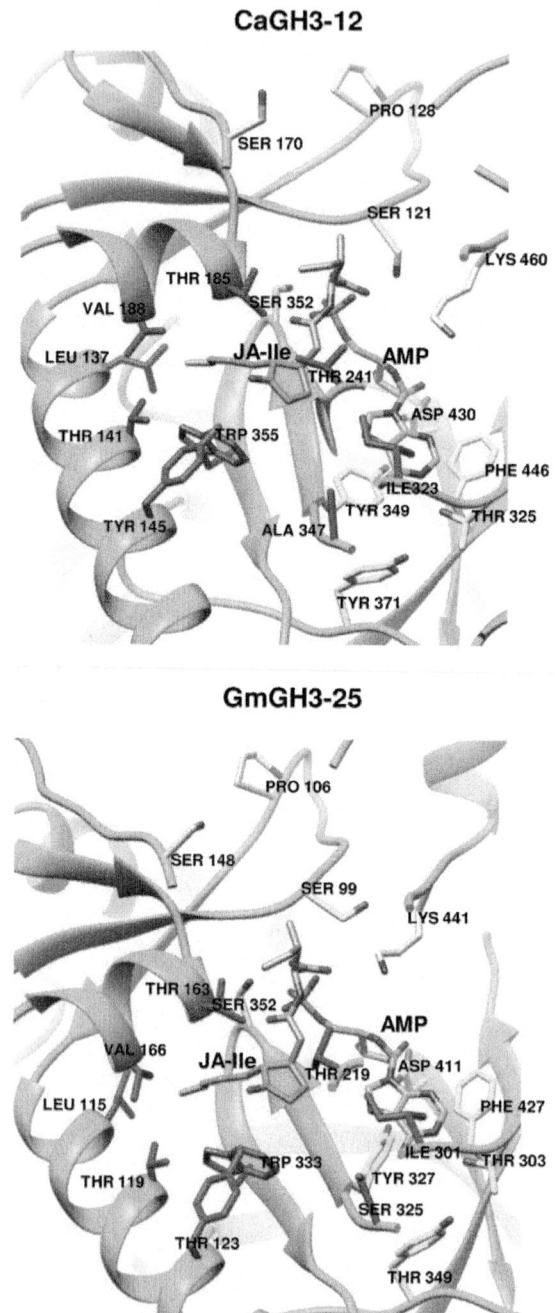

Figure 7. Hormone and nucleotide binding residues in GH3 proteins. Ribbon diagram showing hormone binding residues in magenta, nucleotide (ATP/AMP) binding residues in yellow, and residues in pink determine amino-acid preferences.

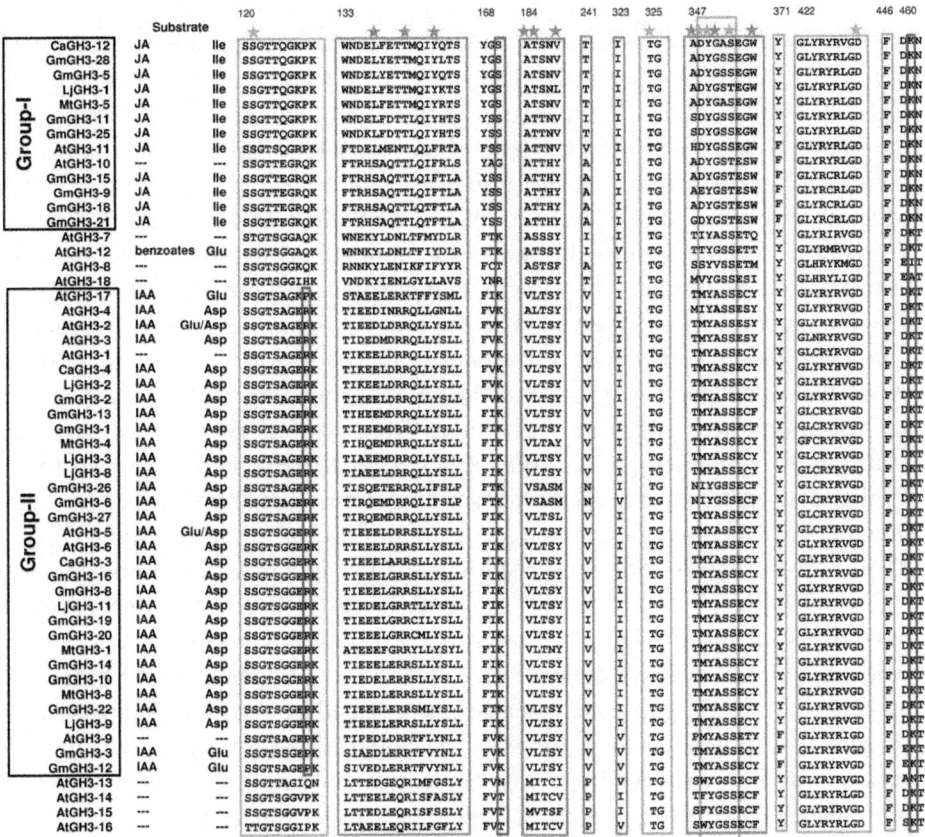

Figure 8. Proposed substrates of GH3 proteins based on conserved amino acid residues. Protein sequences of all the identified *GH3* genes were aligned using MAFFT. Green and blue boxes represent nucleotide (ATP/AMP) and hormone-binding motifs/residues, respectively. Magenta boxes represent residues determining amino-acid preferences. Only sequences with complete *C*- and *N*-terminal domains were included. Star across the top of the alignment indicates conserved residues in pocket forming active site. Numbering at the top corresponds to CaGH3-12.

Further, to determine amino-acid specificities, we identified residues involved in discrimination of apolar (i.e., Ile) and acidic (i.e., Asp/Glu) substrates in the transferase reaction by comparing structures. Within the active site, a lysine residue (Lys450 in CaGH3-3, Lys460 in CaGH3-12, Lys451 in GmGH3-8 and Lys441 in GmGH3-25) was conserved at same position (Figures 7 and 8). This residue was also found to be highly conserved in GH3 proteins with known amino-acid preference (Westfall et al., 2012).

Also, it has been found that Lys428 (AtGH3-12) is conserved in GH3-proteins that accept acidic-amino acid, whereas Ser151 (AtGH3-11) at the same position is conserved in enzymes specific to isoleucine conjugation (Westfall et al., 2012). We also found Lys153 in CaGH3-3 and GmGH3-8, suggesting CaGH3-3 and GmGH3-8 may accept acidic-amino acid (i.e., Asp/Glu) and at the same position Ser170 in CaGH3-12 and Ser148 in GmGH3-25, indicating their preferences for isoleucine (Figures 7 and 8). In addition, conservation of another residue Arg110 in IAA-specific CaGH3-3 and GmGH3-8 (Figures 7 and 8), further specified their Asp-conjugating nature. This was also corroborated by another study in grapevine, where residue Arg115 in VvGH3-1 (IAA-specific) was conserved in all the four Asp-conjugating GH3s, whereas Glu-conjugating GH3s had a Proline at that position (Peat et al., 2012). Also, the same pattern was found in IAA-conjugating GH3 enzymes with known amino-acid substrate preferences from *Arabidopsis* (Staswick et al., 2005) and rice (Zhang et al., 2009; Chen et al., 2010). Next, we also found similar conservation in residues determining amino-acid preferences for other members of GH3 proteins, which led us to propose substrates for them (Figure 8). Group-I proteins with conserved Ser and Lys at similar position as that of CaGH3-12 (Ser170 and Lys461; magenta boxes; Figure 8) are proposed to have Ile as substrate (Figure8). For group-II proteins, Asp will be the substrate when Arg at 128 and Lys at positions 170 and 461 (magenta boxes; Figure 8); and Glu, will be the substrate when Arg is replaced by Pro at the same position (Figure 8). Altogether, the structures presented here showed conservation of residues at hormone-binding site, nucleotide-binding site, and amino-acid preferences determining residues, indicating similar function.

Several previous studies have reported the differential expression of *GH3* genes in various tissues/developmental stages and in response to various stimuli, including auxin, jasmonic acid, salicyclic acid, and abiotic/biotic stresses in different plants (Park et al., 2007; Zhang et al., 2007; Jain and Khurana, 2009;Kumar et al., 2012; Yuan et al., 2013; Wu et al., 2014). Our results also revealed preferential/tissue-specific and stress-responsive expression of many *GH3* genes in different legumes. The knowledge of motifs/residues of GH3 proteins that determine substrate preferences and conjugation to auxin may help modulate their binding efficiency and substrate preferences for engineering plants with desired agronomic traits.

CONCLUSIONS

We performed a genome-wide analysis of GH3 gene family in legumes to reveal gene structure, phylogenetic relationship, and expression profiles during various developmental stages and abiotic stress conditions. Some*GH3* genes exhibited preferential/specific expression in a particular tissue and/or under abiotic stress condition(s). Our analysis revealed that *GH3* genes seem to be involved in biology of various tissues or organs and actively participate in stress responses in legumes. The analysis of protein structures of few members identified key features of substrate recognition, which might help in investigation of their molecular functions in legumes. The data generated in this study will serve as a foundation for functional characterization of GH3 gene family members in legumes.

ACKNOWLEDGMENTS

This work was funded by the Department of Biotechnology, Government of India, under the Next Generation Challenge Programme on Chickpea Genomics (grant number BT/PR12919/AGR/02/676/2009 from 2009 to 2014) to Mukesh Jain. Rohini Garg acknowledges INSPIRE faculty award from the Department of Science and Technology, Government of India. Vikash K. Singh acknowledges the award of research fellowship from the Department of Biotechnology, Government of India.

REFERENCES

1. Benedito, V. A., Torres-Jerez, I., Murray, J. D., Andriankaja, A., Allen, S., Kakar, K.,et al. (2008). A gene expression atlas of the model legume *Medicago truncatula*. *Plant J.* 55, 504–513. doi: 10.1111/j.1365-313X.2008.03519.x

2. Bostock, R. M. (2005). Signal crosstalk and induced resistance: straddling the line between cost and benefit. *Annu. Rev. Phytopathol.* 43, 545–580. doi: 10.1146/annurev.phyto.41.052002.095505

3. Böttcher, C., Dennis, E. G., Booker, G. W., Polyak, S. W., Boss, P. K., and Davies, C. (2012). A novel tool for studying auxin-metabolism: the inhibition of grapevine indole-3-acetic acid-amido synthetases by a reaction intermediate analogue. *PLoS ONE* 7:e37632. doi: 10.1371/journal.pone.0037632

4. Böttcher, C., Keyzers, R. A., Boss, P. K., and Davies, C. (2010). Sequestration of auxin by the indole-3-acetic acid-amido synthetase GH3–1 in grape berry (Vitis vinifera L.) and the proposed role of auxin conjugation during ripening. *J. Exp. Bot.* 61, 3615–3625. doi: 10.1093/jxb/erq174

5. Chen, Q., Westfall, C. S., Hicks, L. M., Wang, S., and Jez, J. M. (2010). Kinetic basis for the conjugation of auxin by a GH3 family indole-acetic acid-amido synthetase. *J. Biol. Chem.* 285, 29780–29786. doi: 10.1074/

jbc.M110.146431

6.   Chen, Q., Zhang, B., Hicks, L. M., Wang, S., and Jez, J. M. (2009). A liquid chromatography-tandem mass spectrometrybased assay for indole-3-acetic acid-amido synthetase. *Anal. Biochem.* 390, 149–154. doi: 10.1016/j.ab.2009.04.027

7.   Ding, X., Cao, Y., Huang, L., Zhao, J., Xu, C., Li, X., et al. (2008). Activation of the indole-3- acetic acid-amidosynthetase GH3-8 suppresses expansin expression and promotes salicylate-and jasmonate- independent basal immunity in rice. *Plant Cell* 20, 228–240. doi: 10.1105/tpc.107.055657

8.   Domingo, C., Andrés, F., Tharreau, D., Iglesias, D. J., and Talón, M. (2009). Constitutive expression of OsGH3.1 reduces auxin content and enhances defense response and resistance to a fungal pathogen in rice. *Mol. Plant Microbe Interact.* 22, 201–210. doi: 10.1094/MPMI-22-2-0201

9.   Du, D., Hao, R., Cheng, T., Pan, H., Yang, W., Wang, J., et al. (2013a). Genome-wide analysis of the AP2/ERF gene family in Prunus mume. *Plant Mol. Biol. Rep.* 31, 741–750. doi: 10.1007/s11105-012-0531-6

10.   Du, H., Liu, H., and Xiong, L. (2013b). Endogenous auxin and jasmonic acid levels are differentially modulated by abiotic stresses in rice. *Front. Plant Sci.* 4:397. doi: 10.3389/fpls.2013.00397

11.   Du, H., Wu, N., Fu, J., Wang, S., Li, X., Xiao, J., et al. (2012). A GH3 family member, OsGH3-2, modulates auxin and abscisic acid levels and differentially affects drought and cold tolerance in rice. *J. Exp. Bot.* 63, 6467–6480. doi: 10.1093/jxb/ers300

12.   Fu, J., Liu, H., Li, Y., Yu, H., Li, X., Xiao, J., et al. (2011). Manipulating broad-spectrum disease resistance by suppressing pathogen-induced auxin accumulation in rice. *Plant Physiol.* 155, 589–602. doi: 10.1104/pp.110.163774

13.   Garg, R., Bhattacharjee, A., and Jain, M. (2014). Genome-scale transcriptomic insights into molecular aspects of abiotic stress responses in chickpea. *Plant Mol. Biol. Rep.* doi: 10.1007/s11105-014-0753-x

14.   Garg, R., Patel, R. K., Jhanwar, S., Priya, P., Bhattacharjee, A., Yadav, G., et al. (2011). Gene discovery and tissue-specific transcriptome analysis in chickpea with massively parallel pyrosequencing and web resource development. *Plant Physiol.* 156, 1661–1678. doi: 10.1104/pp.111.178616

15.   Garg, R., Sahoo, A., Tyagi, A. K., and Jain, M. (2010). Validation of internal control genes for quantitative gene expression studies in chickpea (*Cicer arietinum* L.). *Biochem. Biophys. Res. Commun.* 396, 283–288.

doi: 10.1016/j.bbrc.2010.04.079

16. Gee, M. A., Hagen, G., and Guilfoyle, T. J. (1991). Tissue-specific and organ-specific expression of soybean auxin-responsive transcripts GH3 and SAURs. *Plant Cell* 3, 419–430. doi: 10.1105/tpc.3.4.419

17. Gulick, A. M. (2009). Conformational dynamics in the Acyl-CoA synthetases, adenylation domains of non-ribosomal peptide synthetases, and firefly luciferase. *ACS Chem. Biol.* 4, 811–827. doi: 10.1021/cb900156h

18. Hagen, G., and Guilfoyle, T. (2002). Auxin-responsive gene expression: genes, promoters and regulatory factors. *Plant Mol. Biol.* 49, 373–385. doi: 10.1023/A:1015207114117

19. Hagen, G., Kleinschmidt, A., and Guilfoyle, T. (1984). Auxin-regulated gene expression in intact soybean hypocotyls and excised hypocotyls sections. *Planta* 162, 147–153. doi: 10.1007/BF00410211

20. Hagen, G., Martin, G., Li, Y., and Guilfoyle, T. J. (1991). Auxin-induced expression of the soybean GH3 promoter in transgenic tobacco plants. *Plant Mol. Biol.* 17, 567–579. doi: 10.1007/BF00040658

21. Hsieh, H. L., Okamoto, H., Wang, M., Ang, L. H., Matsui, M., Goodman, H.,et al. (2000). FIN219, an auxin-regulated gene, defines a link between phytochrome A and the downstream regulator COP1 in light control of *Arabidopsis* development. *Genes Dev.* 14, 1958–1970.

22. Jagadeeswaran, G., Raina, S., Acharya, B. R., Maqbool, S. B., Mosher, S. L., Appel, H. M.,et al. (2007). *Arabidopsis* GH3-LIKE DEFENSE GENE 1 is required for accumulation of salicylic acid, activation of defense responses and resistance to *Pseudomonas syringae*. *Plant J.* 51, 234–246. doi: 10.1111/j.1365-313X.2007.03130.x

23. Jain, M., Kaur, N., Tyagi, A. K., and Khurana, J. P. (2006). The auxin-responsive GH3 gene family in rice (*Oryza sativa*). *Funct. Integr. Genomics* 6, 36–46. doi: 10.1007/s10142-005-0142-5

24. Jain, M., and Khurana, J. P. (2009). An expression compendium of auxin-responsive genes during reproductive development and abiotic stress in rice. *FEBS J.* 276, 3148–3162. doi: 10.1111/j.1742-4658.2009.07033.x

25. Jain, M., Misra, G., Patel, R. K., Priya, P., Jhanwar, S., Khan, A. W.,et al. (2013). A draft genome sequence of the pulse crop chickpea (*Cicer arietinum* L.). *Plant J.* 74, 715–729. doi: 10.1111/tpj.12173

26. Kelley, L. A., and Sternberg, M. J. E. (2009). Protein structure prediction on the Web: a case study using the Phyre server. *Nat. Protoc.* 4, 363–371. doi: 10.1038/nprot.2009.2

27. Khan, S., and Stone, J. M. (2007). *Arabidopsis thaliana* GH3.9 influences primary root growth. *Planta* 226, 21–34. doi: 10.1007/s00425-006-0462-2

28. Kuang, J. F., Zhang, Y., Chen, J. Y., Chen, Q. J., Jiang, Y. M., Lin, H. T.,et al. (2011). Two GH3 genes from longan are differentially regulated during fruit growth and development. *Gene* 485, 1–6. doi: 10.1016/j.gene.2011.05.033

29. Kumar, R., Agarwal, P., Tyagi, A. K., and Sharma, A. K. (2012). Genome-wide investigation and expression analysis suggest diverse roles of auxin-responsive GH3 genes during development and response to different stimuli in tomato (*Solanum lycopersicum*). *Mol. Genet. Genomics* 287, 221–235. doi: 10.1007/s00438-011-0672-6

30. Li, Y., Liu, Z. B., Shi, X., Hagen, G., and Guilfoyle, T. J. (1994). An auxin-inducible element in soybean SAUR promoters. *Plant Physiol.* 106, 645–657. doi: 10.1104/pp.106.1.37

31. Libault, M., Farmer, A., Joshi, T., Takahashi, K., Langley, R. J., Franklin, L. D.,et al. (2010). An integrated transcriptome atlas of the crop model Glycine max, and its use in comparative analyses in plants. *Plant J.* 63, 86–99.

32. Liu, S. Q., and Hu, H. F. (2013). Genome-wide analysis of the auxin response factor gene family in cucumber. *Genet. Mol. Res.* 12, 4317–4331. doi: 10.4238/2013.April.2.1

33. Lorenzo, O., and Solano, R. (2005). Molecular players regulating the jasmonate signaling network. *Curr. Opin. Plant Biol.* 8, 532–554. doi: 10.1016/j.pbi.2005.07.003

34. Ludwig-Muller, J. (2011). Auxin conjugates: their role for plant development and in the evolution of land plants. *J. Exp. Bot.* 62, 1757–1773. doi: 10.1093/jxb/erq412

35. Ludwig-Müller, J., Jülke, S., Bierfreund, N. M., Decker, E. L., and Reski, R. (2008). Moss (*Physcomitrella patens*) GH3 proteins act in auxin homeostasis. *New Phytol.* 181, 323–338. doi: 10.1111/j.1469-8137.2008.02677.x

36. Mauch-Mani, B., and Mauch, F. (2005). The role of abscisic acid in plant–pathogen interactions. *Curr. Opin. Plant. Biol.* 8, 409–414. doi: 10.1016/j.pbi.2005.05.015

37. Nakazawa, M., Yabe, N., Ichikawa, T., Yamamoto, Y. Y., Yoshizumi, T., Hasunuma, K.,et al. (2001). DFL1, an auxin-responsive GH3 gene homologue, negatively regulates shoot cell elongation and lateral root formation, and positively regulates the light response of hypocotyl length.

Plant J. 25, 213–221. doi: 10.1046/j.1365-313x.2001.00957.x

38. Nobuta, K., Okrent, R. A., Stoutemyer, M., Rodibaugh, N., Kempema, L., and Wildermuth, M. C.,et al. (2007). The GH3 acyl adenylase family member PBS3 regulates salicylic acid-dependent defense responses in *Arabidopsis. Plant Physiol.* 144, 1144–1156. doi: 10.1104/pp.107.097691

39. Okrent, R. A., Brooks, M. D., and Wildermuth, M. C. (2009). *Arabidopsis* GH3.12 (PBS3) conjugates amino acids to 4-substituted benzoates and is inhibited by salicylate. *J. Biol. Chem.* 284, 9742–9754. doi: 10.1074/jbc. M806662200

40. Park, J. E., Park, J. Y., Kim, Y. S., Staswick, P. E., Jeon, J., Yun, J.,et al. (2007). GH3-mediated auxin homeostasis links growth regulation with stress adaptation response in *Arabidopsis. J. Biol. Chem.* 282, 10036–10046. doi: 10.1074/jbc.M610524200

41. Peat, T. S., Bottcher, C., Newman, J., Lucent, D., Cowieson, N., and Davies, C. (2012). Crystal structure of an indole-3-acetic acid amido synthetase from grapevine involved in auxin homeostasis. *Plant Cell* 24, 4525–4538. doi: 10.1105/tpc.112.102921

42. Prasad, K., Sriram, P., and Usha, V. (2005). OsMADS1, a rice MADS box factor, controls differentiation of specific cell type in lemma and plea and is an early-acting regulator of inner floral organs. *Plant J.* 43, 915–928. doi: 10.1111/j.1365-313X.2005.02504.x

43. Prince, V. E., and Pickett, F. B. (2002). Splitting pairs: the diverging fates of duplicated genes. *Nat. Rev. Genet.* 3, 827–837. doi: 10.1038/nrg928

44. Sani, E., Herzyk, P., Perrella, G., Colot, V., and Amtmann, A. (2013). Hyperosmotic priming of *Arabidopsis* seedlings establishes a long-term somatic memory accompanied by specific changes of the epigenome. *Genome Biol.* 14:R59. doi: 10.1186/gb-2013-14-6-r59

45. Sato, S., Nakamura, Y., Kaneko, T., Asamizu, E., Kato, T., Nakamura, Y.,et al. (2008). Genome structure of the legume, *Lotus japonicus. DNA Res.* 15, 227–239. doi: 10.1093/dnares/dsn008

46. Schmutz, J., Cannon, S. B., Schlueter, J., Ma, J., Mitros, T., Nelson, W.,et al. (2010). Genome sequence of the palaeopolyploid soybean. *Nature* 463, 178–183. doi: 10.1038/nature08670

47. Severin, A. J., Woody, J. L., Bolon, Y. T., Joseph, B., Diers, B. W., Farmer, A. D.,et al. (2010). RNA-Seq Atlas of Glycine max: a guide to the soybean transcriptome. *BMC Plant Biol.* 10:160. doi: 10.1186/1471-2229-10-160

48. Singh, V. K., Garg, R., and Jain, M. (2013). A global view of transcriptome

dynamics during flower development in chickpea by deep sequencing. *Plant Biotechnol. J.* 11, 691–701. doi: 10.1111/pbi.12059

49. Staswick, P. E., Serban, B., Rowe, M., Tiryaki, I., Maldonado, M. T., Maldonado, M. C.,et al. (2005). Characterization of an *Arabidopsis* enzyme family that conjugates amino acids to indole-3-acetic acid. *Plant Cell* 17, 616–627. doi: 10.1105/tpc.104.026690

50. Staswick, P. E., Tiryaki, I., and Rowe M. L. (2002). Jasmonate response locus JAR1 and several related Arabidopsis genes encode enzymes of the firefly luciferase superfamily that show activity on jasmonic, salicylic and indole-3-acetic acids in an assay for adenylation. *Plant Cell* 14, 1405–1415. doi: 10.1105/tpc.000885

51. Takase, T., Nakazawa, M., Ishikawa, A., Kawashima, M., Ichikawa, T., Takahashi, N.,et al. (2004). ydk1-D, an auxin-responsive GH3 mutant that is involved in hypocotyl and root elongation. *Plant J.* 37, 471–483. doi: 10.1046/j.1365-313X.2003.01973.x

52. Tebbji, F., Nantel, A., and Matton, D. P. (2010). Transcription profiling of fertilization and early seed development events in a solanaceous species using a 7.7 K cDNA microarray from *Solanum chacoense* ovules. *BMC Plant Biol.* 10:174. doi: 10.1186/1471-2229-10-174

53. Terol, J., Domingo, C., and Talon, M. (2006). The GH3 family in plants: genome wide analysis in rice and evolutionary history based on EST analysis. *Gene* 371, 279–290. doi: 10.1016/j.gene.2005.12.014

54. Ulmasov, T., Liu, Z. B., Hagen, G., and Guilfoyle, T. J. (1995). Composite structure of auxin response elements. *Plant Cell* 7, 1611–1623. doi: 10.1105/tpc.7.10.1611

55. Varshney, R. K., Song, C., Saxena, R. K., Azam, S., Yu, S., Sharpe, A. G.,et al. (2013). Draft genome sequence of chickpea (*Cicer arietinum*) provides a resource for trait improvement. *Nat. Biotechnol.* 31, 240–246. doi: 10.1038/nbt.2491

56. Verdier, J., Torres-Jerez, I., Wang, M., Andriankaja, A., Allen, S. N., He, J.,et al. (2013). Establishment of the *Lotus japonicus* Gene Expression Atlas (LjGEA) and its use to explore legume seed maturation. *Plant J.* 74, 351–362. doi: 10.1111/tpj.12119

57. Wang, D., Pei, K., Fu, Y., Sun, Z., Li, S., Liu, H.,et al. (2007). Genome-wide analysis of the auxin response factors (ARF) gene family in rice (*Oryza sativa*). *Gene* 394, 13–24. doi: 10.1016/j.gene.2007.01.006

58. Wang, S. K., Bai, Y. H., Shen, C. J., Wu, Y. R., Zhang, S. N., Jiang, D. A.,et al. (2010). Auxin-related gene families in abiotic stress response in*Sorghum bicolor*. *Funct. Integr. Genomics* 10, 533–546. doi: 10.1007/

s10142-010-0174-3

59. Wei, F., Coe, E., Nelson, W., Bharti, A. K., Engler, F., Butler E.,et al. (2007). Physical and genetic structure of the maize genome reflects its complex evolutionary history. *PLoS Genet* 3:e123. doi: 10.1371/journal. pgen.0030123

60. Westfall, C. S., Muehler, A. M., and Jez, J. M. (2013). Enzyme action in the regulation of plant hormone responses. *J. Biol. Chem.* 288, 19304–19311. doi: 10.1074/jbc.R113.475160

61. Westfall, C. S., Zubieta, C., Herrmann, J., Kapp, U., Nanao, M. H., and Jez, J. M. (2012). Structural basis for prereceptor modulation of plant hormones by GH3 proteins. *Science* 336, 1708–1711. doi: 10.1126/science.1221863

62. Woodward, A. W., and Bartel, B. (2005). Auxin: regulation, action, and interaction. *Ann. Bot.* 95, 707–735. doi: 10.1093/aob/mci083

63. Wu, J., Liu, S., Guan, X., Chen, L., He, Y., Wang, J.,et al. (2014). Genome-wide identification and transcriptional profiling analysis of auxin response-related gene families in cucumber. *BMC Res. Notes* 7:218. doi: 10.1186/1756-0500-7-218

64. Wu, J., Wang, F., Cheng, L., Kong, F., Peng, Z., Liu, S.,et al. (2011). Identification, isolation and expression analysis of auxin response factor (ARF) genes in Solanum lycopersicum. *Plant Cell Rep.* 30, 2059–2073. doi: 10.1007/s00299-011-1113-z

65. Young, N. D., Debellé, F., Oldroyd, G. E., Geurts, R., Cannon, S. B., Udvardi, M. K.,et al. (2011). The *Medicago* genome provides insight into the evolution of rhizobial symbioses. *Nature* 480, 520–524. doi: 10.1038/nature10625

66. Yuan, H. Z., Zhao, K., Lei, H. J., Shen, X. J., Liu, Y., Liao, X.,et al. (2013). Genome-wide analysis of the GH3 family in apple (*Malus*×domestica). *BMC Genomics* 14:297. doi: 10.1186/1471-2164-14-297

67. Zhang, S. W., Li, C. H., Cao, J., Zhang, Y. C., Zhang, S. Q., Xia, Y. F.,et al. (2009). Altered architecture and enhanced drought tolerance in rice via the downregulation of indole-3-acetic acid by TLD1/OsGH3.13 activation. *Plant Physiol.* 151, 1889–1901. doi: 10.1104/pp.109.146803

68. Zhang, Z. Q., Li, Q., Li, Z. M., Staswick, P. E., Wang, M. Y., Zhu, Y.,et al. (2007). Dual regulation role of GH3.5 in salicylic acid and auxin signaling during Arabidopsis-Pseudomonas syringae interaction. *Plant Phsiol.* 145, 450–464. doi: 10.1104/pp.107.106021

69. Zhao, Y. (2010). Auxin biosynthesis and its role in plant development. *Annu. Rev. Plant Biol.* 61, 49–64. doi: 10.1146/annurev-arplant-042809-112308

Chapter 10

ENHANCEMENT OF PHOSPHATE ABSORPTION BY GARDEN PLANTS BY GENETIC ENGINEERING: A NEW TOOL FOR PHYTOREMEDIATION

Keisuke Matsui[1], Junichi Togami[2], John G. Mason[3], Stephen F. Chandler[4] and Yoshikazu Tanaka[1]

[1]Research Institute, Suntory Global Innovation Center Limited, 1-1-1 Wakayama-dai, Shimamoto-cho, Mishima-gun, Osaka 618-8503, Japan

[2]Safety Science Institute, Quality Assurance Division, Suntory Business Expert Limited, 57 Imaikami-cho, Nakahara-ku, Kanagawa Kawasaki 211-0067, Japan

[3]Biosciences Research Division, Department of Environment & Primary Industries, AgriBio, Centre for AgriBioscience, 5 Ring Road, La Trobe University, Bundoora, VIC 3083, Australia

[4]School of Applied Sciences, RMIT University, Bundoora, VIC 3083, Australia

ABSTRACT

Although phosphorus is an essential factor for proper plant growth in natural environments, an excess of phosphate in water sources causes serious pollution. In this paper we describe transgenic plants which hyperaccumulate inorganic phosphate (Pi) and which may be used to reduce environmental water pollution by phytoremediation. AtPHR1, a transcription factor for a key regulator of the Pi starvation response in Arabidopsis thaliana, was overexpressed in the ornamental garden plants Torenia, Petunia, and Verbena. The transgenic plants showed hyperaccumulation of Pi in leaves and accelerated Pi absorption rates from hydroponic solutions. Large-scale hydroponic experiments indicated that the enhanced ability to absorb Pi in transgenic torenia (AtPHR1) was comparable to water hyacinth a plant that though is used for phytoremediation causes overgrowth problems.

INTRODUCTION

Water pollution has become a serious problem around the world. Contamination by toxic substances such as endocrine disruptors and heavy metals and excessive inflows of phosphorus, nitrogen and other elements all contribute to water pollution. Eutrophication is one of the major problems associated with water pollution and is caused by inflow of excess amounts of nutrients (especially phosphorus and nitrogen) [1]. The sources of excessive amounts of phosphorus and nitrogen are agricultural run-off, sewage, industrial effluents, and natural erosion from soil and rocks. Eutrophication is due to rapid growth of phytoplankton causing algal blooms or "red tides," the result of which are serious environmental problems such as bad odor and fish death as a result of oxygen depletion and accumulation of toxic cyanotoxins [2].

Phosphorus can be removed by physical, chemical, and biological methods [3–6]. Physical and chemical methods (e.g., electrolytic, crystallization, filtration, and aggregation/separation methods) are superior in terms of removal efficiency and throughput capacity. However, these methods require complicated equipment and large quantities of chemicals, resulting in high cost and environmental burdens. A biological method, the anaerobic-anoxic-oxic method (A2O), is one of the advanced activated sludge methods and has been widely examined in sewage plants. However this method is also very expensive [7], and presently, there are no practically useable technologies to remove inorganic ions such as phosphorus and nitrogen during sewage treatment using activated sludge methods. Thus, though various types of water purification systems have been developed for water and sewage plants [8], these technologies are often difficult to apply directly to aquatic environments due to cost and the need for special equipment. Eutrophication therefore remains a problem.

Concurrently with improving sewage treatment technology, a low-cost and highly efficient method is still needed for sustainable water purification in aquatic environments. A treatment for environmental pollution using plants (phytoremediation) is a possible solution [9, 10]. Since phosphorus is an essential and often limiting nutritive substance for plants, plants actively absorb it from environments through the roots. Phytoremediation of aquatic systems has been attempted using water plants such as water hyacinth and Phragmites, as these plants absorb phosphorus relatively efficiently in comparison to terrestrial plants, and they also grow rapidly [11]. However, the high cost of collection and disposal of water plants (especially water hyacinth) presents difficulties in habitat management, and the impact of the plants on preexisting ecosystems hamper their wide application. In addition, the ability of these water plants to eliminate phosphorus in aquatic ecosystems is still inadequate as an even higher efficiency is needed for effective phytoremediation.

Inorganic phosphate (Pi) transporter is a key component in Pi absorption by plant roots. In Arabidopsis thaliana, 9 high-affinity transporters are known [12]. One of these, AtPHT1, encodes a cell membrane-located Pi transporter with high affinity for Pi. It has been reported that overexpression of AtPHT1 in cultured cells of Nicotiana leads to an acceleration of Pi absorption and an increase in cell growth rate [13]. In contrast, when the same Pi transporter was overexpressed in Hordeum vulgare, an increase in absorption of Pi was not observed [14]. These two contradicting reports suggest that merely increasing the number of Pi transporters does not necessarily lead to enhanced Pi absorption.

Several Pi starvation-related genes have been identified in A. thaliana mutants [15]. One of the known control factors which function when plants enter a state of Pi starvation is the AtPHR1 gene. AtPHR1 gene encodes a transcription factor which activates the transcription of genes in response to states of Pi starvation [16]. Recently, it is reported that overexpression of AtPHR1 in A. thaliana increases the Pi concentration in aerial plant parts [17].

In this study, we introduced the AtPHR1 gene into the garden plants Torenia, Petunia, and Verbena, in order to enhance Pi absorption. Small and large-scale hydroponic trials with transgenic torenia plants expressing the AtPHR1 gene were performed. We demonstrate for the first time that over expression of the AtPHR1 gene results in enhanced Pi absorption rate in different plant species. The AtPHR1 transgenic plants can possibly facilitate effective phytoremediation in polluted aquatic environments.

MATERIALS AND METHODS

Plant Materials

Plants of Torenia hybrida cv. Summer Wave blue, Petunia hybrida cv. Surfinia purple mini, and Verbena hybrida cv. Temari scarlet (Suntory Flowers, Ltd.) were grown in soil and supplied with full nutrients every week in a green house or a growth chamber in controlled conditions (22–25°C, 12 hours light).

Constructs for Expression in Plants and Plant Transformation

Molecular biology techniques were employed according to the methods described by Sambrook et al. [18], unless otherwise specified.

The AtPHR1 gene was amplified by PCR using primers PHRf (5′-ATGGAGGCTCGTCCAGTTCAT-3′) and PHRr (5′-TCAATTATCGATTTGGGACGC-3′) and subcloned into the pCR2.1

vector using a TOPO-TA cloning kit (Life Technologies) according to the manufacturer's instructions. A fragment of the AtPHR1 gene was inserted into binary vector pBinPLUS [19] which contains an enhanced cauliflower mosaic virus 35S promoter [20] and a nopaline synthase (nos) terminator. This plasmid was named pSPB1898.

Transformation with transformation vector pSPB1898 was carried out as described previously for Torenia [21], Petunia [22], and Verbena [23] using Agrobacterium tumefaciens strain AGL0 [24].

RNAs were extracted from leaves of the obtained recombinant plants using the RNeasy Plant Mini Kit (Qiagen). Positive strains were selected by RT-PCR.

Method for Measuring Phosphorus Concentration

Phosphorus concentration was measured according to a modified method of Ames [25]. Leaves were weighed (approximately 100 mg per sample) and inserted into a 2 mL tube for crushing with zirconia beads (4 mm diameter), at $-80°C$. The frozen sample was taken to room temperature, and 500 μL of 1% (v/v) acetic acid was added to each tube. The mixture was then shaken and crushed for 6 minutes using a TissueLyser (Qiagen). After crushing, the mixture was centrifuged at 15,000 rpm for 5 minutes using a desktop centrifuge to obtain 500 μL of supernatant. This Pi extract was diluted with distilled water (from 10 to 100-fold dilution) to a final concentration of 800 μL. To this solution, 160 μL of measuring buffer (1.25 M sulfuric acid, 30 mM ascorbic acid, 0.405 mg/mL antimony potassium tartrate, and 24 mg/mL ammonium molybdate) was added, and the mixture was stirred well and left for 10 minutes. The absorbance was measured at 880 nm using a spectrophotometer BioSpec-mini (Shimadzu, Japan). The amount of phosphorus in 1 g of leaf was calculated from phosphorus concentration and weight of the sample. For calculations on a dry weight basis, samples were dried at 80°C for about 2 days. An independent Student's t-test was used to compare differences between host and transgenic plants. All tests were two-sided, and $P < 0.05$ was considered statistically significant. Data are the mean ± SD from at least three different samples.

Hydroponic Experiment

Wild-type torenia or transgenic torenia was grown on a support made of polystyrene foam with holes to allow the root systems of the plants to grow into the hydroponic solution. Plants were floated on 5 liter of hydroponic solution (0.5 mM KNO_3, 0.2 mM $MgSO_4$, 0.2 mM $Ca(NO_3)_2$, 0.161 mM KPO_4,

5 µM Fe-EDTA, 7 µM H_3BO_3, 1.4 µM $MnCl_2$, 0.05 µM $CuSO_4$, 0.1 mM $ZnSo_4$, 0.02 µM Na_2MoO_4, 1 µM NaCl, and 0.001 µM $CoCl_2$). The initial phosphorus concentration in the hydroponic solution was 5 mg/L. Four plants were used in each support. The Pi concentration in the hydroponic solution was measured each day. Since the fluid volume of the hydroponic solution decreased due to transpiration and evaporation, on every fourth day, deionised water was added to the solution. For large container experiments, the same solution was used, but the volume of hydroponic solution was 400 liter, and 13 plants were used per container. The volume of each container was adjusted with deionised water on a weekly basis.

RESULTS AND DISCUSSION

Overexpression of AtPHR1 Enhances Pi Accumulation and Absorption in Transgenic Plants

It has been shown in A. thaliana that over expression of AtPHR1 causes enhanced Pi accumulation in aerial parts [17]. To examine whether AtPHR1 is effective in other plant species, we transformed torenia, petunia, and verbena with AtPHR1. These plants were transformed with the plasmid pSPB1898, which contains the AtPHR1gene under the control of the constitutive 35S promoter. We screened over 30 transgenic plants for each species for the presence of the transgene with RT-PCR and for leaf Pi concentration 4 weeks after potting up from tissue culture. Concentration of phosphorus per fresh leaf weight was then measured for selected lines. In each of the 3 plant species, phosphorus concentration in the leaves of the transgenic plants was 2 to 3-fold higher than that of control host plants (Figure 1).

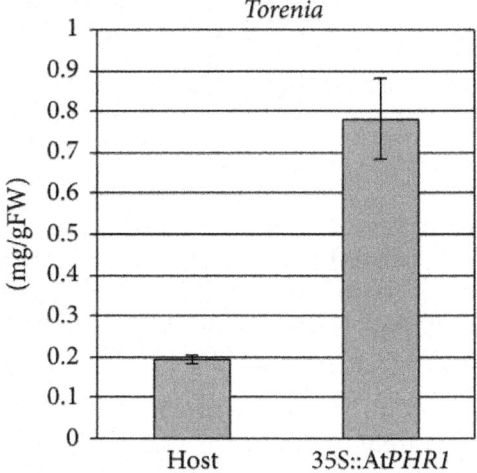

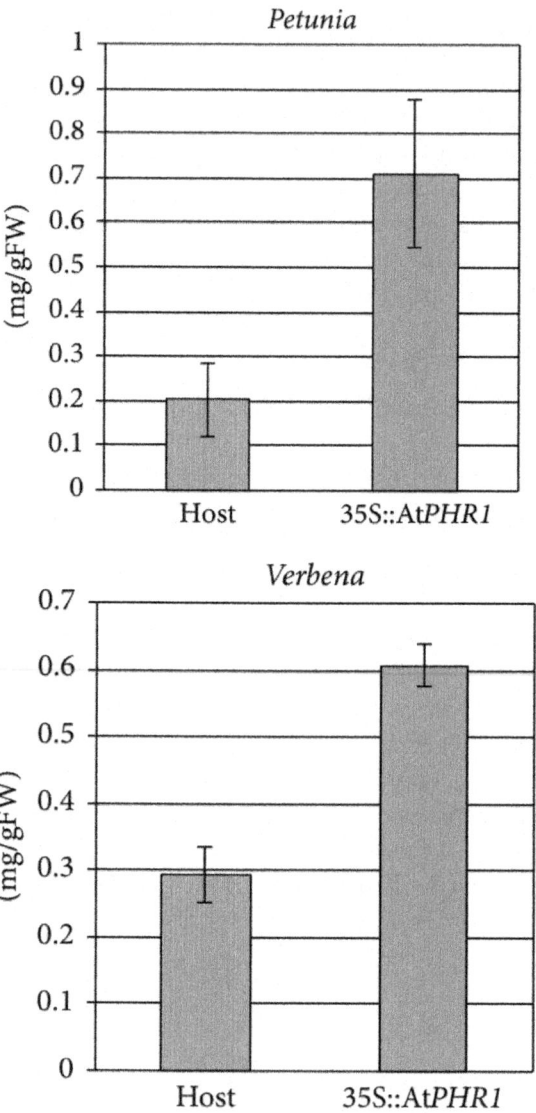

Figure 1: Phosphorus measurements of AtPHR1 transgenic plants. Phosphorus concentrations in the leaves of AtPHR1 transgenic plants of potted torenia, petunia, and verbena were measured. The longitudinal axis shows the phosphorus amounts per gram fresh weight (mg/gFW). Significant differences in means between host and transgenic plants were detected for all three species.

We examined other Pi starvation-related genes (AtPHT1;1, AtPHT1;2, AtIPS1, and AtPHO1) from A. thalianaby constitutively overexpressing them in transgenic torenia and petunia (data not shown). None of these transgenic plants showed enhanced Pi accumulation. This result is consistent with the observation that over-expression of the Pi transporter did not cause any change to Pi accumulation in H. vulgare [14]. Thus, we focused on AtPHR1 in the following experiments.

To confirm that introduction of the AtPHR1 gene accelerates Pi absorption rates, we grew plants of a transgenic torenia line in a hydroponic system. Torenia was chosen as this plant grows luxuriantly and roots tolerate being submerged in water. The torenia plants were grown in 5 liters of hydroponic solution containing 5 mg/L phosphorus for 1 to 2 months in a green house or a growth chamber. The phosphorus concentration of the hydroponic solutions was measured daily. The superior transgenic line expressing AtPHR1 (35S::AtPHR1) showed enhanced Pi absorption from the hydroponic solution (Figure 2(a)). Enhanced accumulation of Pi in the transgenic leaves was also confirmed by measurements of leaf phosphorus concentration (Figure 2(b)). The phosphorus concentration of the hydroponic solution in which 35S::AtPHR1 was grown decreased during the two weeks of the experiment. The Pi absorption rate observed for 35S::AtPHR1 was up to 0.091 mgP/day/plant in this experiment compared to 0.056 mgP/day/plant for the host (Figure 2(a)). This result suggests that the enhanced Pi accumulation observed in the potted AtPHR1 transgenic torenia plants is mainly due to enhanced Pi absorption rate.

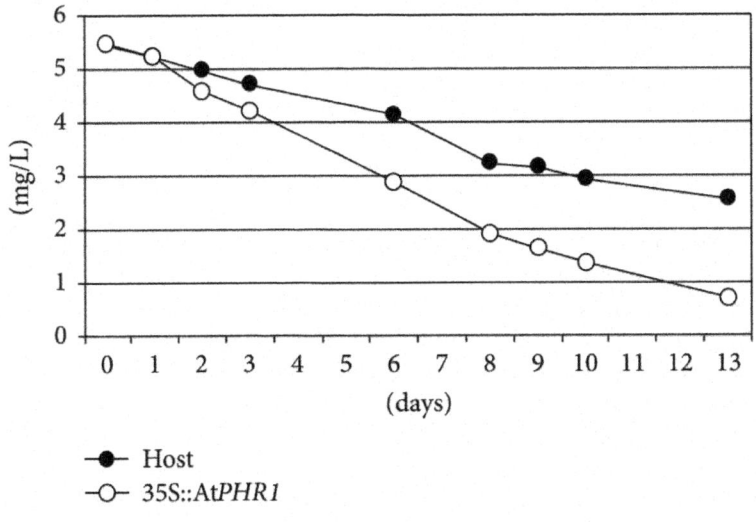

(a)

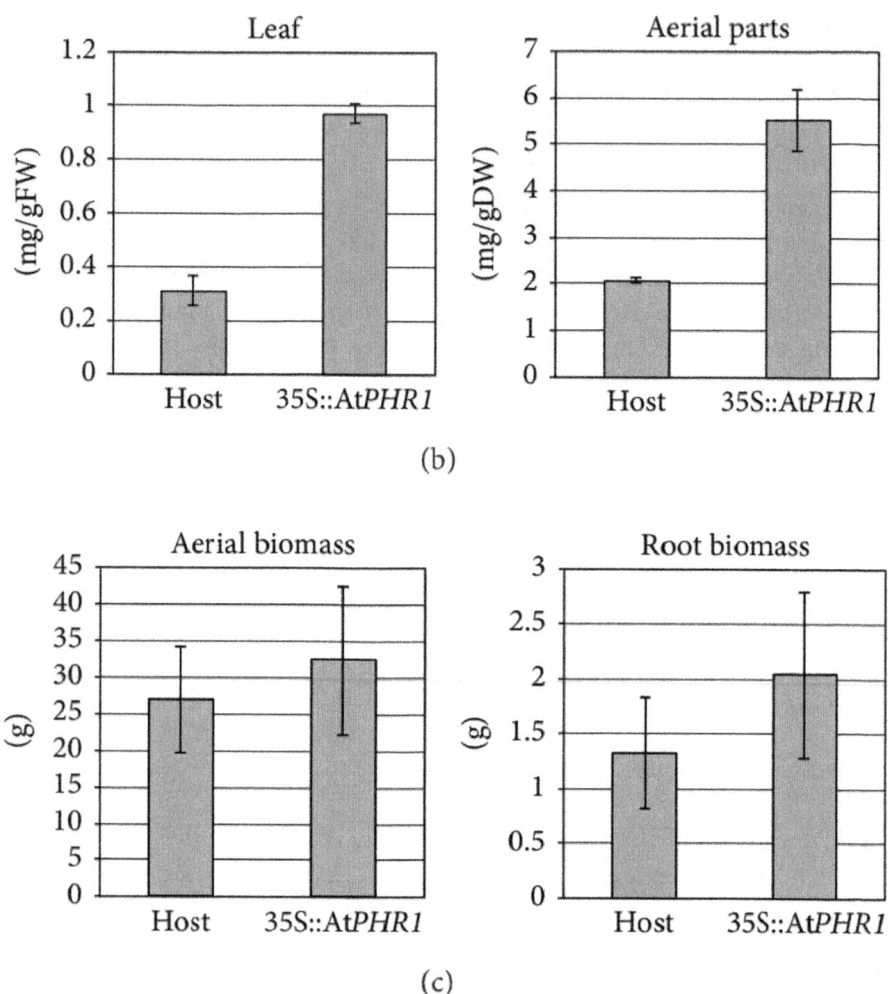

(b)

(c)

Figure 2: Pi accumulation and growth properties of AtPHR1 transgenic torenia. (a) Changes of Pi concentration in hydroponic solutions. The phosphorus concentration in a hydroponic solution in which host (filled circle) and AtPHR1 transgenic torenia (empty circle) were cultured was measured. The longitudinal axis shows the phosphorus concentration (mg/L), and the horizontal axis shows the number of days after exchange of the hydroponic solution. (b) Pi concentration in the leaves and aerial parts of hydroponically-cultivated torenia. The longitudinal axis shows the phosphorus concentration per gram fresh weight of samples (mg/gFW) (left) and the phosphorus concentration per gram dry weight of samples (mg/gDW) (right). There were significant differences in means between host and transgenic plants. (c) Comparison of growth rate. Weights of aerial parts and root parts of the torenia plants were measured at the end of hydroponic experiments. There was no statistically significant difference between transgenic and host.

To see if the decrease of Pi concentration in the hydroponic solution was also reflected in an increase in Pi accumulation in the plant, Pi accumulation in the aerial parts of the plants was measured. Three plants each of the transgenic and the host torenia were hydroponically cultivated in the solution containing 5 mg/L phosphorus for about 2 months. The aerial parts of those plants were collected and dried on the phosphorus concentration measured (Figure 2(b)). The Pi concentration in the transgenic plants was approximately 2.5-fold that of the host.

We weighed aerial and root parts of the tested plants after each hydroponic experiment. Even though slightly less weight was measured in the host, there was no statistically significant difference between the transgenic and host (Figure 2(c)). This suggests that excessively absorbed Pi is not used for plant growth but is accumulated and stored in the aerial part of the plants. As a result, overexpression of AtPHR1 does not retard plant growth. Since the transgenic plants did not show any morphological or reproductive abnormalities, over-expression of the AtPHR1 gene can enhance Pi accumulation with no negative effects on plant growth.

Limitation of Pi Capacity

Sections of dead tissues in the leaves were often observed in transgenic torenia during the 4 weeks of the hydroponic experiments (Figures 3(a)–3(c)). We collected the dead sections and compared them to the unaffected areas of the leaves from the same plants. The harvested leaves were dried and then measured for phosphorus concentration. The phosphorus concentration in the dead sections was slightly higher than that of unaffected portions of leaves (Figure 3(d)). Since excess Pi may cause cell toxicity [26], the death may have been the result of exceeding a critical limit of Pi concentration in the torenia leaf cells. It thus appears that the critical limit of Pi accumulation level in AtPHR1 transgenic torenia is approximately 20 mg/gDW. One possible way to overcome the death of leaf tissues due to high Pi accumulation is to convert Pi to a nontoxic form of phosphorus that is phytic acid. Genetic modification could be used to achieve this, resulting in transgenic plants accumulating even more Pi than reported here.

(a)

(b)

(c)

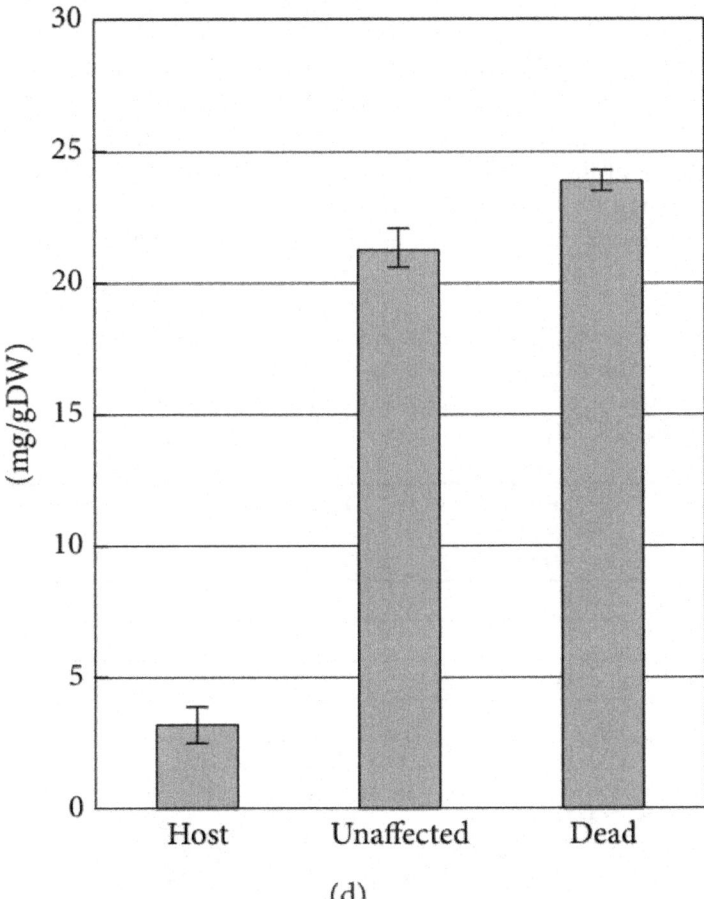

(d)

Figure 3: Dead tissue in AtPHR1 transgenic torenia. (a) Host plant at the end of hydroponic experiment. (b) AtPHR1 transgenic plant after 4 weeks of hydroponic experiment. (c) Magnified image of (b). Arrowheads indicate partially dead sections. (d) Phosphorus concentration in unaffected and dead areas from leaves of host and AtPHR1transgenic plants. The longitudinal axis shows the phosphorus concentration per gram dry weight of sample (mg/gDW).

Large-Scale Hydroponic Experiment

To access the potential for phytoremediation using the transgenic torenia at a larger scale, we performed longer term hydroponic experiments. Thirteen torenia plants were put each into 400-liter tub and incubated for approximately 2 months (Figure 4). There was no significant difference in average biomass between transgenic and host plants after 65 days incubation (Figure 4 and Table 1). However, approximately 3-fold more Pi accumulation was seen in the transgenic plant when compared to the host. This confirmed that transgenic torenia shows the accelerated absorption as well as accumulation of Pi in the leaves when grown on a larger scale. From the daily calculation of Pi accumulation of the transgenic torenia plant, Pi accumulation rates were able to be compared to water hyacinth (Table 1). The AtPHR1 transgenic torenia showed an equivalent efficiency of Pi accumulation to that of water hyacinth [27, 28].

Table 1: Comparison of phosphate absorption performances. Phosphorus content, total biomass, and absorption rate after 65 days of the hydroponic experiment are indicated. Data are the mean ± SD from 13 plants. Values of water hyacinth were calculated from values listed in [27, 28]

	Phosphorus in leaf (mg/gFW)	Total biomass (g/ plant)	Absorption rate (mg/plant/day)
Host			1.08
35S::AtPHR1			4.15
Water hyacinth	0.38		1.79

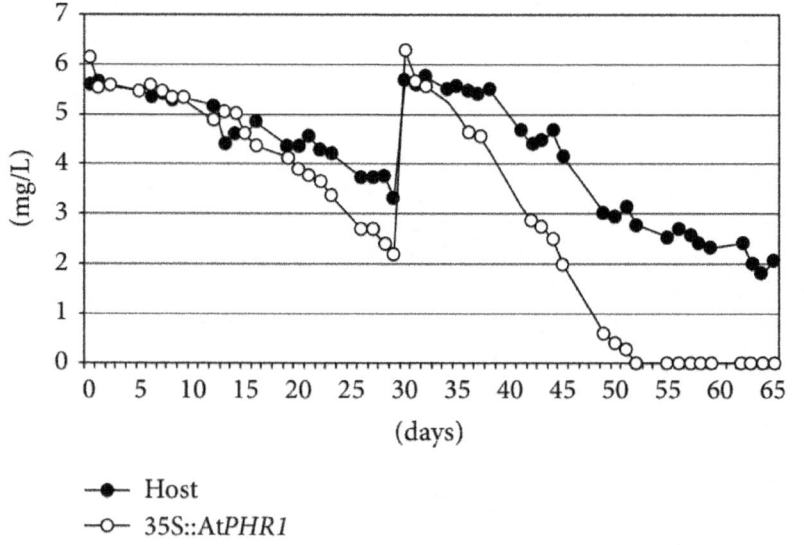

(a)

(b)

(c)

Figure 4: Large-scale hydroponic experiments. (a) Changes in Pi concentration in hydroponic solutions. The phosphorus concentration in a hydroponic solution in which host (filled circle) and AtPHR1 transgenic torenia (empty circle) were cultured was measured. The longitudinal axis shows the phosphorus concentration (mg/L), and the horizontal axis shows the number of days. Hydroponic solutions were fully exchanged 30 days after starting the experiment. (b) Large-scale experiment (0 day). (c) Large-scale experiment (65 days).

Overexpression of AtPHR1 gene might drive a Pi starvation response in the transgenic plants. As a result, excessive amounts of Pi accumulated in transgenic leaves. In A. thaliana, AtPHR1 gene is not transcriptionally regulated even under Pi starvation condition [17]. Since the key mechanism of the Pi starvation response is still debatable in Arabidopsis thalinana [17, 29], it is difficult to postulate why overexpression of AtPHR1 is effective for Pi uptake in other species. We have isolated orthologous Pi starvation-related genes (AtPHR1, AtIPS1, AtPHT1;1, and AtPHO2) in torenia and examined expression pattern of these genes (data not shown). We could not detect any differences between transgenic torenia and host plants. Overexpression of AtPHR1may interfere with the proper posttranscriptional modification of the endogenous AtPHR1 counterpart, possibly through competitive inhibition.

Since phosphorus is expected to be exhausted as a natural resource within a hundred year [30], it is necessary to recover phosphorus from the environment, especially in polluted areas. Currently, over 90% of the produced phosphorus in the world is used as fertilizers. Therefore, it is most reasonable to recover phosphorus from fertilized soils and agricultural run-offs. Phytoremediation is a suitable method for such a recycling process, in addition to cleaning up phosphorus from the aquatic environment. One of the critical problems of phytoremediation is the cost of the disposal of the plant [31]. The plant used for phytoremediation was in many cases simply discarded without being used as a source of Pi. Ideally, plants containing high accumulation of Pi can be returned to soils of agricultural land without processing and can be directly used as fertilizer. However, at present, absorbing ability of the existing plants used for phytoremediation is not efficient enough to be used as Pi sources for agriculture in this way. In this study, the AtPHR1 transgenic plants accumulated a high level of Pi. Therefore, applications of AtPHR1 transgenic plants for phytoremediation of water could be cost-effective. Moreover, the Pi recycling ability of flowers and ornamental plants for gardening can be increased by means of AtPHR1 gene introduction, and thereby purifying water with plants having both ornamental beauty and high purification ability.

CONCLUSIONS

In this study, we prove the feasibility of using AtPHR1 as an enhancer of Pi uptake in transgenic plants. By introducing AtPHR1 to garden plants, amounts of Pi accumulation and absorption of Pi were increased to rates approximately 3-fold higher than host plant. There was no significant reduction in biomass or morphology of the transgenic plant expressing AtPHR1. Taken together, these observations indicate that the AtPHR1 gene will be valuable for production of hyperaccumulator plants for the purification of waters polluted with Pi. In addition, an improved appearance of purification sites can be provided by using ornamental plants with many flowers, as shown in Figure 4(c).

ACKNOWLEDGMENTS

The authors thank Mses. Keiko Takeda, Masumi Taniguchi, and Sarah Parsons for producing the transgenic plants and Mses. Chika Shimadzu, Kumi Takemura, Miyuki Ogawa, and Kim Stevenson for their technical assistance. The authors thank Dr. Robert A. Ludwig for providing A. tumefaciens Agl0 and appreciate Mr. Masayasu Yoshikawa for his critical reading of the paper.

REFERENCES

1. V. H. Smith, G. D. Tilman, and J. C. Nekola, "Eutrophication: impacts of excess nutrient inputs on freshwater, marine, and terrestrial ecosystems," Environmental Pollution, vol. 100, no. 1–3, pp. 179–196, 1998. View at Publisher · View at Google Scholar · View at Scopus

2. V. H. Smith and D. W. Schindler, "Eutrophication science: where do we go from here?" Trends in Ecology and Evolution, vol. 24, no. 4, pp. 201–207, 2009. View at Publisher · View at Google Scholar ·View at Scopus

3. C. Vohlaa, M. Koiva, H. J. Bavor, et al., "Filter materials for phosphorus removal from wastewater in treatment wetlands—A review," Ecological Engineering, vol. 37, no. 1, pp. 70–89, 2011. View at Google Scholar

4. J. QU, "Research progress of novel adsorption processes in water purification: a review," Journal of Environmental Sciences, vol. 20, no. 1, pp. 1–13, 2008. View at Publisher · View at Google Scholar · View at Scopus

5. D. W. de Haas, M. C. Wentzel, and G. A. Ekama, "The use of simultaneous chemical precipitation in modified activated sludge systems exhibiting biological excess phosphate removal part 1: literature review," Water SA, vol. 26, no. 4, pp. 439–452, 2000. View at Google Scholar · View at Scopus

6. F. Y. Wang, V. Rudolph, and Z. H. Zhu, "Sewage Sludge technologies," Encyclopedia of Ecology, pp. 3227–3242, 2008. View at Publisher · View at Google Scholar

7. Y. Peng, X. Wang, W. Wu, J. Li, and J. Fan, "Optimisation of anaerobic/anoxic/oxic process to improve performance and reduce operating costs," Journal of Chemical Technology and Biotechnology, vol. 81, no. 8, pp. 1391–1397, 2006. View at Publisher · View at Google Scholar · View at Scopus

8. M. A. Shannon, P. W. Bohn, M. Elimelech, J. G. Georgiadis, B. J. Marĩas, and A. M. Mayes, "Science and technology for water purification in the coming decades," Nature, vol. 452, no. 7185, pp. 301–310, 2008.View at

Publisher · View at Google Scholar · View at Scopus

9.   E. Pilon-Smits, "Phytoremediation," Annual Review of Plant Biology, vol. 56, pp. 15–39, 2005. View at Publisher · View at Google Scholar · View at Scopus

10. M. Luqman, T. M. Batt, A. Tanvir, et al., "Phytoremediation of polluted water by trees: a review,"African Journal of Agricultural Research, vol. 8, no. 17, pp. 1591–1595, 2013. View at Google Scholar

11. P. Gupta, S. Roy, and A. B. Mahindrakar, "Treatment of water using water hyacinth, water lettuce and vetiver grass—a review," Resources and Environment, vol. 2, no. 5, pp. 202–215, 2012. View at Google Scholar

12. C. Rausch and M. Bucher, "Molecular mechanisms of phosphate transport in plants," Planta, vol. 216, no. 1, pp. 23–37, 2002. View at Publisher · View at Google Scholar · View at Scopus

13. N. Mitsukawa, S. Okumura, Y. Shirano et al., "Overexpression of an Arabidopsis thaliana high-affinity phosphate transporter gene in tobacco cultured cells enhances cell growth under phosphate-limited conditions," Proceedings of the National Academy of Sciences of the United States of America, vol. 94, no. 13, pp. 7098–7102, 1997. View at Publisher · View at Google Scholar · View at Scopus

14. A. L. Rae, J. M. Jarmey, S. R. Mudge, and F. W. Smith, "Over-expression of a high-affinity phosphate transporter in transgenic barley plants does not enhance phosphate uptake rates," Functional Plant Biology, vol. 31, no. 2, pp. 141–148, 2004. View at Publisher · View at Google Scholar · View at Scopus

15. C. A. Ticconi and S. Abel, "Short on phosphate: plant surveillance and countermeasures," Trends in Plant Science, vol. 9, no. 11, pp. 548–555, 2004. View at Publisher · View at Google Scholar · View at Scopus

16. V. Rubio, F. Linhares, R. Solano et al., "A conserved MYB transcription factor involved in phosphate starvation signaling both in vascular plants and in unicellular algae," Genes and Development, vol. 15, no. 16, pp. 2122–2133, 2001. View at Publisher · View at Google Scholar · View at Scopus

17. L. Nilsson, R. MÜller, and T. H. Nielsen, "Increased expression of the MYB-related transcription factor,PHR1, leads to enhanced phosphate uptake in Arabidopsis thaliana," Plant, Cell and Environment, vol. 30, no. 12, pp. 1499–1512, 2007. View at Publisher · View at Google Scholar · View at Scopus

18. J. Sambrook, E. F. Fritsch, and T. Maniatis, Molecular Cloning: A Laboratory Manual, Cold Spring Harbor Laboratory Press, 1989.

19. F. A. van Engelen, J. W. Molthoff, A. J. Conner, J. P. Nap, A. Pereira, and W. J. Stiekema, "pBINPLUS: an improved plant transformation vector based on pBIN19," Transgenic Research, vol. 4, no. 4, pp. 288–290, 1995. View at Google Scholar · View at Scopus

20. I. Mitsuhara, M. Ugaki, H. Hirochika et al., "Efficient promoter cassettes or enhanced expression of foreign genes in dicotyledonous and monocotyledonous plants," Plant and Cell Physiology, vol. 37, no. 1, pp. 49–59, 1996. View at Google Scholar · View at Scopus

21. R. Aida and M. Shibata, "Agrobacterium-mediated transformation of torenia (Torenia fournieri),"Breeding Science, vol. 45, no. 1, pp. 71–74, 1995. View at Google Scholar · View at Scopus

22. R. B. Horsch, J. E. Fry, N. L. Hoffmann, D. Eichholtz, S. G. Rogers, and R. T. Fraley, "A simple and general method for transferring genes into plants," Science, vol. 227, no. 4691, pp. 1229–1231, 1985.View at Google Scholar · View at Scopus

23. M. Tamura, J. Togami, K. Ishiguro et al., "Regeneration of transformed verbena (Verbena × hybrida) byAgrobacterium tumefaciens," Plant Cell Reports, vol. 21, no. 5, pp. 459–466, 2003. View at Google Scholar · View at Scopus

24. G. R. Lazo, P. A. Stein, and R. A. Ludwig, "A DNA transformation-competent Arabidopsis genomic library in Agrobacterium," Bio/Technology, vol. 9, no. 10, pp. 963–967, 1991. View at Publisher · View at Google Scholar · View at Scopus

25. B. N. Ames, "Assay of inorganic phosphate, total phosphate and phosphatases," Methods in Enzymology, vol. 8, pp. 115–118, 1966. View at Publisher · View at Google Scholar · View at Scopus

26. D. T. Clarkson and C. B. Scattergood, "Growth and phosphate transport in barley and tomato plants during the development of, and recovery from, phosphate-stress," Journal of Experimental Botany, vol. 33, no. 5, pp. 865–875, 1982. View at Publisher · View at Google Scholar · View at Scopus

27. W. T. Haller and D. L. Sutton, "Effect of pH and high phosphorus concentrations on growth of water hyacinth," Hyacinth Control Journal, pp. 59–61, 1973. View at Google Scholar

28. M. Morii, Y. Doyama, and J. Katayama, "On the absorption of nitrogen and phosphorus from water by water hyacinth, Eichhornia crassipes (Mart.) Solms," Bulletin of the Osaka Agricultural Research Center, vol. 26, pp. 11–15, 1990. View at Google Scholar

29. R. Bustos, G. Castrillo, F. Linhares et al., "A central regulatory system largely controls transcriptional activation and repression responses to phosphate starvation in arabidopsis," PLoS Genetics, vol. 6, no. 9, Article ID e1001102, 2010. View at Publisher · View at Google Scholar · View at Scopus

30. A. Maggio, J.-P. Malingreau, A.-K. Bock, et al., "NPK: Will there be enough plant nutrients to feed a world of 9 billion in 2050?" Publication Office of the European Union, 2012,http://publications.jrc.ec.europa.eu/repository/handle/111111111/25770.

31. A. Sas-Nowosielska, R. Kucharski, E. Małkowski, M. Pogrzeba, J. M. Kuperberg, and K. Kryński, "Phytoextraction crop disposal—an unsolved problem," Environmental Pollution, vol. 128, no. 3, pp. 373–379, 2004. View at Publisher · View at Google Scholar · View at Scopus

Chapter 11

A GENE STACKING APPROACH LEADS TO ENGINEERED PLANTS WITH HIGHLY INCREASED GALACTAN LEVELS IN ARABIDOPSIS

Vibe M Gondolf[1,2,3], Rhea Stoppel[1,2], Berit Ebert[1,2,3], Carsten Rautengarten[1,2], April JM Liwanag[1,2], Dominique Loqué[1,2] and Henrik V Scheller[1,2,4]

[1]Feedstocks Division, Joint BioEnergy Institute, Emeryville, California 94608, USA

[2] Physical Biosciences Division, Lawrence Berkeley National Laboratory, Berkeley, California 94720, USA

[3] Department of Plant and Environmental Sciences, University of Copenhagen, DK-1871 Frederiksberg C, Denmark

[4] Department of Plant and Microbial Biology, University of California, Berkeley, California 94720, USA

ABSTRACT

Background

Engineering of plants with a composition of lignocellulosic biomass that is more suitable for downstream processing is of high interest for next-generation biofuel production. Lignocellulosic biomass contains a high proportion of pentose residues, which are more difficult to convert into fuels than hexoses. Therefore, increasing the hexose/pentose ratio in biomass is one approach for biomass improvement. A genetic engineering approach was used to investigate whether the amount of pectic galactan can be specifically increased in cell walls of Arabidopsis fiber cells, which in turn could provide a potential source of readily fermentable galactose.

Results

First it was tested if overexpression of various plant UDP-glucose 4-epimerases (UGEs) could increase the availability of UDP-galactose and thereby increase the biosynthesis of galactan. Constitutive and tissue-specific expression of a poplar UGE and three Arabidopsis UGEs in Arabidopsis plants could not significantly increase the amount of cell wall bound galactose. We then investigated co-overexpression of *At*UGE2 together with the β-1,4-galactan synthase GalS1. Co-overexpression of *At*UGE2 and GalS1 led to over 80% increase in cell wall galactose levels in Arabidopsis stems, providing evidence that these proteins work synergistically. Furthermore, *At*UGE2 and GalS1 overexpression in combination with overexpression of the NST1 master regulator for secondary cell wall biosynthesis resulted in increased thickness of fiber cell walls in addition to the high cell wall galactose levels. Immunofluorescence microscopy confirmed that the increased galactose was present as β-1,4-galactan in secondary cell walls.

Conclusions

This approach clearly indicates that simultaneous overexpression of *At*UGE2 and GalS1 increases the cell wall galactose to much higher levels than can be achieved by overexpressing either one of these proteins alone. Moreover, the increased galactan content in fiber cells while improving the biomass composition had no impact on plant growth and development and hence on the overall biomass amount. Thus, we could show that the gene stacking approach described here is a promising method to engineer advanced feedstocks for biofuel production.

BACKGROUND

Plant cell walls are complex structures composed of polysaccharides influencing plant morphology, defense, growth, and signaling. They also constitute the most abundant biomaterial on earth and have the potential to provide a source of cheap sugars for industrial biotechnology. In lignocellulosic biomass, the cell wall polysaccharides comprise mostly cellulose and glucuronoxylan, a hemicellulose, embedded in highly cross-linked lignin polymers, which protect the polysaccharides from chemical and enzymatic degradation. The hemicellulosic fraction is mostly composed of pentoses (such as xylose and arabinose), which unlike hexoses cannot be easily fermented by yeast into fuels. Two main goals of engineering plants with an altered cell wall composition in order to lower costs and improve efficiency of biofuel production is to decrease recalcitrance by decreasing the lignin content or altering the lignin composition [1,2] or to reduce the content of glucuronoxylan and at the same time increasing the content of polysaccharides composed of a larger proportion of fermentable hexoses [3].

β-1,4-galactan is found as sidechains attached to rhamnogalacturonan I and is generally not highly abundant in lignocellulosic biomass. However, since β-1,4-galactan is composed entirely of galactose residues, which can be easily fermented by yeast, an increased content of this polysaccharide would potentially improve the biomass composition for biofuel purposes. In this study we used a genetic engineering approach to specifically increase the amount of β-1,4-galactan in stem cell walls.

Cell wall polysaccharides are synthesized by glycosyltransferases, which catalyze the formation of glycosidic linkages to form glycosides. During this process monosaccharides from activated sugar substrates are transferred onto glycosyl acceptors. Donor sugars are usually nucleotide sugars, while acceptors can be oligo- or polysaccharides, lipids, proteins, nucleic acids or other small molecules [4]. Recently, the Arabidopsis glycosyltransferase GALACTAN SYNTHASE 1 (GalS1) was shown to be a β-1,4-Galactan synthase. Constitutive overexpression of GalS1 in Arabidopsis wild-type plants led to a 50% increase in cell wall bound galactose in leaves [5]. Loss-of-function mutants in *Gals1* or its homologs *Gals2* and *Gals3* had a larger decrease in cell wall bound galactose in leaves than in stems [5], which suggested that the supply of UDP-galactose might be limiting in stems.

Nucleotide sugars are synthesized by different types of interconverting enzymes such as epimerases, decarboxylases and dehydrogenases. Most of these enzymes are located in the cytosol but some are found within the Golgi lumen [6]. Changes in nucleotide sugar pools can affect the biosynthesis of cell wall polysaccharides, as shown for example for the UDP-glucose dehydrogenase (UGD) double mutant *ugd2/ugd3,* which exhibits significantly reduced cell wall arabinose, xylose, apiose, and galacturonic acid levels [7]. Similarly, the UDP-xylose 4-epimerase (UXE) mutant *mur4* has a 50% decrease in cell wall arabinose [8].

The nucleotide sugar UDP-galactose is formed from UDP-glucose by UDP-glucose 4-epimerase (UGE). Five UGE isoenzymes exist in Arabidopsis (*At*UGEs), all of which have been functionally characterized *in vivo* [9-11]. Differences in the expression pattern, kinetics and amino acid sequences of the five *At*UGEs suggest that these isoenzymes have an overlapping, but not identical function in plants. UGE-overexpressing plants and knockout mutants have been generated, but specific roles for each of the UGEs could not be unambiguously concluded from those experiments. Only the *AtUGE4* knockout mutant ROOT HAIR DEFICIENT 1 *(rhd1/UGE4 [rhd1])* produces a visible phenotype. Roots of *uge4* mutants are shorter as compared to the wild type, and the root epidermis cells are swollen due to a defective synthesis of xyloglucan and type II arabinogalactan [10,12,13]. All five *At*UGEs can rescue this phenotype when expressed under the control of the constitutive cauliflower mosaic virus 35S promoter [9]. In double, triple and quadruple *At*UGE mutants, Rösti et al. [10] observed growth defects and cell wall compositional changes that suggested a partial functional overlap of the five UGE isoenzymes. The authors concluded that *At*UGE2 and *At*UGE4 affect vegetative growth and cell wall carbohydrate biosynthesis whereas*At*UGE1 and *At*UGE5 act in stress situations, and *At*UGE3 seems to be important for pollen development. Analysis of global co-expression profiles led to the conclusion that *At*UGE1 and *At*UGE3 are co-expressed with putative trehalose-6-phosphate synthase genes, whereas *At*UGE2, −4, and −5 are co-expressed with various known glycosyltransferases and other cell wall biosynthetic enzymes, suggesting that *At*UGE1 and *At*UGE3 might preferentially act in the UDP-galactose (UDP-Gal) to UDP-glucose (UDP-Glc) direction, while *At*UGE2, −4, and −5 might act in the UDP-Glc to UDP-Gal direction *in vivo* [9].

All UGE isoenzymes can interconvert UDP-Glc and UDP-Gal *in vitro*, although they show differences in the substrate affinity and reaction requirements [9,11]. Interestingly, two isoenzymes, *At*UGE1 and *At*UGE3 have been shown to also interconvert UDP-xylose (UDP-Xyl) and UDP-arabinose (UDP-Ara) [11]. This bifunctionality is reflected in the amino acid sequence of the different *At*UGEs. Phylogenetic analysis of UGE homologs in different organisms revealed that UGEs in plants are distributed in two plant-specific clades with *At*UGE1 and *At*UGE3 grouping together in clade I, while *At*UGE2, −4, and −5 group together in clade II (Figure 1). Besides *At*UGE1 and *At*UGE3, the pea *Ps*UGE1 has also been shown to interconvert UDP-Xyl and UDP-Ara [11]. *Ps*UGE1 is also located in UGE clade I, indicating that this clade contains additional bifunctional UGEs. Overexpression of two potato UGEs (*St*UGE45 and *St*UGE51) led to an increase of galactose in potato tuber cell walls [14] consistent with the hypothesis that the amount of available UDP-Gal rather than that of galactosyltransferases can be the limiting factor for the accumulation of cell wall galactose.

In the present study, we overexpressed different UGEs from Arabidopsis and poplar, in order to increase the level of cell wall galactose and thereby improve the C_6/C_5 sugar ratio. In addition to constitutive overexpression of one poplar and three Arabidopsis UGEs, we expressed one of the Arabidopsis UGEs (*At*UGE2) under the control of the secondary cell wall specific promoter pIRX5 together with the master transcription factor NST1. IRX5 is one of the catalytic subunits of the cellulose synthase complex in secondary cell walls [15] and its expression is induced by the transcription factor NST1 [16].

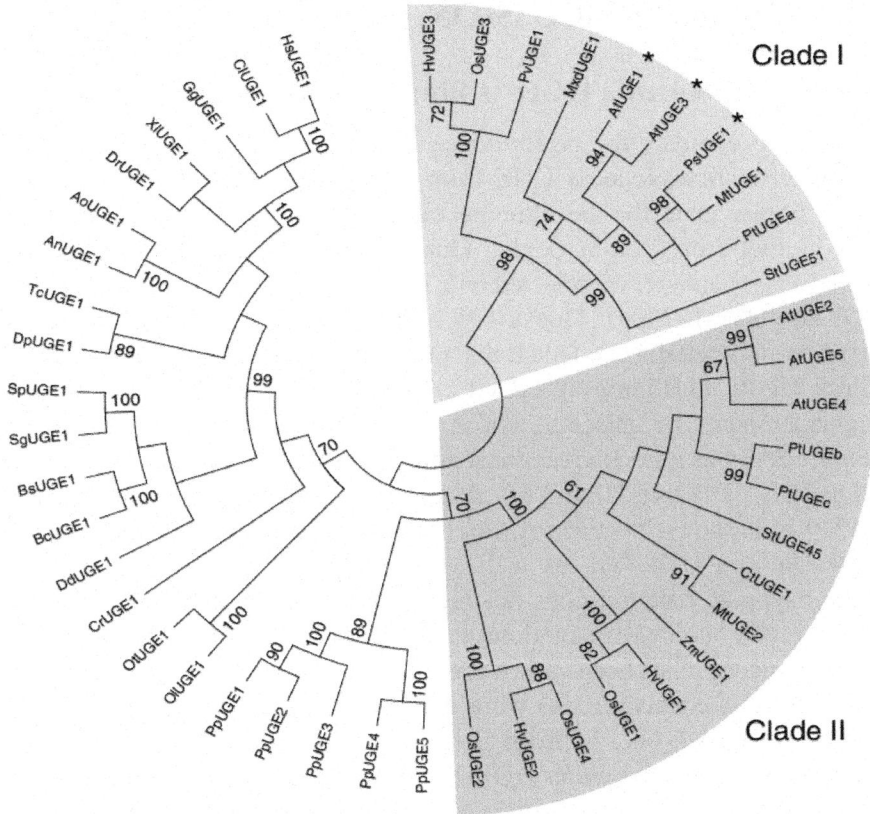

Figure 1: Cladogram of UDP-glucose 4-epimerases. The phylogenetic relationships of UGEs from different organisms are shown. Two vascular plant UGE families described by Kotake et al. [11] are highlighted. UGEs that have documented UDP-Xyl 4-epimerase activity are marked with asterisks. The tree was generated using the Neighbor-Joining method, the bootstrap consensus tree inferred from 1000 replicates (MEGA6 program). Bootstrap values greater than 60% are indicated. Species and accession numbers are described in the Methods section.

Expression of NST1 under the pIRX5 promoter has been shown to create a transcriptional positive feedback loop enhancing overall expression of secondary cell wall biosynthesis genes and thereby increasing secondary cell wall deposition in fiber cells [2]. Finally, constitutive and fiber-specific promoter *At*UGE2 constructs were expressed in the background of transgenic plants overexpressing the galactan synthase GalS1 [5].

While expression of any of the four UGEs alone did not alter the galactose content significantly, regardless of the promoter used, co-overexpression of *At*UGE2 and GalS1 led to an increase in the cell wall galactose content of stems of up to 80%.

RESULTS AND DISCUSSION

Populus trichocarpa UGEc is Bifunctional *in vitro*

In order to compare the performance of different UGEs for the engineering purpose, we first cloned a UGE from poplar since it might be preferable to use a poplar gene for the ultimate translation on the engineering approach to a biofuel crop such as poplar. Out of the UGEs encoded in the *Populus trichocarpa* genome, we selected one, referred to as *Pt*UGEc (XP_002299469, POPTR_0001s10700g). Due to its higher sequence similarity to the non-bifunctional *At*UGEs of Clade II that preferentially act in interconverting UDP-Glc to UDP-Gal (Figure 1), *Pt*UGEc seemed a good candidate for generating higher amounts of galactan. The *in vitro* activity of purified His-*Pt*UGEc (Figure 2A) was tested in reactions with different nucleotide sugars (UDP-Glc, UDP-Gal, UDP-Xyl, UDP-Ara). Proportions of UDP-sugars after incubation for 30 min show that the *Pt*UGEc enzyme is bifunctional, interconverting UDP-Glc and UDP-Gal, as well as UDP-Xyl and UDP-Ara (Figure 2B). The enzymatic reaction does not require the addition of NAD^+, most likely because of non-covalent binding and co-purification of NAD^+ together with the enzyme, as it has been shown for the barley *Hv*UGE1 [17]. Final UDP-Gal and UDP-Glc concentrations were the same in reactions starting with UDP-Gal or with UDP-Glc, indicating that equilibrium levels were reached. The reactions starting with either UDP-Xyl or UDP-Ara did not completely reach the equilibrium level, suggesting that the reaction rates for UDP-Xyl/UDP-Ara interconversion are slower than for UDP-Gal/UDP-Glc. The bifunctional enzymatic characteristics of *Pt*UGEc were somewhat unexpected for a UGE belonging to Clade II since other UGEs from this clade have been found to be specific for UDP-Glc/UDP-Gal interconversion. However, structural features that determine bifunctionality vs. mono-functionality are not known yet.

A

Total protein   Flow through   Wash I   Wash II   Elution I   Elution II

39 kDa

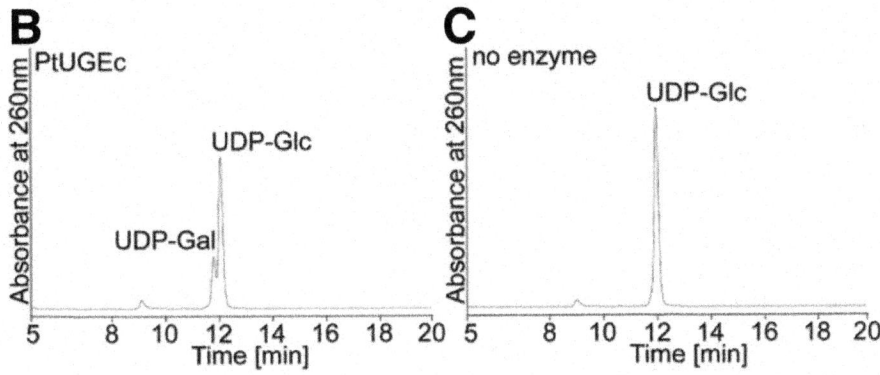

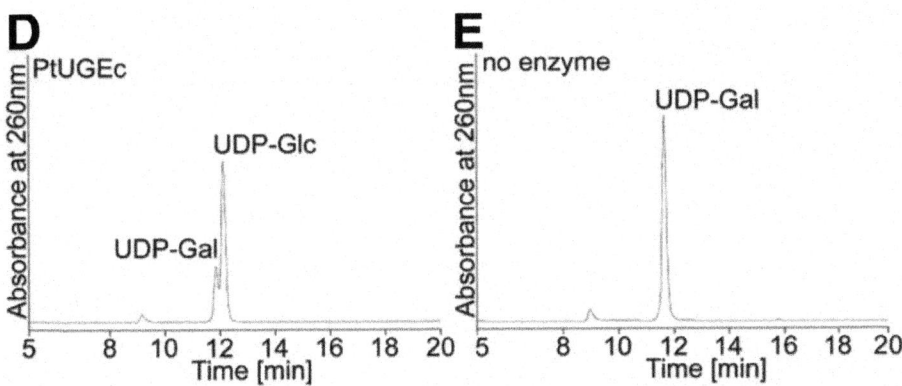

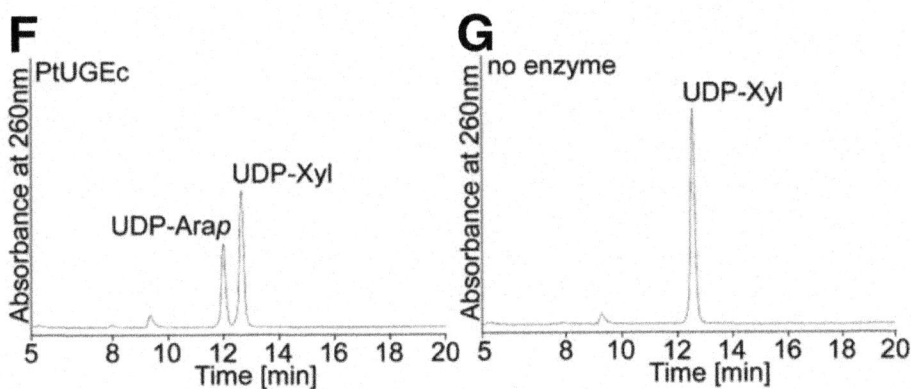

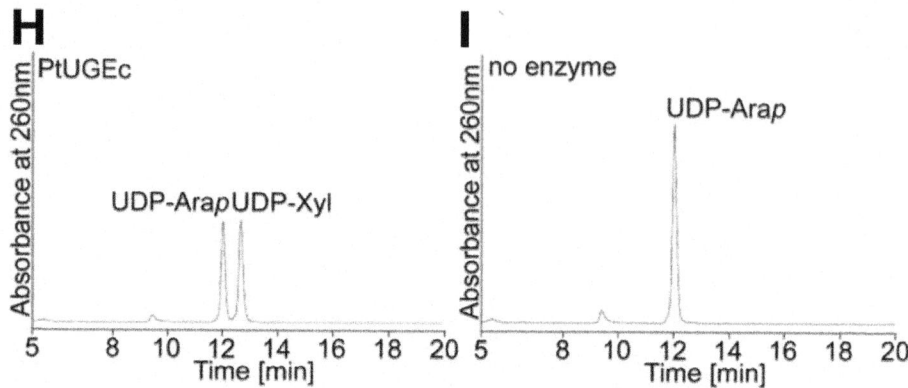

Figure 2: UDP-glucose 4-epimerase activity of recombinant *Pt* UGEc protein. (A) SDS-PAGE analysis of different fractions from the purification of His-*Pt*UGEc. **(B-I)** High-performance liquid chromatography (HPLC) chromatograms of nucleotide sugars incubated for 30 min with (left panels) or without (right panels) recombinant *Pt*UGEc protein.

All UGE Constructs Complement the *UGE4* ^{rhd1} Phenotype

In order to verify the functionality of UGE constructs, *Pt*UGEc and the Arabidopsis *At*UGE2, *At*UGE4, and *At*UGE5 proteins were overexpressed under the control of the constitutive cauliflower mosaic virus 35S promoter in the Arabidopsis *uge4*mutant background. Loss of function of *UGE4* results in a reduced root elongation rate and swelling of root epidermal cells probably as a result of defective cell wall matrix carbohydrate biosynthesis [12,18]. Thus, a simple visual screen can confirm complementation of the wild-type phenotype and thereby not only expression but also functionality of the UGE proteins. The four different UGE constructs all suppressed the root epidermal swelling and the reduced root length confirming previous UGE complementation results published by Barber et al. [9] and demonstrating that the poplar *Pt*UGEc is functional *in planta* (Figure 3). The slight root length decrease in our UGE overexpressor plants as compared to wild type indicates however, that complementation is not complete. Expression of *At*UGE2 and *At*UGE5 resulted in almost complete complementation.

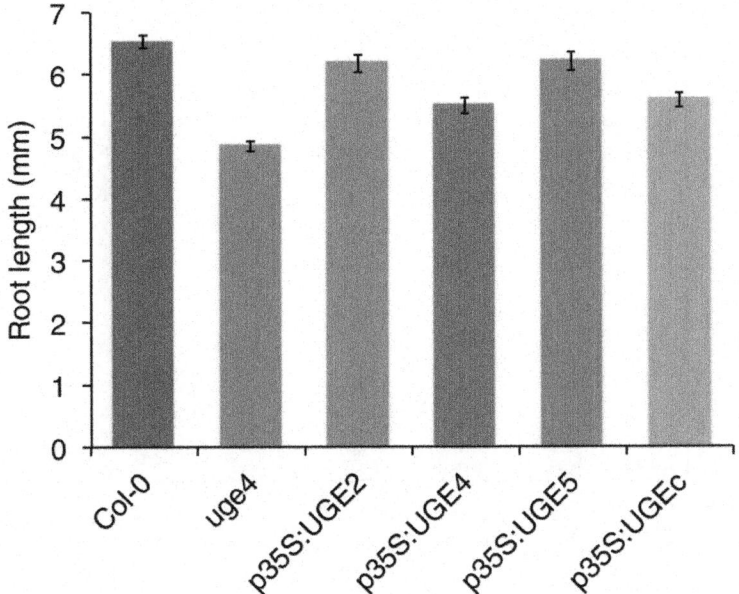

Figure 3: Overexpression of *At* UGE2, *At* UGE4, *At* UGE5, or *Pt* UGEc rescues the *uge4* mutant phenotype. The UGE expressing plants rescue the *uge4* phenotype to various extent. Average root length of Col-0, *uge4* and transgenic seedlings in *uge4* background was determined. Data show mean ± SD (n = 50).

Plants Overexpressing UGE Show no Increase in Cell Wall Bound Galactose in Leaf or Stem Cell Walls

All *At*UGE2, *At*UGE4, *At*UGE5, or *Pt*UGEc overexpressing Arabidopsis plants had a growth phenotype similar to wild-type Col-0 and the empty vector control plants. The monosaccharide composition of non-cellulosic polysaccharides from leaves and stems was analyzed by high-performance anion exchange chromatography with pulsed amperometric detection (HPAEC-PAD). Two to four independent transgenic lines were analyzed for each construct in the T2 generation. Data for one representative line for each construct is shown in Figure 4. The transgenic Arabidopsis UGE lines had no significant changes in sugar composition compared to the empty vector control plants in leaf cell walls (Figure 4A) or in stem cell walls (Figure 4B).

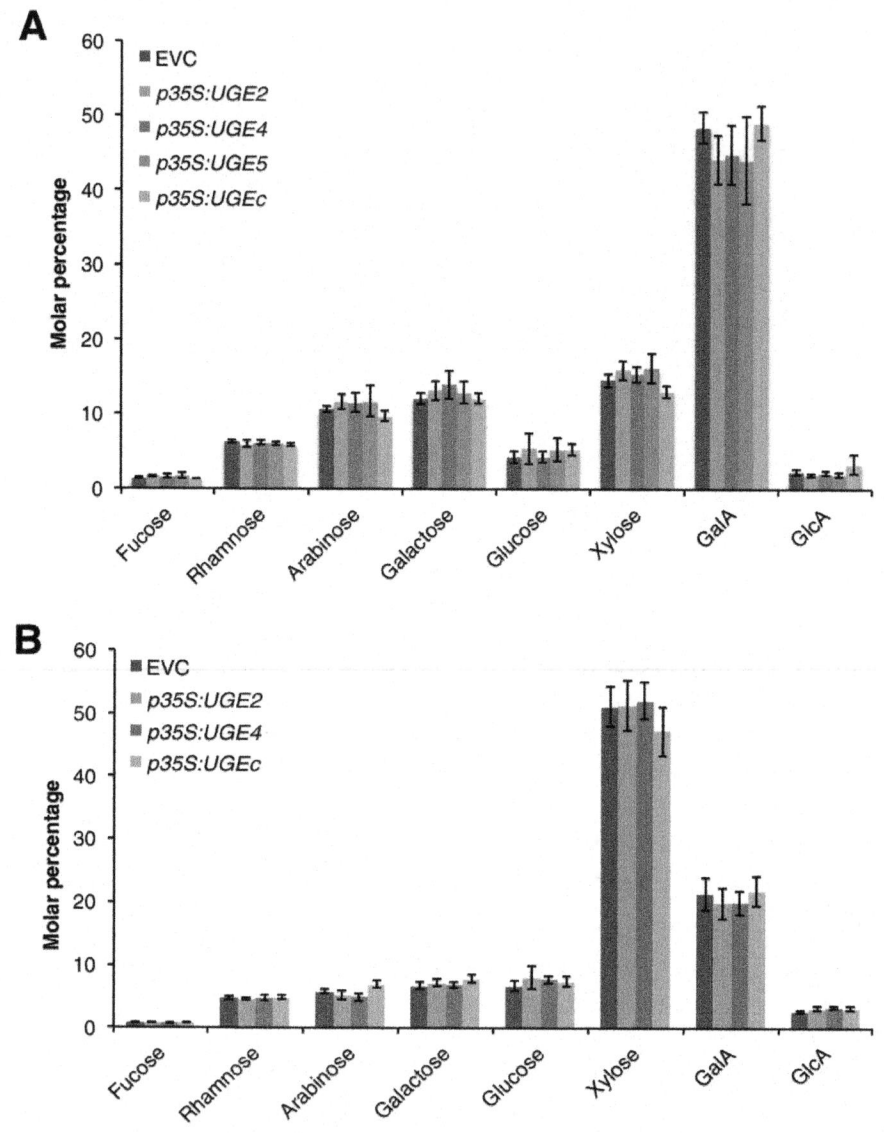

Figure 4: Cell wall monosaccharide composition analysis of leaf (A) and stem (B) material from UGE overexpression plants. Monosaccharide composition analysis of leaf (A) and stem (B) cell walls of plants constitutively overexpressing *At*UGE2, *At*UGE4,*At*UGE5 did not show any significant differences compared to the empty vector control (EVC) ($p > 0.05$, pairwise t-test with Holm-Bonferroni correction for family-wise error rate). Monosaccharide levels are shown as molar percentage ± SD (n = 6). GalA, α-d-galacturonic acid; GlcA, α-d-glucuronic acid.

Recently, we have shown that overexpression of GalS1 *in planta* can lead to a significant 40% increase in total cell wall galactose in leaves [5]. Thus, the UDP-Gal substrate does not seem to be limiting for galactose incorporation into the cell wall in leaves and therefore an increase in interconverting UGE enzymes is not necessarily expected to lead to an increase in cell wall galactan.

Cell Wall Bound Galactose Levels are Increased in Stem Cell Walls of Co-Overexpressers

Since overexpression of *At*UGE2, *At*UGE4, *At*UGE5, or *Pt*UGEc alone did not result in a significant increase of total galactose in stems, we designed a gene stacking approach by co-expressing *At*UGE2 together with the galactan synthase GalS1. AtUGE2 was chosen because it showed efficient complementation of the *uge4* root phenotype (Figure 3). Although we had initially preferred to use a poplar UGE, PtUGEc was not a good choice because of its bispecificity and incomplete ability to complement *uge4*. We designed two different constructs for *At*UGE2 expression. One under the control of the constitutive 35S promoter (*p35S:UGE2*) and a second fiber-specific construct (*pIRX5:NST1-UGE2*). The fiber-specific construct is under control of the *IRX5* promoter and in addition expresses the transcription factor NST1 leading to a positive-artificial feedback loop and increased wall thickness in fiber cells, as previously reported [2]. In this construct *UGE2* is expressed from the same *pIRX5* promoter, separated from *NST1* with the 2A sequence from foot-and-mouth disease virus allowing coordinate expression of multiple proteins [19]. The stem cell wall composition was analyzed in the T2 generation for three independent lines for each construct. For each construct there was no difference between the independent lines and one line was selected for confirmation of the results in the T3 generation (Figure 5). Expression of *p35S:UGE2* and *pIRX5:NST1-UGE2*, respectively, in the background of plants constitutively overexpressing GalS1 (*p35S:GalS1*) led to significantly increased galactose levels (p≤0.01) as shown by analysis of the monosaccharide composition of cell walls from stems (Figure 5). While *p35S:UGE2/p35S:GalS1* plants showed a galactose increase of more than 80%, the *pIRX5:NST1-UGE2/p35S:GalS1* galactose levels were only increased by 44% as compared to empty vector control (EVC) plants (Figure 5). Plants expressing *p35S:GalS1* alone had only a slight increase in stem wall bound galactose, and no galactose increase was observed when the *pIRX5:NST1-UGE2* construct was incorporated in the wild-type background. The apparent increase in xylose in plants containing the *pIRX5:NST1-UGE2* construct for overexpression of the feedback-loop construct with NST1 could be expected in plants with increased fiber cell wall density and more xylan. However, the xylose content in these plants is not significantly different from the control (p>0.05 even without Holm-Bonferroni correction). These results show that GalS1 is limiting for galactan synthesis in both leaves and stems while UGEs are not. However, when GalS1 is overexpressed, *At*UGE2 or other UGEs seem to become limiting for how much galactan can be accumulated in cell walls of Arabidopsis plant stems. The 80% increase in cell wall galactose may not be the limit for what can be achieved. UDP-Gal formed by UGE2 must be transported into the Golgi lumen to be used by GalS1, and it is possible that the transport becomes limiting when both UGE2 and GalS1 are overexpressed. Recently, a UDP-Gal transporter URGT1 has been characterized, the overexpression of which results in increased cell wall galactan in leaves [20]. We are currently investigating the effect of overexpressing the transporter together with UGE2 and GalS1.

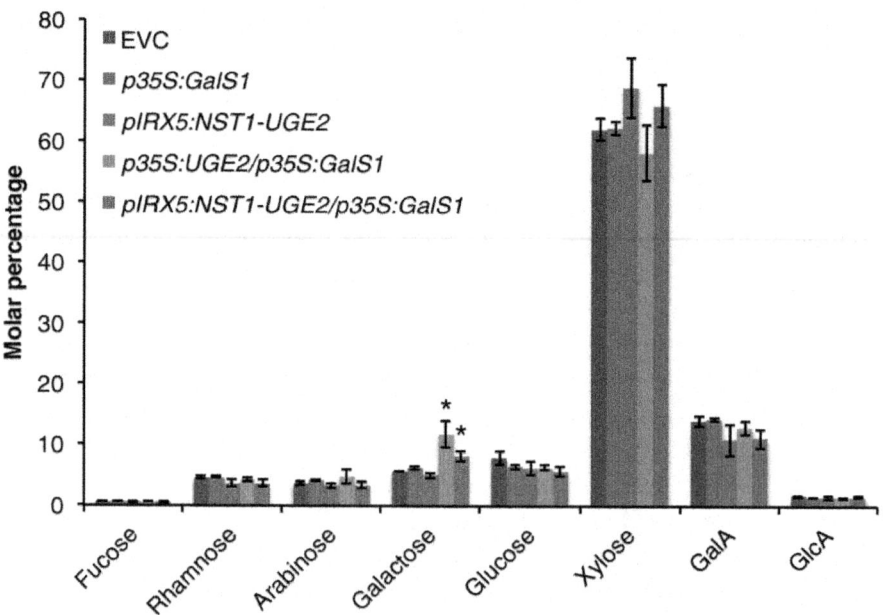

Figure 5: Cell wall monosaccharide composition analysis of stems from plants co-overexpressing *At* UGE2 and GalS1. The monosaccharide composition of stems from plants in the T3 generation expressing either *p35S:GalS1* or *pIRX5:NST1-UGE2* or co-overexpressing either *pIRX5:NST1-UGE2/p35S:GalS1* or *p35S:UGE2/p35S:GalS1* was determined. Monosaccharide levels are shown as molar percentage±SD (n =5). Significantly increased levels of galactose (p≤0.01, indicated with asterisks) were found with the two co-expressing constructs *pIRX5:NST1-UGE2/p35S:GalS1* and *p35S:UGE2/p35S:GalS1*, while no other sugars were different from empty vector control plants (EVC) in any of the transgenic plants (p>0.05). Analysis in the T2 generation of three independent lines for each construct showed the same

results. The data were analyzed by pairwise t-test with Holm-Bonferroni correction for family-wise error rate. GalA, α-d-galacturonic acid; GlcA, α-d-glucuronic acid.

The increase in galactose of *p35S:UGE2/p35S:GalS1* and *pIRX5:NST1-UGE2/p35S:GalS1* co-overexpressor plants was further investigated by immunofluorescence microscopy of stem sections (Figure 6). Although immunofluorescence microscopy is not easily quantified, the detection of the LM5 galactan epitope was strongly increased specifically in the secondary cell walls of top and bottom stem sections of *p35S:UGE2/p35S:GalS1* and *pIRX5:NST1-UGE2/p35S:GalS1* co-overexpressors, as compared to overexpressor lines of *p35S:GalS1* or *pIRX5:NST1-UGE2* alone, or the empty vector control. The fiber-specific pIRX5 constructs in addition resulted in highly thickened cell walls as an effect of overexpressing NST1 under the IRX5 promoter and as visualized by lignin autofluorescence using a confocal microscope (Figure 6). On the one hand this increase in biomass density is highly desirable to improve the cost-effectiveness of lignocellulosic bioenergy production. On the other hand however, the increase in xylan rather results in more recalcitrance and is counteracting the increase in the C6/C5 sugar ratio, which is obtained due to increased galactan deposition. Therefore, future implementation of galactan overexpression could be improved by simultaneously downregulating xylan deposition in fiber cells. This could be achieved by using mutants deficient in xylan that have been complemented by reintroducing xylan biosynthesis specifically into the xylem vessels in order to restore wild type-like growth of the plants, as recently described [3]. Conceivably, the lignin content in fiber cells also needs to be decreased.

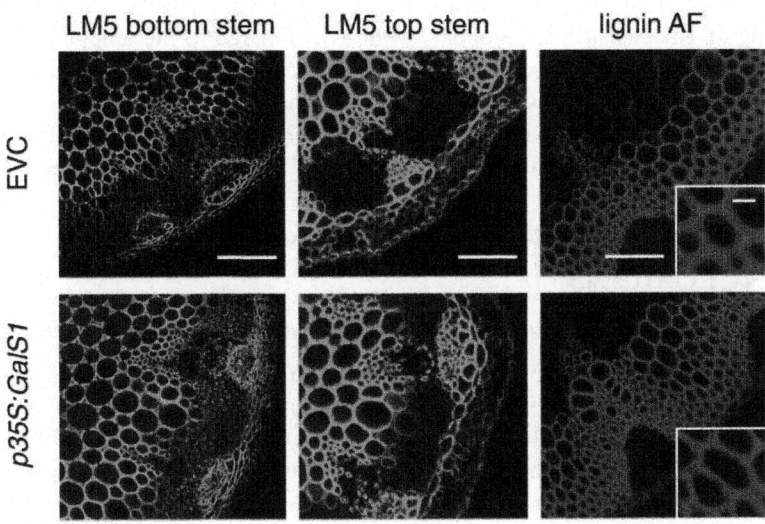

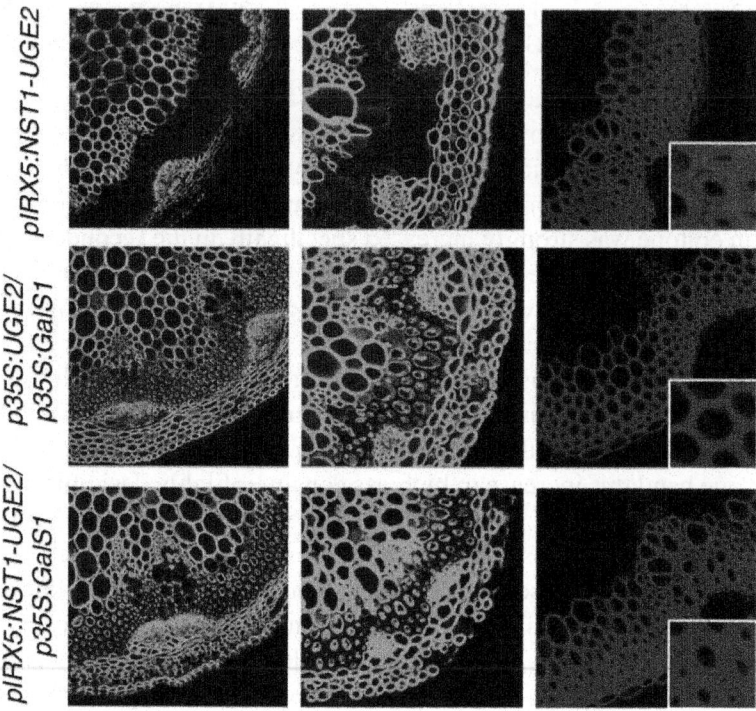

Figure 6: Galactan and lignin detection in stem sections. β-1,4-galactan was detected by immunofluorescence microscopy using the LM5 antibody. Stem sections from the top and bottom of inflorescence stems were analyzed. Plants co-expressing either *pIRX5:NST1-UGE2/ p35S:GalS1* or *p35S:UGE2/p35S:GalS1* show a very strong labeling of galactan in the secondary wall as compared to the empty vector control or plants expressing only *p35S:GalS1* or *pIRX5:NST1-UGE2*. Visualization of lignin autofluorescence using a confocal microscope under UV light shows the increase in fiber cell wall density with constructs using the *pIRX5* promoter. Bars are 100 μm for bottom stem and lignin autofluorescence pictures, 50 μm for top stems and 10 μm for lignin close ups.

CONCLUSIONS

Rescuing of the *uge4* root phenotype proved functionality of the constitutively expressed poplar *p35S:UGEc* and the three Arabidopsis *p35S:UGE2*, *p35S:UGE4* and *p35S:UGE5* constructs. Overexpression of any of the four UGE proteins alone did not increase the total cell wall galactose content in Arabidopsis leaves or stems. However, our gene stacking approach, combining overexpression of *AtUGE2* and *GalS1*, clearly showed that it is possible to engineer plants with an even higher galactose content than *GalS1* overexpressing plants by combining the overexpression of multiple genes into one plant. The Arabidopsis plants obtained have a more than 80% increase in stem galactose levels as compared to wild-type or empty vector control plants. Importantly, these transgenic plants exhibit no impairment of growth and development. Our study shows the promise of the gene stacking approach for engineering plants with improved properties for biofuel applications.

METHODS

Phylogenetic Analysis

Phylogenetic analyses were conducted in MEGA6 [21]. The bootstrap consensus tree inferred from 1000 replicates was taken to represent the evolutionary history of the taxa analyzed using the Neighbor-Joining method. The percentages of replicate trees in which the associated taxa clustered together in the bootstrap test are shown next to the branches retaining only groups with a frequency ≥60%. Species and Genbank accession numbers are: An, *Aspergillus niger*[XP_001401007]; At, *Arabidopsis thaliana* [AEE28928, UGE1; AEE84827, UGE2; AEE34065, UGE3; AEE34241, UGE4; AEE82951, UGE5]; Ao, *Aspergillus oryzae* [XP_001827449]; Bc, *Bacillus cereus* [ZP_01180393]; Bs, *Bacillus subtilis* [P55180]; Cl, *Canis lupus* [XP_544499]; Cr, *Chlamydomonas reinhardtii* [XP_001698706]; Ct, *Cyamosis tetragonoloba* [O65781]; Dd,*Dictyostelium discoideum* [XP_643834]; Dp, *Drosohila pseudoobscura* [XP_001352806]; Dr, *Danio rerio* [NP_001035389]; Gg,*Gallus gallus* [XP_417833]; Hs, *Homo sapiens* [Q14376]; Hv, *Hordeum vulgare* [AAX49504, UGE1; AAX49505, UGE2; AAX49503, UGE3]; Mt, *Medicago truncatula* [ACJ85116, UGE1; ACJ84690, UGE2]; Mxd, *Malus x domestica* [BAF51705]; Ol,*Ostreococcus lucimarinus* [XP_001419325]; Os, *Oryza sativa* [BAF18426, UGE1; BAF23582, UGE2; BAF25641, UGE3; BAF24783, UGE4]; Ot, *Ostreococcus tauri* [CAL54894]; Pp, *Physcomitrella patens* [XP_001768301, UGE1; XP_001777464, UGE2; XP_001775163, UGE3; XP_001751529, UGE4; XP_001771084, UGE5]; Ps, *Pisum sativum* [AB381885]; Pt, *Populus trichocarpa* [XP_002304478, UGEa: XP_002303653, UGEb; XP_002299469, UGEc;] Pv, *Paspalum vaginatum* [BAE92559]; Sg,*Streptococcus gordonii* [AAN64559]; St, *Solanum tuberosum* [AAP42567, UGE45; AAP97493, UGE51]; Sp, *Streptococcus pneumonia* [ZP_01825231]; Tc, *Tribolium castaneum* [XP_968616]; Xl, *Xenopus laevis* [NP_001080902]; Zm, *Zea mays*[AAP68981].

Plant Material and Vectors

All *Arabidopsis thaliana* (L.) Heynh. wild-type and mutant plant lines used were of ecotype Columbia-0 (Col-0). The *At*UGE4 mutant *uge4/rhd1-1* (At1g64440, CS2257) was obtained from the Arabidopsis Biological Resource Center (ABRC, http://www.arabidopsis.org). Plants overexpressing YFP-GalS1 (*p35S:GalS1*) have been previously described [5]. *Populus trichocarpa* Nisqually-1 leaf tissue was kindly donated by Dr. Lee Gunter (Oak Ridge National Laboratory). Entry vectors containing *AtUGE2*, *AtUGE4* and *AtUGE5* cDNA (At4g23920, At1g64440, At4g10960) and the Gateway-compatible plant transformation vectors pMDC32 and pMDC43 were obtained from ABRC. The destination vector pTKan-pIRX5-GWR3R2 was made as described [1] except that it had attR2 and attR3 recombination sites. Gene constructs used to generate transgenic plants are shown in Figure 7.

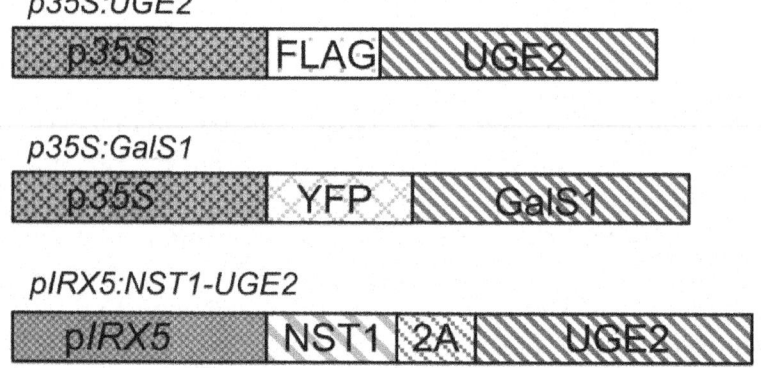

Figure 7: Schematic presentation of the gene constructs used. The constructs used to generate the transgenic plants used for Figures 4, 5and 6 incorporated either the *p35S* or the Arabidopsis *pIRX5* promoter. The UGE2 and GalS1 proteins were expressed with tags as indicated in the two upper constructs. The *pIRX5:NST1-UGE2* construct had the two open reading frames separated by a 2A sequence. For additional details see Methods and Table 2.

Generation of Vectors and Transformation of Plants

Flag-tagged UGE constructs were generated by PCR using Phusion Polymerase (Thermo Scientific) with primers fU2-F, fU2-R (*At*UGE2); fU4-F, fU4-R (*At*UGE4); fU5-F, fU5-R (*At*UGE5); and fUc-F, fUc-R (*Pt*UGEc) (Table 1), and cloned into vectors pMDC32 and pMDC43 using Gateway technology (Life Technologies). The fiber expressed *At*UGE2 construct was generated in 2 steps. The sequence encoding for the NST1-2A-AtUGE2 fusion protein was first assembled in a gateway pDON-P3P2 vector (Life Sciences) to generate pDON-NST1-2A-AtUGE2-L3L2 plasmid. pDON-NST1-2A-AtUGE2-L3L2 vector was built using In-Fusion HD Cloning System (Clontech) to assemble PCR products of *At*UGE2 and pDON-L3L2 vector containing the encoding sequence of NST1-2A (synthesized by GenScript, Piscatway, NJ) such that the encoding sequence of *At*UGE2 was inserted in frame with the 3′end of NST1-2A and at 5′end of the attL2 sequences. *At*UGE2 and pDON-NST1-2A-L3L2 PCR products were generated using Phusion Polymerase with ntU2-F/ntU2-R primer pair and F-pDON-attL2/R-pDON-NST1-2A primer pair respectively. The NST1-2A-AtUGE2 gene fusion was then transferred from into pTKan-pIRX5-GWR3R2 by LR recombination (Life technologies) to express NST1-2A-AtUGE2 under the control of the fiber-specific pIRX5 promoter. *Pt*UGEc cDNA was generated from RNA isolated from Poplar leaf tissue. RNA was extracted using the RNeasy Plant Minikit (Qiagen), treated with DNase I (Sigma) and cDNA was generated using iScript™ Reverse Transcription Supermix (Biorad). The cloned UGE open reading frames were confirmed by sequencing to be identical to the published sequences (see Figure 1 for accession numbers). For stable transformation of *A. thaliana* wild-type Columbia-0, *uge4/rhd1-1* mutants or GalS1 overexpressing lines, constructs were transformed into *A. tumefaciens* strain C58-1 pGV3850 and plants were transformed using the floral dip method [22] (Table 2). Transformants were selected on MS medium containing hygromycin and transferred to soil. Plants confirmed to express the transgene were propagated and further characterized in the T2 and T3 generations.

Table 1: Primer sequences used in this work

Primer name	Primer sequence
fU2-F	5′-CACCATGGATTACAAGGATGACGATGACAAGGC-GAAGAGTG-3′
fU2-R	5′-TTATGAAGAGGAGCCATTGGAGGAGGA-3′
fU4-F	5′-CACCATGGATTACAAGGATGACGAT-GACAAGGTTGGGAATATT-3′
fU4-R	5′-TTATGTTGAGTTTGGTGAAGAACCGTAACC-3′
fU5-F	5′-CACCATGGATTACAAGGATGACGATGACAAGATGGC-TAGAAACGT-3′
fU5-R	5′-TTTAATGAGAGTTGTCTTCAGAAGAGG-3′

fUc-F	5′-CACCATGGATTACAAGGATGACGATGACAAGATG-GCCTATAATATTC TGGTTACCG-3′
fUc-R	5′-TCAGTTTGTGCCGTCAGGAGATC-3′
ntU2-F	5′-TCTAATCCAGGACCTATGGCGAAGAGTGTTTTGGT-TAC-3′
ntU2-R	5′-CAAGAAAGCTGGGTCTGAAGAGGAGCCATTGGAG-GAG-3′
F-pDON-attL2	5′-GACCCAGCTTTCTTGTACAAAGT-3′
R-pDON-NST1-2A	5′-AGGTCCTGGATTAGACTCAACG-3′

Table 2: Constructs used for plant expression

Name	Construct	Expression vector	Plant background
p35S:UGE2	*p35S:FLAG-AtUGE2*	pMDC32	Col-0, *uge4,* GalS1-OE
p35S:UGE4	*p35S:FLAG-AtUGE4*	pMDC32	Col-0, *uge4*
p35S:UGE5	*p35S:FLAG-AtUGE5*	pMDC32	Col-0, *uge4*
p35S:UGEc	*p35S:FLAG-PtUGEc*	pMDC32	Col-0, *uge4*
p35S:GalS1	*p35S:YFP-GalS1*	pEarleyGate104	Col-0
pIRX5:NST1-UGE2	*pIRX5:NST1-2A-AtUGE2*	pTKan-pIRX5-GWR3R2	Col-0, GalS1-OE
EVC	*35S:pvu2* (non coding)	pMDC32	Col-0, *uge4*

Plant Growth Conditions and Measurements

All plants were grown in a 14–16 h photoperiod at 120 µmol m^{-2} s^{-1} photon flux density. Plants for root length measurements were grown vertically on MS plates and scanned at 6 days after germination. The root length of about 50 individual seedlings was measured using ImageJ [23].

Monosaccharide Composition Analysis of Cell Walls

Bottom sections (5 cm length of stems) or whole leaves from 5-week-old plants were ground in liquid nitrogen and alcohol-insoluble residue (AIR) was prepared and enzymatically destarched as described [24]. Destarched AIR samples (1 mg) were hydrolyzed with 2 M triflouroacetic acid (TFA) and monosaccharide composition was determined by HPAEC-PAD as described [25,26].

Expression and Purification of His-*Pt*UGEc

*Pt*UGEc was introduced into pDEST17 expression vector (Invitrogen) containing an N-terminal 6xHis-tag and an IPTG inducible promoter. Gene expression in BL21 Star cells (Invitrogen) was induced by adding IPTG to a final concentration of 1 mM, and cultures were grown at 18°C overnight. *Pt*UGEc was purified from the supernatant of lysed cell pellets using HIS-Select Nickel Affinity Gel purification (Sigma-Aldrich). Lysis of bacterial cells took place using CelLytic B 2X containing 0.2 mg/ml lysozyme, 50 U/ml benzonase (all Sigma-Aldrich) and proteinase inhibitor cocktail (Roche). HIS-tagged *Pt*UGEc was desalted with PD-10 desalting columns (Amersham Biosciences). Samples of 30 µg His-*Pt*UGEc protein were separated on a Novex 8-16% Tris Glycine Gradient gel (Invitrogen) and stained with Coomassie Brilliant Blue.

His-*Pt*UGEc Activity Assay

5 µg of purified His-*Pt*UGEc was mixed with 1 mM UDP-glucose in 50 mM Tris–HCl (pH 8.0), 1 mM DTT for 30 min at 30°C. Reactions were terminated by incubating for 10 min at 90°C and filtration through 0.45 µm filters (Millipore) prior to quantification. To test more nucleotide sugar substrates, the same reaction mixture as described above was used, substituting UDP-Glc with UDP-Gal, UDP-Xyl or UDP-Ara. Negative controls of purified protein boiled for 10 min at 90°C were used for all reactions. Separation and quantification of UDP-sugars in terminated reactions was performed by HPAEC analysis using a Dionex Ultimate 3000 system (Thermo Fisher) with detection at 262 nm. Samples were separated on a CarboPac PA20 column (Thermo Fisher) and eluted with an ammonium formate gradient according to Rautengarten *et al.* [27]. Standard solutions containing UDP-Glc, UDP-Gal (both Sigma Aldrich), UDP-Ara and UDP-Xyl (both Carbosource Service) were run as references.

Immunofluorence Microscopy

The top and base 3 cm of main stems from 6-week-old plants (3 stems/line) were harvested and fixed overnight at 4°C in fixative solution (4% paraformaldehyde in 50 mM piperazine-N-N'-bis(2-ethanesulphonic acid), 5 mM EGTA, pH 6.9). Fixed stem sections were embedded in 7% agarose and sectioned using a Leica VT1000S vibratome. Stem sections were labeled with monoclonal LM5 rabbit antibody (PlantProbes, Leeds, UK), which recognizes 1,4-linked β-galactan [28]. The labeling was performed according to Verhertbruggen et al. [29]. Sections were mounted on slides and pictures were taken using a LSM 710 confocal Microscope (Carl Zeiss). Lignin autofluorescence was monitored using a 405 nm Diode laser. Images were acquired with the Zen software package (Carl Zeiss) and analyzed with ImageJ [23].

ABBREVIATIONS

GalA: α-D-galacturonic acid

GlcA: α-D-glucuronic acid

GalS1: GALACTAN SYNTHASE1

GT: Glycosyltransferase

NST: NAC secondary wall thickening promoting factor

NAC: NAM ATAF1/2 and CUC2

UGE: UDP-glucose 4-epimerase

UDP: Uridine diphosphate

ACKNOWLEDGEMENTS

This work conducted by the Joint BioEnergy Institute was supported by the Office of Science, Office of Biological and Environmental Research, through contract DE-AC02-05CH11231 between Lawrence Berkeley National Laboratory and the U. S. Department of Energy. RS was supported by a research fellowship of the German Research Foundation (Deutsche Forschungsgemeinschaft). Funding for BE was provided by the Danish Council for Strategic Research. We thank Ms. Sherry Chan for assistance with plant growth and Dr. Lee Gunter for providing poplar material for cDNA purification.

COMPETING INTERESTS

Parts of the strategy described in this paper have been included in a patent application with HVS as inventor. The remaining authors declare that they have no competing interests.

AUTHORS' CONTRIBUTIONS

HVS designed and coordinated the study; VMG, RS, AJML, BE, DL and CR conducted the experiments; VMG, RS, BE, CR and HVS conducted data analysis; VMG, RS and HVS wrote the manuscript, and all authors read and approved the final manuscript.

REFERENCES

1. Eudes A, George A, Mukerjee P, Kim JS, Pollet B, Benke PI, Yang F, Mitra P, Sun L, Cetinkol OP, Chabout S, Mouille G, Soubigou-Taconnat L, Balzergue S, Singh S, Holmes BM, Mukhopadhyay A, Keasling JD, Simmons BA, Lapierre C, Ralph J, Loque D: Biosynthesis and incorporation of side-chain-truncated lignin monomers to reduce lignin polymerization and enhance saccharification. *Plant Biotechnol J* 2012, 10:609–620.

2. Yang F, Mitra P, Zhang L, Prak L, Verhertbruggen Y, Kim JS, Sun L, Zheng K, Tang K, Auer M, Scheller HV, Loque D: Engineering secondary cell wall deposition in plants. *Plant Biotechnol J* 2013, 11:325–335.

3. Petersen PD, Lau J, Ebert B, Yang F, Verhertbruggen Y, Kim JS, Varanasi P, Suttangkakul A, Auer M, Loque D, Scheller HV: Engineering of plants with improved properties as biofuels feedstocks by vessel-specific complementation of xylan biosynthesis mutants.*Biotechnol Biofuels* 2012, 5:84.

4. Lairson LL, Henrissat B, Davies GJ, Withers SG: Glycosyltransferases: structures, functions, and mechanisms. *Annu Rev Biochem*2008, 77:521–555.

5. Liwanag AJ, Ebert B, Verhertbruggen Y, Rennie EA, Rautengarten C, Oikawa A, Andersen MC, Clausen MH, Scheller HV: Pectin biosynthesis: GALS1 in Arabidopsis thaliana is a beta-1,4-galactan beta-1,4-galactosyltransferase. *Plant Cell* 2012, 24:5024–5036.

6. Bar-Peled M, O'Neill MA: Plant nucleotide sugar formation, interconversion, and salvage by sugar recycling. *Annu Rev Plant Biol*2011, 62:127–155.

7. Reboul R, Geserick C, Pabst M, Frey B, Wittmann D, Lutz-Meindl U, Leonard R, Tenhaken R: Down-regulation of UDP-glucuronic acid biosynthesis leads to swollen plant cell walls and severe developmental defects associated with changes in pectic polysaccharides.*J Biol Chem* 2011, 286:39982–39992.

8. Burget EG, Verma R, Molhoj M, Reiter WD: The biosynthesis of

L-arabinose in plants: molecular cloning and characterization of a Golgi-localized UDP-D-xylose 4-epimerase encoded by the MUR4 gene of Arabidopsis. *Plant Cell* 2003, 15:523–531.

9. Barber C, Rosti J, Rawat A, Findlay K, Roberts K, Seifert GJ: Distinct properties of the five UDP-D-glucose/UDP-D-galactose 4-epimerase isoforms of Arabidopsis thaliana. *J Biol Chem* 2006, 281:17276–17285.

10. Rösti J, Barton CJ, Albrecht S, Dupree P, Pauly M, Findlay K, Roberts K, Seifert GJ: UDP-glucose 4-epimerase isoforms UGE2 and UGE4 cooperate in providing UDP-galactose for cell wall biosynthesis and growth of *Arabidopsis thaliana*. *Plant Cell* 2007, 19:1565–1579.

11. Kotake T, Takata R, Verma R, Takaba M, Yamaguchi D, Orita T, Kaneko S, Matsuoka K, Koyama T, Reiter WD, Tsumuraya Y: Bifunctional cytosolic UDP-glucose 4-epimerases catalyse the interconversion between UDP-D-xylose and UDP-L-arabinose in plants. *Biochem J* 2009, 424:169–177.

12. Seifert GJ, Barber C, Wells B, Dolan L, Roberts K: Galactose biosynthesis in Arabidopsis: genetic evidence for substrate channeling from UDP-D-galactose into cell wall polymers. *Curr Biol* 2002, 12:1840–1845.

13. Nguema-Ona E, Andeme-Onzighi C, Aboughe-Angone S, Bardor M, Ishii T, Lerouge P, Driouich A: The reb1-1 mutation of Arabidopsis. Effect on the structure and localization of galactose-containing cell wall polysaccharides. *Plant Physiol* 2006, 140:1406–1417.

14. Oomen RJFJ, Dao-Thi B, Tzitzikas EN, Bakx EJ, Schols HA, Visser RGF, Vincken JP: Overexpression of two different potato UDP-Glc 4-epimerases can increase the galactose content of potato tuber cell walls. *Plant Science* 2004, 166:1097–1104.

15. Taylor NG: Cellulose biosynthesis and deposition in higher plants. *New Phytol* 2008, 178:239–252.

16. Mitsuda N, Iwase A, Yamamoto H, Yoshida M, Seki M, Shinozaki K, Ohme-Takagi M: NAC transcription factors, NST1 and NST3, are key regulators of the formation of secondary walls in woody tissues of Arabidopsis. *Plant Cell* 2007, 19:270–280.

17. Zhang Q, Hrmova M, Shirley NJ, Lahnstein J, Fincher GB: Gene expression patterns and catalytic properties of UDP-D-glucose 4-epimerases from barley (Hordeum vulgare L.). *Biochem J* 2006, 394:115–124.

18. Andeme-Onzighi C, Sivaguru M, Judy-March J, Baskin TI, Driouich A: The reb1-1 mutation of Arabidopsis alters the morphology of trichoblasts, the expression of arabinogalactan-proteins and the organization of cortical microtubules. *Planta* 2002, 215:949–958.

19. Halpin C, Cooke SE, Barakate A, Amrani A, Ryan MD: Self-processing 2A-polyproteins – a system for co-ordinate expression of multiple proteins in transgenic plants. *Plant J* 1999, 17:453–459.

20. Rautengarten C, Ebert B, Moreno I, Temple H, Herter T, Link B, Doñas D, Moreno A, Saéz-Aguayo S, Blanco MF, Mortimer J, Schultink A, Reiter WD, Dupree P, Pauly M, Heazlewood JL, Scheller HV, Orellana A: The Golgi UDP-Rhamnose / UDP-Galactose transporter family in Arabidopsis. *Proc Natl Acad Sci USA* 2014, 111:11563–11568.

21. Tamura K, Stecher G, Peterson D, Filipski A, Kumar S: MEGA6: Molecular Evolutionary Genetics Analysis version 6.0. *Mol Biol Evol* 2013, 30:2725–2729.

22. Clough SJ, Bent AF: Floral dip: a simplified method for Agrobacterium-mediated transformation of Arabidopsis thaliana. *Plant J* 1998, 16:735–743.

23. Abramoff MD, Magalhaes PJ, Ram SJ: Image processing with ImageJ. *Biophotonics Int* 2004, 11:36–42.

24. Harholt J, Jensen JK, Sorensen SO, Orfila C, Pauly M, Scheller HV: Arabinan deficient 1 is a putative arabinosyltransferase involved in biosynthesis of pectic arabinan in Arabidopsis. *Plant Physiol* 2006, 140:49–58.

25. Øbro J, Harholt J, Scheller HV, Orfila C: Rhamnogalacturonan I in Solanum tuberosum tubers contains complex arabinogalactan structures. *Phytochemistry* 2004, 65:1429–1438.

26. Yin L, Verhertbruggen Y, Oikawa A, Manisseri C, Knierim B, Prak L, Jensen JK, Knox JP, Auer M, Willats WG, Scheller HV: The cooperative activities of CSLD2, CSLD3, and CSLD5 are required for normal Arabidopsis development. *Mol Plant* 2011, 4:1024–1037.

27. Rautengarten C, Ebert B, Herter T, Petzold CJ, Ishii T, Mukhopadhyay A, Usadel B, Scheller HV: The interconversion of UDP-arabinopyranose and UDP-arabinofuranose is indispensable for plant development in Arabidopsis. *Plant Cell* 2011, 4:1373–90.

28. Jones L, Seymour GB, Knox JP: Localization of pectic galactan in tomato cell walls using a monoclonal antibody specific to (1->4)-beta-D-galactan. *Plant Physiol* 1997, 113:1405–1412.

29. Verhertbruggen Y, Marcus SE, Haeger A, Verhoef R, Schols HA, McCleary BV, McKee L, Gilbert HJ, Knox JP: Developmental complexity of arabinan polysaccharides and their processing in plant cell walls. *Plant J* 2009, 59:413–425.

CITATION

CHAPTER 1

Boyle et al.: A BioBrick compatible strategy for genetic modification of plants. Journal of Biological Engineering 2012 6:8. doi:10.1186/1754-1611-6-8.

CHAPTER 2

Laura C Roden, Berthold Göttgens and Effie S Mutasa-Göttgens, "Protocol: Precision engineering of plant gene loci by homologous recombination cloning in Escherichia coli," Plant Methods20051:6, DOI: 10.1186/1746-4811-1-6.

CHAPTER 3

Singh, H. and Singh, B. (2014) Genetic Engineering of Field, Industrial and Pharmaceutical Crops. American Journal of Plant Sciences, 5, 3974-3993. doi: 10.4236/ajps.2014.526416.

CHAPTER 4

M. K. Mishra and A. Slater, "Recent Advances in the Genetic Transformation of Coffee," Biotechnology Research International, vol. 2012, Article ID 580857, 17 pages, 2012. doi:10.1155/2012/580857.

CHAPTER 5

Wang H, Wang H, Shao H and Tang X (2016) Recent Advances in Utilizing Transcription Factors to Improve Plant Abiotic Stress Tolerance by Transgenic Technology. *Front. Plant Sci.* 7:67. doi: 10.3389/fpls.2016.00067.

CHAPTER 6

A. Karthikeyan, R. Valarmathi, S. Nandini and M.R. Nandhakumar, 2012. Genetically Modified Crops: Insect Resistance. Biotechnology, 11: 119-126. DOI: 10.3923/biotech.2012.119.126.

CHAPTER 7

Asis Datta, "Genetic engineering for improving quality and productivity of crops," Agriculture & Food Security20132:15, DOI: 10.1186/2048-7010-2-15.

CHAPTER 8

Afonso-Grunz F, Molina C, Hoffmeier K, Rycak L, Kudapa H, Varshney RK, Drevon J-J, Winter P and Kahl G (2014) Genome-based analysis of the transcriptome from mature chickpea root nodules. *Front. Plant Sci.* 5:325. doi: 10.3389/fpls.2014.00325.

CHAPTER 9

Singh VK, Jain M and Garg R (2015) Genome-wide analysis and expression profiling suggest diverse roles of *GH3* genes during development and abiotic stress responses in legumes. *Front. Plant Sci.* 5:789. doi: 10.3389/fpls.2014.00789.

CHAPTER 10

Keisuke Matsui, Junichi Togami, John G. Mason, Stephen F. Chandler, and Yoshikazu Tanaka, "Enhancement of Phosphate Absorption by Garden Plants by Genetic Engineering: A New Tool for Phytoremediation,"BioMed Research International, vol. 2013, Article ID 182032, 7 pages, 2013. doi:10.1155/2013/182032.

CHAPTER 11

Vibe M Gondolf, Rhea Stoppel, Berit Ebert, Carsten Rautengarten, April JM Liwanag, Dominique Loqué and Henrik V Scheller, "A gene stacking approach leads to engineered plants with highly increased galactan levels in Arabidopsis," BMC Plant Biology201414:344, DOI: 10.1186/s12870-014-0344-x.

INDEX

A

abotic stress 119
accumulation 232, 235, 237, 238, 239, 242, 244
Agrobacterium 3, 4, 8, 11, 14, 17
alcohol-insoluble residue (AIR) 267
allergenicity 168
amplification 178
amplified fragment length polymorphism (AFLP) 80
Animal and Plant Health Inspection Service (APHIS) 53
anthocyanins 169
auxin 197, 198, 199, 207, 214, 215, 222, 223, 224, 225, 226, 227, 228, 229
auxin-response factor (ARF) 198

B

Bacillus thuringiensis 151, 154, 155, 157, 161, 163
bacterial artificial chromosome (BAC) 19, 81
bacteroidines 175
BioBrick 1, 2, 3, 4, 5, 6, 7, 8, 9, 10, 11, 12, 13, 273
Brazzein 4, 7, 8, 11, 16

breeding 2, 15

C

calcium-dependent protein kinases (CD-PKs) 120
caroteniod 44
Chrysanthemum lavandulifolium 127, 134
cinnamyl alcohol dehydrogenase (CAD) 49
Cloning 10, 11, 14
Collybia velutipes 167, 171
Cry proteins 155
curculin 4
cyanotoxins 232

D

denaturation 178

E

element-binding factor (ERF) 123
Environmental Impact Quotient (EIQ) 158
expressed sequence tag derived simple sequence repeats (EST-SSR) 80
expressed sequence tag (EST) 81